Frame and Slab Structures

edited by

G. S. T. Armer
Building Research Establishment, UK

D. B. Moore
Building Research Establishment, UK

Butterworths
London   Boston   Singapore   Sydney   Toronto   Wellington

 PART OF REED INTERNATIONAL P.L.C.

All rights reserved. No part of this publication may be reproduced or transmitted in any form or by any means (including photocopying and recording) without the written permission of the copyright holder except in accordance with the provisions of the Copyright Act 1956 (as amended) or under the terms of a licence issued by the Copyright Licensing Agency Ltd, 33–34 Alfred Place, London, England WC1E 7DP. The written permission of the copyright holder must also be obtained before any part of this publication is stored in a retrieval system of any nature. Applications for the copyright holder's written permission to reproduce, transmit or store in a retrieval system any part of this publication should be addressed to the Publishers.

Warning: The doing of an unauthorised act in relation to a copyright work may result in both a civil claim for damages and criminal prosecution.

This book is sold subject to the Standard Conditions of Sale of Net Books and may not be re-sold in the UK below the net price given by the Publishers in their current price list.

First published 1989

© Butterworth & Co. (Publishers) Ltd, 1989

The cover illustration is a model Yield Locus for a doubly reinforced, two-way spanning slab.
The illustration and the papers "The stability of tall buildings" and "The composite action of brick panel walls supported on reinforced concrete beams" are reproduced by permission of the Controller HMSO, Crown Copyright.

British Library Cataloguing in Publication Data

Frame and slab structures
1. Structures. Design
I. Armer, G. S. T.   II. Moore, D. B.
624.1'771

ISBN 0–408–03669–9

Library of Congress Cataloging in Publication Data applied for

Printed and bound in Great Britain by
Anchor Press Ltd, Tiptree, Essex

Preface

This book is dedicated to the memory of Randal H Wood. Dr Wood spent most of his distinguished career within the British Scientific Civil Service and achieved the rare distinction of gaining an individual merit promotion to Deputy Chief Scientific Officer. His international reputation resulted in visiting professorships at the Universities of Berkeley in California and the University of Mexico and in an Associate Chair of Engineering Science at the University of Warwick.

The target for Dr Wood's research was always the engineer in the design office. If serving this target required a study of, for example, existence theorems, then he was the first to venture into the minefield. His great enthusiasm for the world about him brought joy to those who had the good fortune to work with him, and his legacy of technical writings will be required reading for many years to come.

The contents of this work comprise nine invited papers from authors working in fields associated with Dr Wood's areas on interest. The papers represent the state-of-the-art in frame analysis, the design of steel connections, lightweight steel construction, composite construction and plastic design of reinforced concrete slabs. Dr Wood was an early advocate of numerical methods and the use of the theory of plasticity in structural analysis. He was also however, an avid supporter of good experimental work and the papers report recent test results where appropriate. They also contain practical design guidance to aid the designer in the application of the latest theoretical developments.

GSTA   Building Research Establishment
DBM   Garston, Herts

Editors   December, 1988

Acknowledgements

The enthusiastic support of the authors for this project is gratefully acknowledged. Special thanks are due to Mrs. Jenny Alexander whose skill and patience converted the original manuscripts into the printed form of this book.

CONTENTS

Preface

		Page
1.	An historical review on the interaction of plasticity and structural stability in theory and its application to design. **M R Horne**	1
2.	The stability of framed structures. **U Vogel**	29
3.	Frame analysis and the link between connection behaviour and frame performance. **D A Nethercot**	57
4.	The design of end-plate connections. **D B Moore**	75
5.	Developments in lightweight decking and cladding. **E R Bryan**	109
6.	New developments in composite construction. **R P Johnson**	135
7.	Hillerborg's advanced strip method - a review and extensions. **L L Jones**	145
8.	Plastic flow rules for use in the analysis of compressive membrane action in concrete slabs. **K O Kemp, J R Eyre and H M Al-Hassani**	175
9.	Nodal forces in slabs and the 'Equilibrium Method'. **C T Morley**	195
10.	R H Wood. Selected published works 1948-1986.	221
11.	The stability of tall buildings. **R H Wood**	225
12.	The composite action of brick panel walls supported on reinforced concrete beams. **R H Wood**	263

Index   295

1

AN HISTORICAL REVIEW ON THE INTERACTION OF PLASTICITY AND STRUCTURAL STABILITY IN THEORY AND ITS APPLICATION TO DESIGN

M R Horne O.B.E, F.R.S.
Formerly of Manchester University, UK

SYNOPSIS

The post-war development of the theory of plasticity to produce practical design methods for multi-storey frame structures is described. Special attention is paid to the foundations laid by the Steel Structures Research Committee and to those aspects of frame stability in which R H Wood made exemplary contributions. The problems of elastic-plastic design of both sway and non-sway frames are discussed together with Wood's devices to simplify the theories and yet still produce rational and economic design procedures for steel building frames.

INTRODUCTION

The giving of anything like a complete historical review of this subject is beyond the scope of a single paper - it would occupy a complete volume. The present history is bound to be selective, and it seems appropriate for a paper given at this conference that attention should be centred on those aspects which attracted most of Randal Wood's interest and attention, concentrating the presentation on the work of Randal Wood himself and on the work of other investigators which most influenced and interacted with his work. I must therefore claim the indulgence of any who may feel that much important work in the subject has been scantily dealt with or not even mentioned.

Wood's work in the field of steel structures was entirely dominated by his aim to produce rational, economic design procedures for steel building frames, especially multi-storey frames. In dealing with such frames, it is important to distinguish between two types. In no-sway frames, sway forces are resisted by bracing or by interaction with shear wall systems, so that the vertical members can be described as 'no-sway' columns. In 'sway' frames on the other hand, resistance to sway in either or both horizontal directions (assuming the usual rectangular grid-based layout) is taken by the flexural resistance of the columns interacting with beams that are connected to them by either virtually rigid connections, or by connections with a high degree of rigidity. Wood made outstanding contributions to our understanding of the behaviour of both types of frame. In the case of non-sway frames especially, Wood spent many years of devoted effort on the formulation of design procedures directed towards their inclusion in Codes of Practice. It is true to say that the fertility of his mind, and the determination with which he pursued his aims, were in both these fields quite remarkable.

THE FAILURE OF SIMPLY LOADED COLUMNS

In that 'tour de force' containing his most advanced thoughts on the design of no-sway frames ("A new approach to column design"(1)), Wood enumerated some 50 or so independent variables which determined column behaviour and failure loads (variables describing material properties, structural arrangements, dimensions and geometric properties, loading conditions, initial stress states and imperfections). We will come later to a consideration of how Wood dealt with these variables in the design of continuous frames, but even when dealing with the failure of a pin-ended rolled I-section strut subjected to axial loading and failing in the plane perpendicular to the web, one can elaborate at least 10 important independent variables of significance for the magnitude of the failure load.

Material properties   Yield stress, elastic modulus, strain-hardening modulus.

Geometric properties Length, flange width, flange thickness, web depth, web thickness.

Irregularities   Initial lack of straightness and initial stresses, the characteristics of which can in both cases only be described statistically.

The historically dominant basis for the design of steel struts has been the well known Perry-Robertson formula (Robertson (2) 1925), which derives the theoretical axial load which will produce first yield in an initially stress-free column with a sinusoidally shaped imperfection. The magnitude of the assumed imperfection is chosen empirically so that the theoretical load at first yield so calculated is a good fit to the failure loads of tested struts. Moreover, by expressing the initial imperfection in the form

$$\delta = e \frac{Lr_y}{a_y} \text{ where}$$

L = length of strut, r_y = minor axis radius of gyration and
a_y = distance of the extreme fibre from the minor centroidal axis,

the Perry-Robertson formula enables all variables to be eliminated except the

An historical review

minor axis slenderness ratio $\dfrac{L}{r_y}$ and the statistically determined constant e.

It is important to recognise that, although this highly successful formula depends on an elastically calculated maximum stress, it allows empirically for the elastic-plastic behaviour of ductile struts at failure via the statistical fit to test results. Formulae based on the Perry-Robertson type of approach have continued to play an important role in practical design methods in which the aim is to make the criterion the actual failure load, necessarily an elastic-plastic phenomenon.

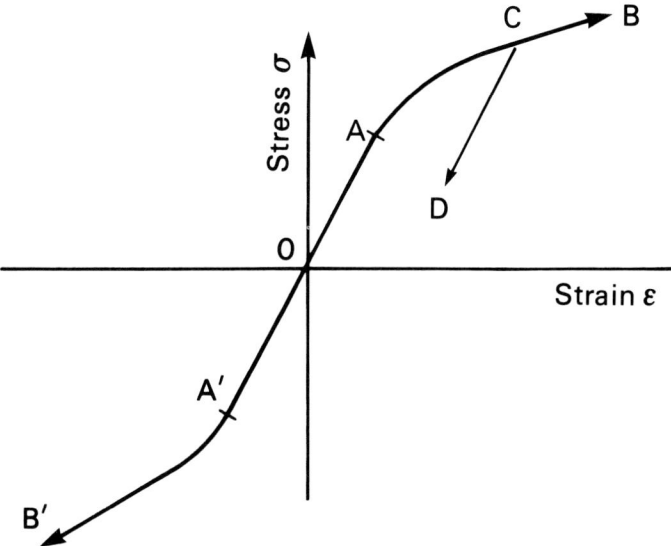

Fig. 1. Stress-strain curve for strain-hardening material

The first major development in stability theory for compression members stressed beyond the elastic limit, taking account of actual non-linear material behaviour, had been Engesser's double modulus theory (3) published in 1895. In this well-known early theory, the flexural stiffness to be used in the stability calculation for a member stressed uniformly in compression to a point C on the stress-strain curve (Fig. 1) is obtained from the reduced modulus calculated by allowing for the following of a tangent to the curve at C on the compression side of the axis of zero incremental strain, but for an elastic unloading along CD for the rest of the cross-section. Later, Shanley (4) pointed out that there could be a progressive increase of buckling deformations, starting at a load less than the reduced modulus load. All stress increments would then follow the tangent to the curve at C in Fig. 1, and Shanley therefore proposed his tangent modulus load, derived from the stiffness given by substituting the slope of the tangent to the stress/strain curve in place of the elastic modulus in the expression for the buckling load.

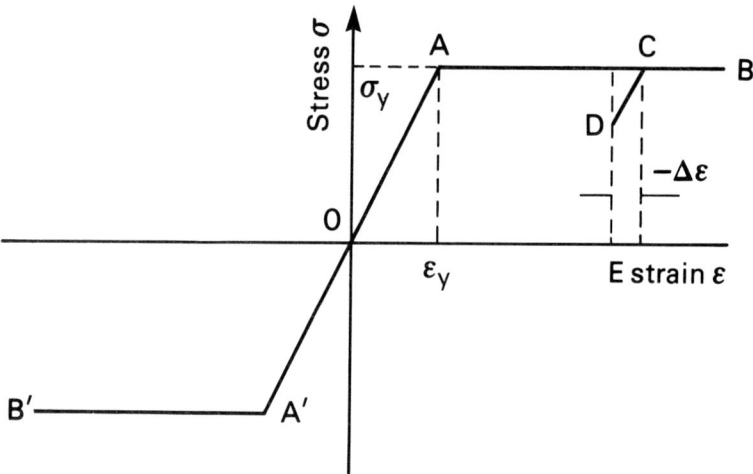

Fig. 2. Stress strain relation for elastic-plastic material

The application of either the reduced or the tangent modulus approach to mild steel columns, for which the elastic/pure plastic stress/strain relation OAB in Fig. 2 is a close representation of material behaviour, leads simply to the prediction that buckling takes place at either the elastic buckling load or at the squash load (all fibres reaching the yield stress) whichever is the lesser.

Neither the reduced nor the tangent modulus load can be regarded as representing a theoretically correct treatment for the buckling of axially loaded struts in practice, since correct treatment necessitates taking into account inevitable out-of-straightness imperfections and initial internal stress states. However, for ideally straight struts buckling about axes of symmetry which are also axes of symmetry for the initial stress states, both the reduced and the tangent modulus approaches may be modified to take account of the stress state at any given axial load. Interestingly enough, both approaches can then be applied meaningfully to struts of elastic/pure plastic material. At any given axial force, some of the material will have yielded, while some will remain elastic. If buckling is assumed to occur according to the reduced modulus theory, material which has been strained to some point C (Fig. 2) will obey CB under continuing compressive strain and CD under reversed strain increments. According to the tangent modulus concept, only those areas of the cross-section which remain elastic will contribute to stiffness. These concepts have been used, particularly by researchers in the United States (5,6), as a basis for design.

As applied to the derivation of design data, no one of the above approaches can be regarded as inherently more accurate than the others - their use for this purpose is in each case empirical. In view of the scatter of test results, and the known variability of geometric imperfections and residual stresses, extended debate claiming superior accuracy for the one or the other is probably misguided. In British practice, preference has been maintained for the Perry-Robertson approach, and in terms of simplicity, there would seem to be much to be said in favour, since empirical adjustment to test results or sophisticated theoretical studies is easily made by the adjustment of the out-of-straightness coefficient.

COLUMNS IN BUILDING FRAMES

The analysis of the behaviour of columns in multi-storey building frames is enormously more complicated than the analysis of isolated struts. Designers have for many years used methods which depend on mutually contradictory assumptions regarding the interactions between beams and columns. The familiar 'simple design' procedure of the British Code BS449 outstandingly lies in this category. Here, beams are designed on the assumption that they are simply supported on brackets attached to the webs and flanges of the columns, and the moments induced by the beams in the columns are calculated on this basis. However, when it comes to the actual design of the columns, effective lengths for buckling of the columns about their minor axes down to only 70% of the actual between-floor lengths are allowed, thus implying the existence not of positive moments applied by the beams to the columns, but negative restraint moments provided to the columns by the beams.

The work of the Steel Structures Research Committee, carried out before the Second World War, was aimed at producing a more rational approach, taking account of the true semi-rigidity of typical beam/column connections. A substantial part of the work which went into the First Interim Report, Second Interim Report and Final Report of the Committee(7) was carried out by the late Lord Baker, then acting as the Technical Officer to the Committee. The work of the Steel Structures Research Committee has also been described and commented upon in some detail by Baker in his own book (8). Despite the relative complexity of the derived elastic design procedure, very little economy was achieved compared with designs using the contemporary "simple" Code procedure. The substantial economies obtained in the design of the beams because of the advantages for them of semi-rigid connection behaviour were off-set by the heavier columns that were required. This result provides an example, in hind-sight, of the pitfalls which can await when one attempts to design rationally against instability effects using a "safe stress" approach. Such considerations strongly influenced the direction of the research pioneered under Baker, first at Bristol and then much more ambitiously at Cambridge, into the ultimate load behaviour of continuous steel structures.

Baker had concluded, from his experience with the Steel Structures Research Committee, that in the design of columns in continuous beam/column building frames, it was most important to allow for the directional restraint provided to the ends of the columns by the beams,

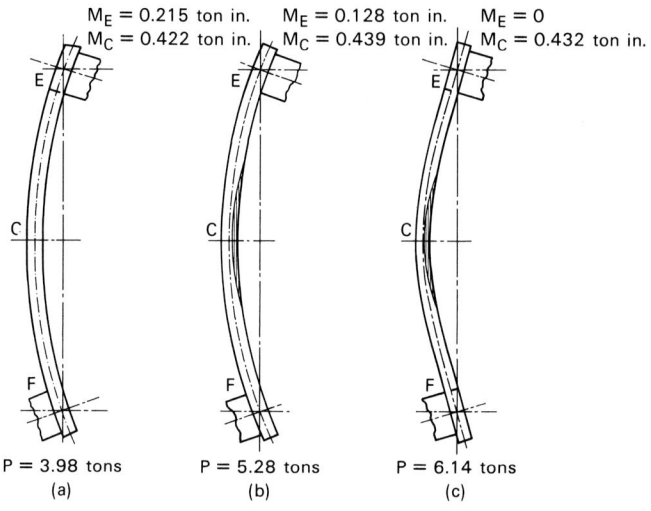
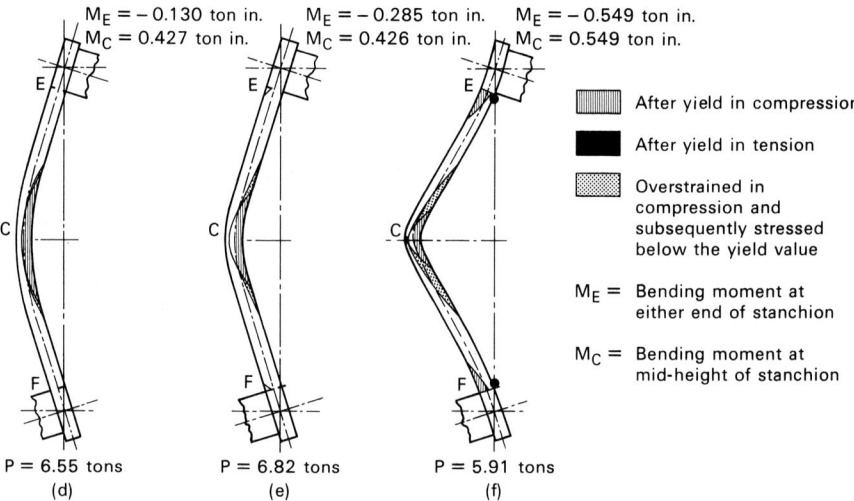

Fig. 3. Theoretical development of plastic zones in a Cambridge single curvature column test (ref 12)

An historical review

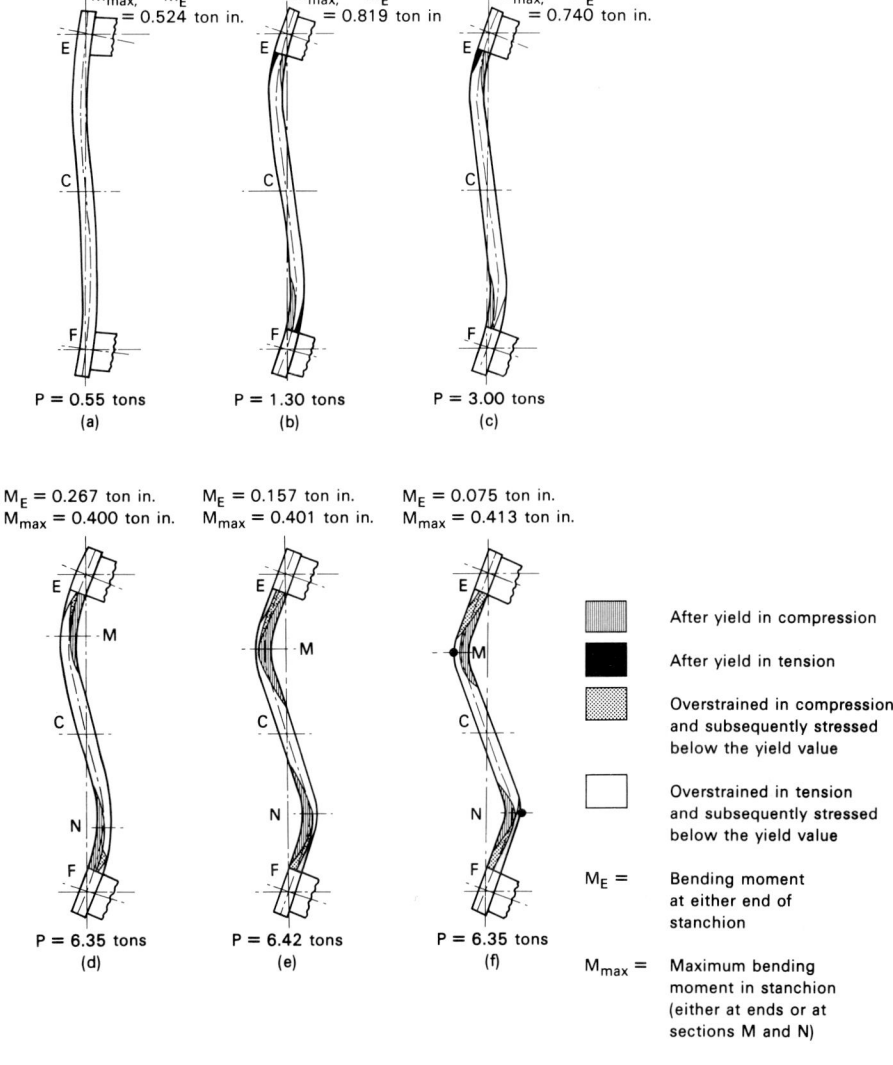

Fig. 4. Theoretical development of plastic zones in a Cambridge double curvature column test (ref 12)

the behaviour of the columns being pursued well beyond the stage at which yield first occurs in them. In a very extensive series of tests reported by Baker and Roderick (9,10,11), and also summarised by Baker et al (12), tests were carried out on model steel columns connected to beams made of high tensile steel to ensure that they remained elastic. In the first series, the beams were first loaded so that the ends of the columns were bent by terminal moments equal and opposite in direction (thus bending the column in a "C" curve, i.e. in "single curvature"), and in the second series, so that the ends of the column were bent by equal terminal moments acting in the same direction (thus bending the column in an "S" curve, i.e. in double curvature"). Direct axial load was then added until the column failed.

Theoretical analyses, based on elastic-plastic theory(13), gave good correlation with test results. Figs. 3 and 4 show the development of plastic zones for typical columns of rectangular cross-section bent in single and double curvature respectively. Under single curvature loading, yield first occurred at an axial load of 3.98 tons (Fig. 3(a)). The addition of further axial load caused plastic zones to develop at mid-height on the compression side, accompanied by decreases in the terminal moments, which ultimately reversed in sign (Figs. 3(b) to (f)). At peak load (Fig. 3(e)), the restraint moments just caused the formation of plastic compression zones at the ends. By the time when (theoretically only, since in the tests all loads were applied as dead loads) plastic hinges had formed at ends and centre, the load capacity had markedly decreased below the peak value (6.82 tons down to 5.91 tons).

In the example of double curvature loading (with the beams carrying the same loads as in the single curvature case above), plastic hinges had already formed at the ends by the time beam loading was complete (Fig. 4(b)), but this did not prevent the axial load being able to rise from 1.30 tons to 6.42 tons (Figs. 4 (b) to (e)) before two plastic hinges formed at a slightly decreased load of 6.35 tons (Fig. 4(f)). For the double curvature case, the analysis assumed that perfect symmetry would persist throughout, whereas in all the double curvature tests, lack of perfect symmetry caused the column to unwrap and fail finally in single curvature. However, the column apparently reached this state of instability at just about the peak load derived in the symmetrical theoretical analysis, since the experimental collapse load was in good agreement with the theoretical load at that stage.

While the above experimental and theoretical results were of considerable interest, they did not contribute directly to the derivation of any economic design procedure, although it was found possible to devise relatively simple methods of estimating the failure loads of the columns so loaded (14,15). The reason was that, for economic design, the primary load-carrying beams needed to frame into the flanges of I- or H-section columns so that they could be designed plastically with three hinged mechanisms, thus applying moments of known value about the major axes of the columns. Following solutions treating the requirements for the stability of I- and H-section members bent about the major axis by

An historical review

terminal moments M_1 and M_2 varying in ratio from

$$\beta = \frac{M_2}{M_1} = 1 \text{ (symmetrical single curvature) to } \beta = \frac{M_2}{M_1} = -1$$

(equal double curvature) (16,17), a design procedure based on a Perry-Robertson treatment for columns so loaded was derived (18). This treatment was later extended (19,20) to allow for the formation of a plastic hinge within one end or (for equal double curvature bending, i.e. $\beta = -1$) within both ends of the column, enabling the envisaged participation of the column in overall failure mechanisms in sway frames. The criterion of failure as the attainment of first yield near the mid-height of the column prevented the introduction of anything approaching full plasticity as the ratio of end moments defined by β approached unity, but a lower "safe curve", based on the study of partially plastic post-buckling behaviour, allowed the application of the design procedure to cases of nearly uniform major axis column bending where lateral supports reduced the slenderness of the column about its minor axis to sufficiently low values between such supports.

The main shortcoming of this treatment of the stability problem in the ultimate load, elastic-plastic design of steel structures lay in its exclusion of the beneficial effects of the restraint of columns against failure about their minor axes as provided by the beams in the planes of their major axes. In BS 449, this had been allowed for in the concept of effective length, although in an arbitrary and inconsistent manner. Wood played a major part in efforts that were mounted to move towards a solution of this design problem, efforts which came to fruition in the publication of two reports under the aegis of the Institution of Structural Engineers and the Welding Institute (21). This was again based on ostensibly elastic design procedures for the columns, although there was hidden use of the fact that high stresses and limited yielding induced near the ends of columns in double curvature bending (see Fig. 4) had much less effect than the lower stresses caused by similar beam loading inducing single curvature bending (Fig. 3). This conclusion, influenced by the Cambridge tests already described, meant that the most severe floor loading pattern for the design of the columns was that producing the nearest approach to uniform single curvature bending, so that in effect an unspecified degree of plasticity at the ends of columns for floor loading patterns which would induce the worst double curvature bending was allowed in the Committee's proposals.

Tests on full scale frames were carried out at the Building Research Station (22,23) to provide a check on certain aspects of the Joint Committee's design proposals.

Despite the considerable effort devoted to the formulation of the Joint Committee design procedure, including notable contributions by Wood aimed at simplifying the application of the column stability design criteria to continuous frames, the Joint Committee method showed only very modest savings compared with BS 449. Wood, with apparently undiminished enthusiasm, set about analysing the reasons for this lack of success. He remained convinced that a rational design procedure, significantly more economic than BS 449, must be possible, since as has been argued above, it is impossible to regard BS 449 as being rational and self-consistent.

Unfortunately, the elastic-plastic behaviour of typically I- or H- section columns subjected to loads and restraints coming from members framing into them about both axes is an impossibly complicated problem to cover comprehensively in any simply conceived design procedure. However, Wood found a clue to a completely new approach to this subject by adapting concepts that had proved fruitful in dealing with the elastic-plastic failure loads of frames free to sway, to which he had himself made significant contributions, and we now turn to this topic.

THE ELASTIC-PLASTIC FAILURE LOADS OF PLANE FRAMES

The effect of plasticity on the stability of a frame may be approached by considering a total energy term given by

$$U_N = U_W + U_E + U_p \tag{1}$$

where U_W, U_E and U_p have significance as follows.

U_W = total potential energy of the external load system,

U_E = elastic strain energy (reversible) stored in the structure,

U_p = energy stored in plastic deformation (non-reversible) absorbed during the deformation of the structure on its load-deformation path.

Taking any monotonically loaded structure, let the load-deflection curve for elastic-plastic behaviour be OAFD in Fig. 5. The horizontal axis represents a dominant deflection parameter μ (e.g. the sway deflection at the top of the structure), while the vertical axis is the load parameter λ. Up to some point A, the structure behaves elastically, and if the elastic limit for the material were indefinitely high, the load-deflection curve would rise to some point C.

The stability criterion for the structure while it is elastic is simply stated in terms of the total energy U given by

$$U = U_W + U_E \tag{2}$$

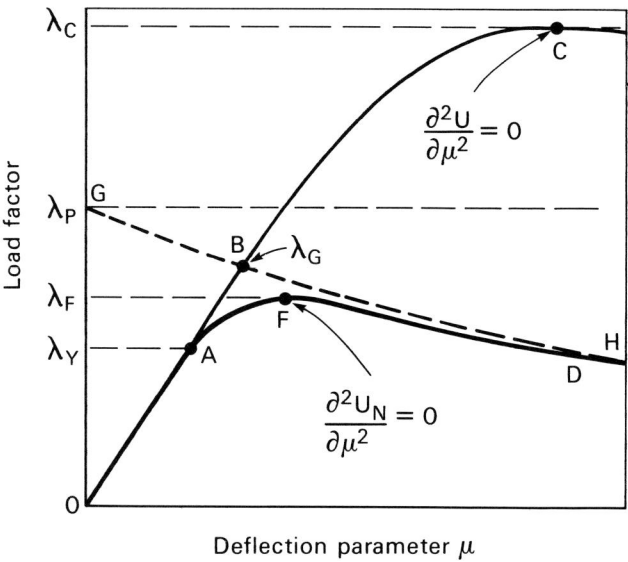

Fig. 5. Load-deflection curves for elastic and elastic-plastic structures

On the rising part of the load-deflection curve OAC, U is a minimum with respect to small deviations from the equilibrium state, whence

$$\frac{\partial^2 U}{\partial \mu^2} > 0$$

If the load-deflection curve falls beyond C, then $\frac{\partial^2 U}{\partial \mu^2} < 0$, while at the maximum load, $\frac{\partial^2 U}{\partial \mu^2} = 0$, and the structure is in neutral equilibrium for small displacements. The structure is then at its elastic critical load.

Considering now the elastic-plastic load-deflection curve OAFD, we see that

$$\frac{\partial^2 U_N}{\partial \mu^2} > 0 \text{ before the peak load factor } \lambda_F \text{ is reached,}$$

$$\frac{\partial^2 U_N}{\partial \mu^2} < 0 \text{ on the falling part of the curve and}$$

$$\frac{\partial^2 U_N}{\partial \mu^2} = 0 \quad \text{at the peak load.}$$

However, in the plastic zones of the structure, the stress is constant at the yield value, whence

$$\frac{\partial^2 U_p}{\partial \mu^2} = 0$$

and so the condition defining the attainment of the peak load becomes

$$\frac{\partial^2 (U_W + U_E)}{\partial \mu^2} = 0 \tag{3}$$

Hence the condition for having reached the load at which deflections can increase without any further increase of loading is the same as the instability condition for the "deteriorated" elastic structure, consisting of the original structure modified by the removal of material which, in the actual structure at failure, is undergoing plastic deformation.

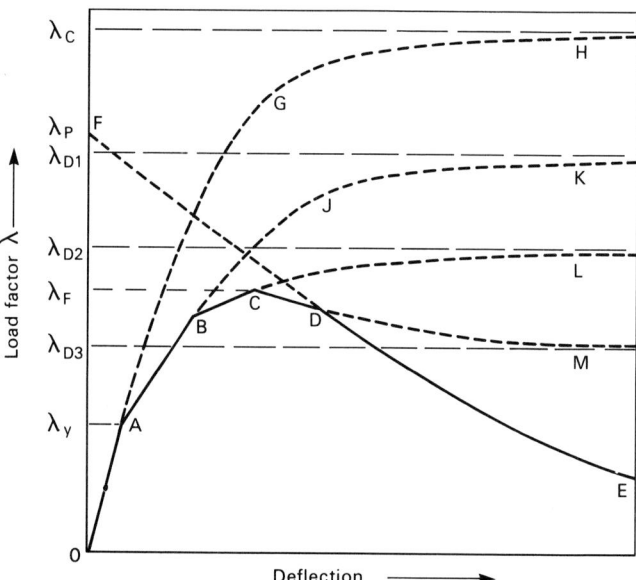

Fig. 6. The concept of deteriorated critical loads

The concept of deteriorated critical loads as applied to rigid frames collapsing by overall sway type mechanisms was first elaborated by Merchant (24) and by Wood (25). Under increasing load factor λ, the deflections in the elastic range (Fig. 6) first follow the

An historical review

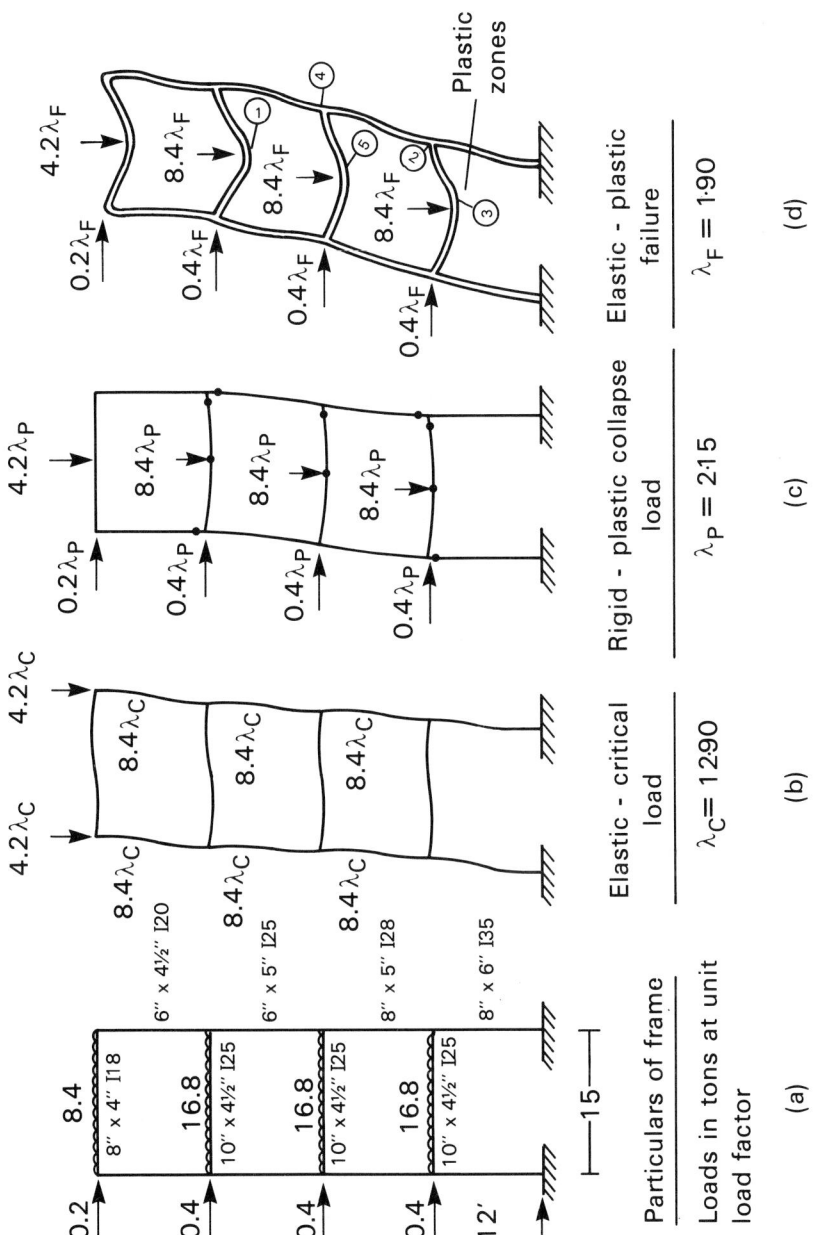

Fig. 7. Four-storey frame analysed by Wood (ref 26)

curve OAGH which, in the absence of plasticity at higher load factors, would rise towards an elastic critical load λ_c at which (ignoring the effects of gross deformations), deflections would tend to become large. However, if a plastic hinge forms somewhere in the structure at point A, the subsequent curve follows instead ABJK, becoming asymptotic to the deteriorated elastic critical load λ_{D1}, which corresponds to the elastic critical load of the same frame identically loaded but containing a structural hinge at the plastic hinge position. The failure load is attained when the formation of an additional plastic hinge (in the case illustrated, the third plastic hinge) causes the deteriorated critical load to fall below the current load value.

In his cardinal paper on the subject published in 1958, Wood (26) gave results for the elastic-plastic analysis of the four-storey frame shown in Fig. 7(a). The elastic-plastic analysis was carried out with the help of a differential analyser, allowance being made for the spread of plastic zones along the members (i.e. plasticity not assumed to be confined to the plastic hinges themselves). Failure was found to occur at a load factor λ_F of 1.90 (Fig. 7(d)), compared with a rigid-plastic collapse load factor λ_p (with an eight-hinge mechanism, Fig. 7(c)) of 2.15. The elastic critical load factor λ_c was 12.90. At the failure load, plastic hinges had formed at positions 1,2,3 and 4 in Fig. 7(d), and a fifth plastic hinge had almost formed at position 5.

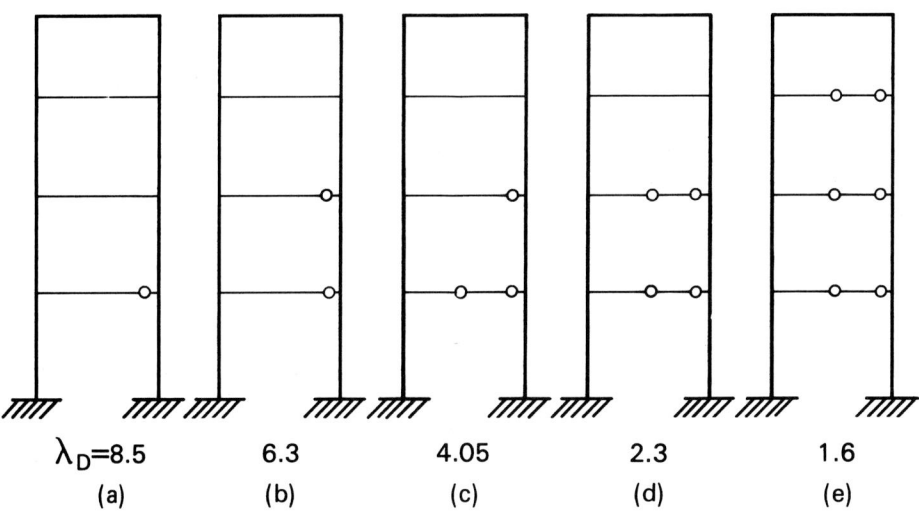

Fig. 8. Deteriorated critical loads of frames analysed by Wood (ref 26)

An historical review

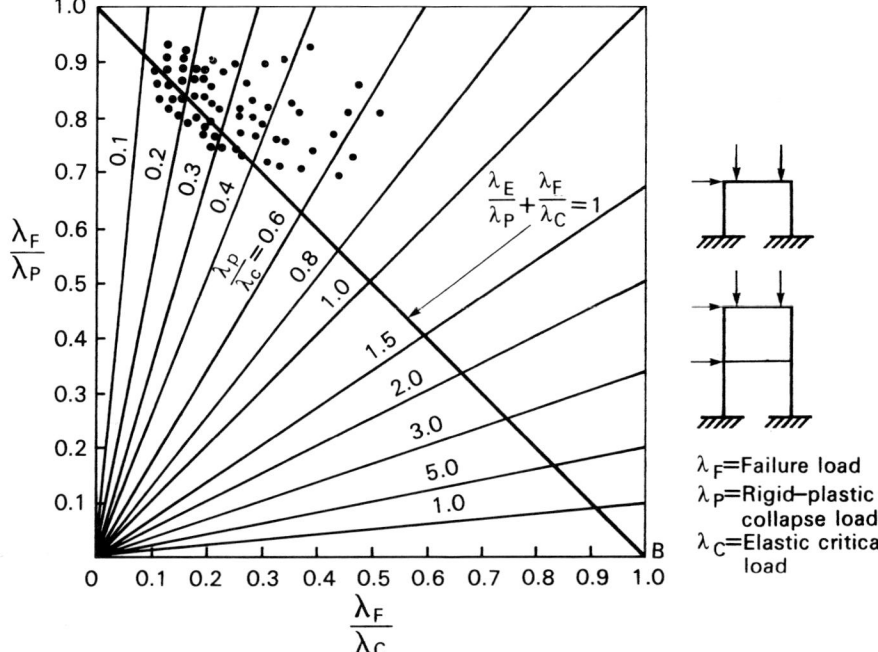

Fig. 9. Theoretical failure loads of frames analysed by Salem (ref 31)

To illustrate the concept of deteriorated elastic critical loads, Wood gave results for the elastic critical loads of his frame with pin joints assumed at various sets of positions. Some of his results are shown in Fig. 8, the circles representing the positions of pin joints. None of the deteriorated structures correspond exactly to the deteriorated structure at the failure load in the full analysis ($\lambda_F = 1.90$, Fig. 7(d)). A deteriorated structure intermediate between those in Fig. 8(d) and (e) is however seen to be appropriate, and the deteriorated critical load factors 2.30 and 1.60 lie appropriately either side of the failure load factor 1.90.

Wood's work showed that, in calculating the elastic-plastic failure loads of plane frames collapsing in overall sway modes, very little error is introduced if the spread of plastic zones along the members is ignored, so that members can be assumed to remain elastic until a full plastic hinge has formed ("unit form factor analysis"). This justified the later production of numerous elastic-plastic computer programs, using stiffness matrices successively modified with increase of load as plastic hinges are formed. The effect of a plastic hinge can be allowed for by introducing at that stage a frictionless pin joint with the addition of equal and opposite moments, equal to the appropriate plastic moment value, applied to the structure as external loads either side of the plastic hinge position. Details of such analyses and examples of their application have been given by Cranston (27), Jennings and Majid (28) and Majid and Anderson (29).

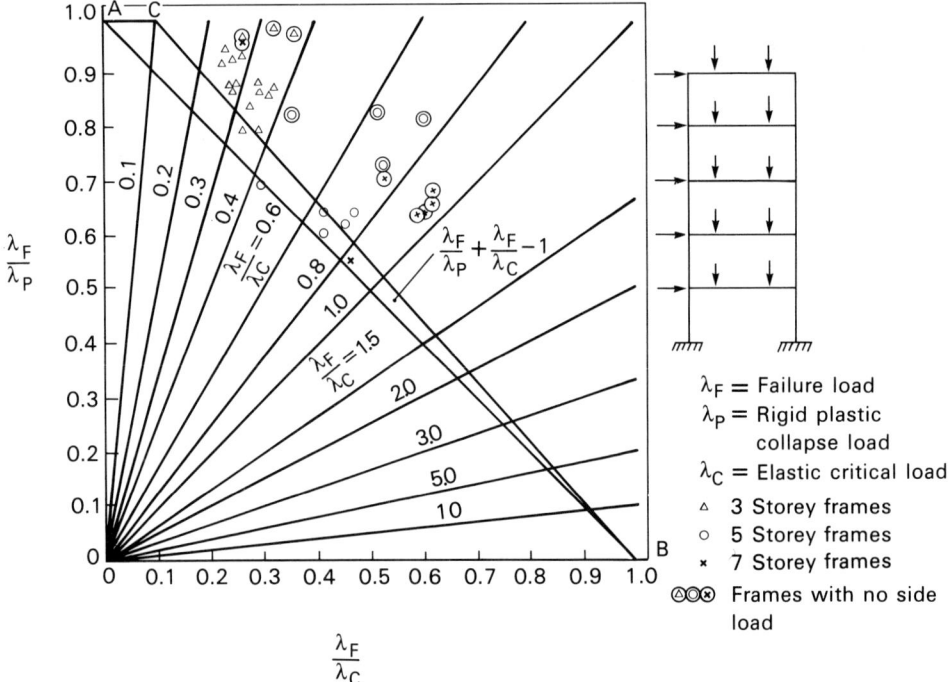

Fig. 10. Experimental failure loads of frames tested to Low (ref 33)

Before Wood published his paper (26), Merchant (30) had suggested that it might be possible to consider the failure load λ_F of an elastic-plastic structure as some function of various idealised load factors, including the load factor at first yield, the elastic critical load factor λ_c and the rigid-plastic collapse load factor λ_p. In particular, Merchant explored the use of the simple empirical formula

$$\frac{\lambda_F}{\lambda_p} + \frac{\lambda_F}{\lambda_c} = 1 \qquad (4)$$

This formula satisfies the requirement that, for very stocky structures for which λ_c is very large and $\lambda_c \gg \lambda_p$, $\lambda_F \to \lambda_p$, while for very slender structures for which $\lambda_c \ll \lambda_p$, $\lambda_F \to \lambda_c$. Working with Merchant, Salem (31) calculated the rigid-plastic, elastic critical and elastic-plastic failure loads of a number of one and two storey frames. Merchant et al (32) plotted pairs of the resulting values of

An historical review

$$\frac{\lambda_F}{\lambda_p} \quad \text{and} \quad \frac{\lambda_F}{\lambda_c} \quad \text{(Fig.9)}$$

so that they could be compared with the relationship given by what has come to be called the "Rankine-Merchant" straight line formula, equation (4) above (AB in Fig. 9). Results of tests carried out by Low (33) on model frames loaded to collapse are shown in Fig. 10. Merchant suggested that the formula (4) represented an <u>approximate</u> lower bound value for λ_F, and the above theoretical and experimental results provide some support for this conjecture.

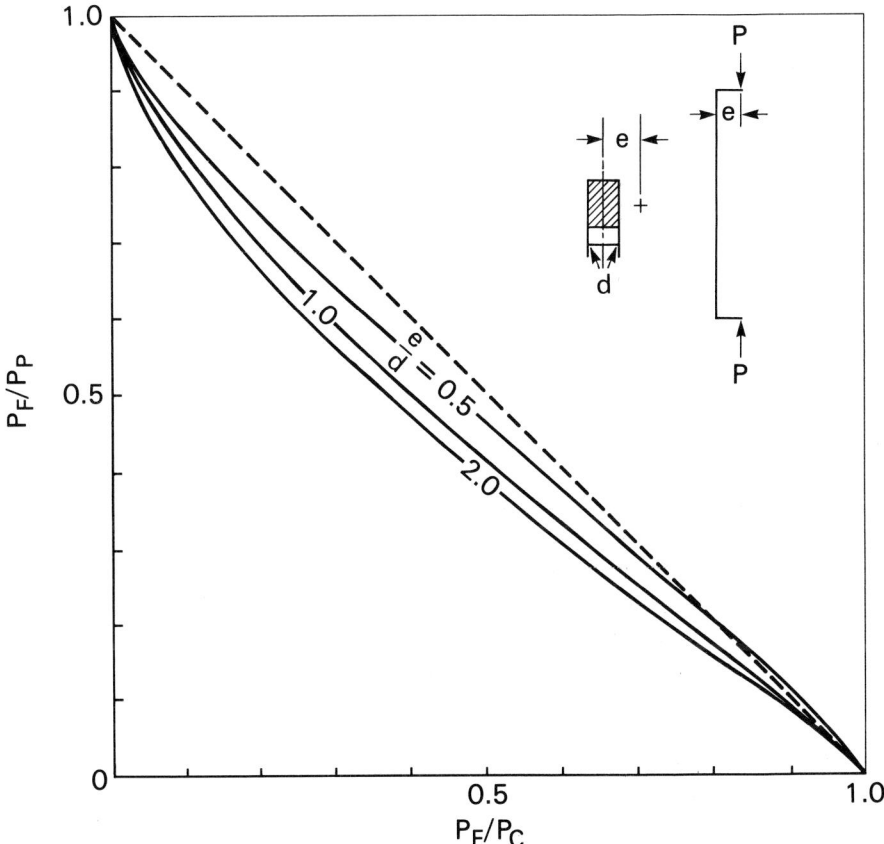

Fig. 11. Comparison between theoretical collapse loads of eccentrically loaded struts and Rankine-Merchant loads (ref 36)

Later investigations, including some carried out by Anderson, Lok and Wood (34) and by Anderson and Lok (35) showed that the Rankine-Merchant load is indeed an approximate, but not a strict, lower bound. The failure load tends to fall slightly below the Rankine-Merchant load when the deflected shapes corresponding to elastic deflections at low loads, the rigid-plastic failure mechanism and the lowest elastic critical load are all similar. The shortfall below the Rankine-Merchant load is most pronounced for very simple structures, such as eccentrically loaded struts, Fig. 11 (see Horne (36)), while for multi-storey frames with predominantly vertical loading, the Rankine-Merchant load does seem to be a safe lower bound (see Fig. 10). Anderson and Lok (35) conclude that, provided the bay width is not less than the storey height (as is always the case for practical building frames), the Rankine-Merchant load does provide a safe estimate of the "exact" failure load obtained by computer. When the beneficial effects of strain-hardening and minimal composite action with cladding are allowed for, it would certainly seem safe to use the Rankine-Merchant load as a safe lower bound for practical frames. In fact, Wood went further, and argued (37) that these factors justified taking for design purposes, the relationship ACB in Fig. 10, in which no allowance is made for any reduction of load factor below the rigid-plastic collapse load when

$$\frac{\lambda_c}{\lambda_p} > 10, \quad \text{while when} \frac{\lambda_c}{\lambda_p} < 10$$

it is assumed that

$$\lambda_F = \frac{\lambda_F}{0.9 + \frac{\lambda_F}{\lambda_c}} \qquad (5)$$

This suggestion has been incorporated in the British Standard "Structural Use of Steelwork in Building" (BS 5950), Part 1(30).

ELASTIC-PLASTIC DESIGN PROCEDURES FOR SWAY FRAMES

The Rankine-Merchant load factor as proposed by Merchant and modified by Wood is simply a means of making an intended "safe" check against sway instability effects for a frame once it has been designed by some means or other. Various proposals have been made for the design procedure itself, i.e. for arriving at suitable member sizes, given the loading and required load factors.

Heyman (39,40) proposed a pattern of plastic hinges for regular multi-storey plane frames, using which the beams and columns sections could be derived directly. The method treated the beam plastic hinge moments as required minimum plastic moments of resistance, but required the columns to sustain the "plastic" values within the limits of their elastic behaviour. There was no direct allowance for frame stability requirements, but the conservatism in the column design represented an undetermined allowance to cope with instability effects. The range of frames for which the method would be satisfactory was not determined.

An historical review

Holmes and Gandhi (41) took a step forward by proposing a method for regular plane frames in which, again, a pattern of plastic hinges was assumed, this time confined to the beams. The columns were designed elastically, allowance for sway instability being made by the application of specially derived stability functions which depended on the elastic stiffnesses of both the beams and the columns and the slenderness of the columns. The necessity of making assumptions about the stiffness ratios introduced an element of iteration into the design procedure.

While the method of Holmes and Gandhi was a step forward in the development of design procedures for multi-storey frames, it had two major defects. First it applied only to regular frames. Secondly, no guarantee existed that at collapse, plastic hinges would form in the pattern assumed. In the presence of instability effects, the uniqueness theorem of plasticity ceases to be valid, and the failure loads of frames designed by the method of Holmes and Gandhi could theoretically be either above or below the stated design requirement.

The development of computer procedures for the elastic-plastic analysis of plane frames led to proposals for their application in producing more rational design methods. Horne and Majid (44) first proposed such an application. They adopted the design criteria that no plastic hinge should form in any beam below the "working load factor" level, while no plastic hinge should occur in a column below the required "ultimate load factor" level (or levels, where alternative load factors for differing load combinations were required). A modification of Heyman's design procedure, mentioned above, was used to obtain a preliminary design. Elastic-plastic analysis was then pursued in steps, linear extrapolation to the next load increment being used systematically to modify the design, using the stated design criteria, before each full matrix computer analysis was performed. The whole automated procedure proved to be capable of dealing with both large and highly irregular frames.

Because of their extensive use, particular studies have been made of the possible incidence of stability problems in plastically designed pitched roof portal frames. Competitive design using plastic theory has led to the emergence of frames of such slenderness about the major axes of the columns and the rafters that deflections and overall stability have become of importance. Strain-hardening can often compensate for modest reductions of load capacity due to instability, and charts for portal frames taking account of this effect have been published (43).

The above design procedures for sway frames are conceived in terms of behaviour in one plane with members assumed to be bending about their major axes only. It is assumed that, in the direction at right angles, away is prevented by bracing or by interaction with shear walls or other stabilising features. Moreover, it is necessary to ensure that columns do not become unstable between floor levels in planes at right angles to the sway plane. When the columns contain plastic hinges at one or both ends, this may be done by using the treatment already described (19,20), which however does not allow for any restraint to the columns by beams framing into their minor axes. This restriction has been necessary because of the interference with continuity restraint when a full plastic hinge forms. While ignoring restraint about the minor axes can be acceptable in the design of low rise frames, the restriction renders this approach uneconomic if applied to multi-storey frames in which axial loads become of greater significance. This has been the reason for the restriction of plastic action to the beams in the above procedures for the design of multi-storey sway frames.

THE ELASTIC-PLASTIC DESIGN OF NON-SWAY FRAMES

The success of the concept of the progressive deterioration of critical loads due to plasticity as applied to sway frames encouraged Wood to investigate the possibility of applying the same concept to the design of columns in non-sway frames. Because of the enormous complexity of the problem of a column subjected to moments and restraints about both axes, with both the columns and the associated beams in developing states of plasticity with increasing level of load, Wood suggested that progress could only be made by treating the stability in generalised, approximate terms.

In "A new approach to column design", Wood discussed the likely spread of plastic zones in an I-section column which was on the point of failure. A column length is assumed to be acted upon by bending moments in the plane of the web by beams some of which, themselves economically designed, have reached a state of plastic collapse, although depending on the pattern of floor loading assumed, some of these beams will still be elastic. In planes at right angles, the columns are rigidly connected to beams which are assumed to remain elastic. Figs. 12 and 13, taken from Wood's paper, show the spread of plastic zones in a typical column when the bending moments applied by the beams act, respectively, in opposite senses (producing single curvature bending) and in the same sense (producing double curvature bending). In single curvature (Fig 12) even if one flange becomes fully plastic, the other flange remains predominantly elastic, so that the effective minor axis flexural rigidity of the column cannot fall much below 50% of the elastic value EI_y. Wood therefore assigned to this case a safe lower limit for the effective minor axis flexural rigidity of the column of 40% of the elastic value, i.e. flexural rigidity = $R.EI_y$ where $R = 0.4$. In double curvature (Fig.13) plasticity is largely confined to the vicinity of the ends, and Wood concluded that one could for this case take $R = 0.8$. While these approximations appear crude, Wood found that the estimates to which they led of the deteriorated critical load of the column was not excessively sensitive to the exact value of R, since with diminishing R, the stiffness of the restraining minor axis beams relative to the stiffness of the column correspondingly increased, thereby decreasing the effective length of the column. This argument reflected the finding, revealed by full theoretical

An historical review

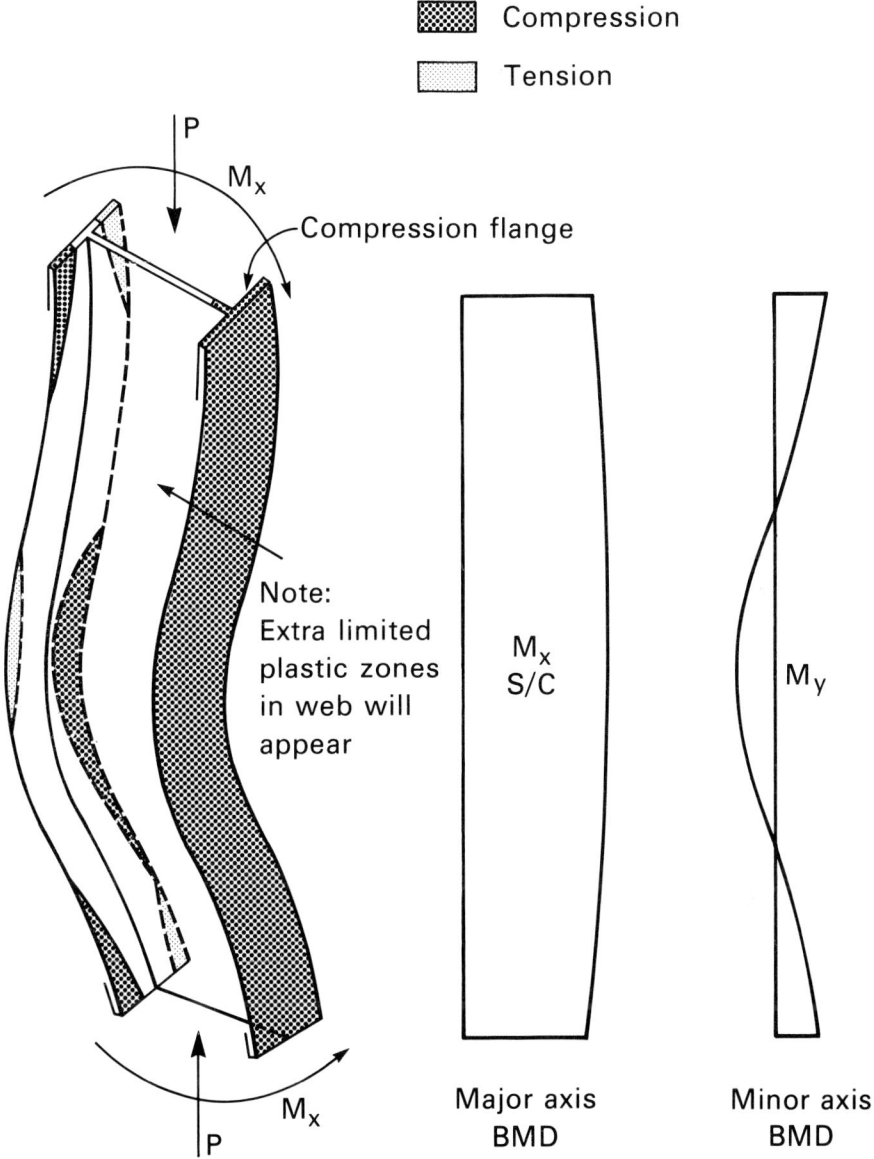

Fig. 12. Wood's assessment of probable plastic zones in the flanges of a restrained slender column at collapse under low axial load, the column bent in single curvature about its major axis (ref 1)

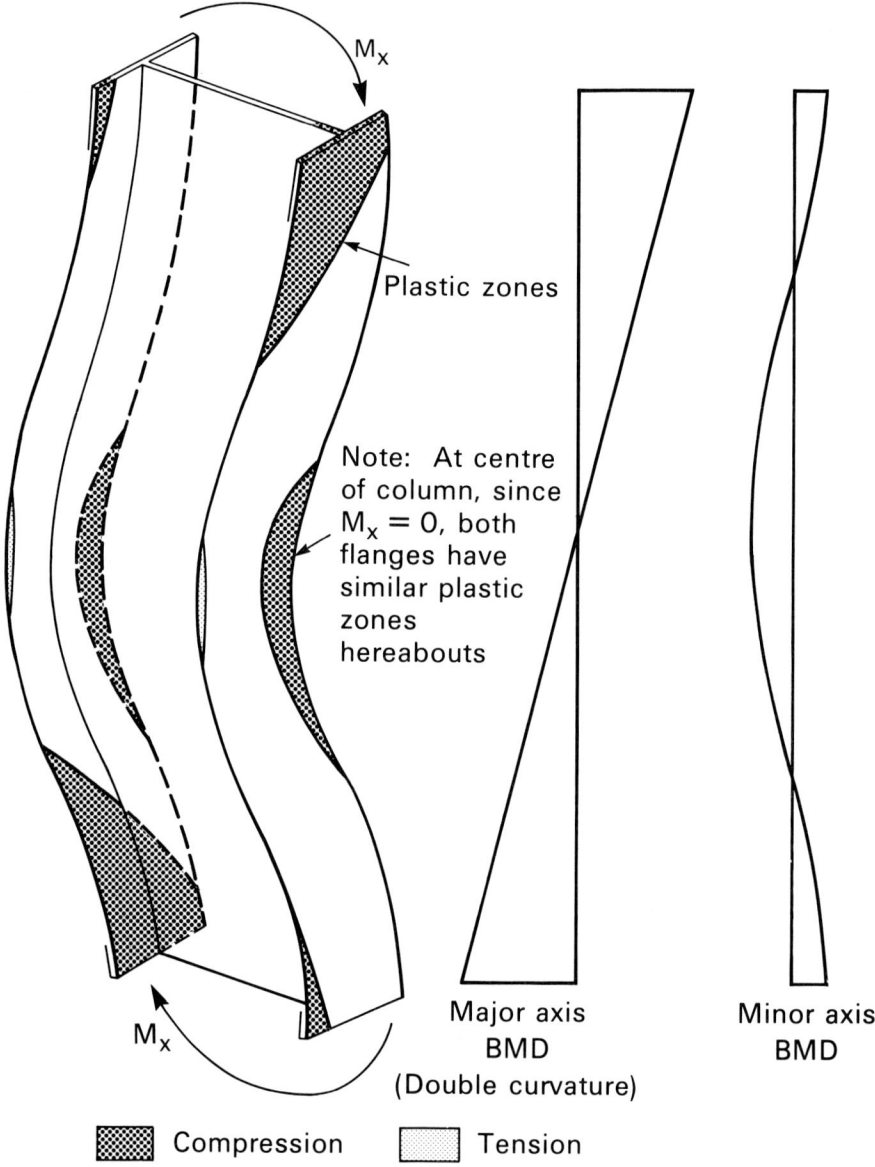

Fig. 13. Wood's assessment of probable plastic zones in the flanges of a restrained slender column at collapse under low axial load, the column bent in double curvature about its major axis (ref 1)

An historical review 23

analyses, that the load-deflection curves for columns in fully continuous frames were distinctly flat-topped, so that a moderately large variation in the assumed state of deformation and plasticity would not correspond, in real behaviour, to any great variation in the derived axial load in the column. The terminal moments acting on the columns were, in Wood's design approach, derived from the pattern loading which produced the nearest approach to single curvature bending since (as for the Joint Committee design method (21)), investigations had shown that this was the more severe loading condition for the column. The actual ratio of terminal bending moments was allowed for by an interpolation formula, taken as being linear in the value of R with respect to that ratio between the extreme values of 0.4 and 0.8, this interpolated value being used when the greater applied column moment just produced plasticity in one flange. When maximum terminal moments were insufficient to produce complete plasticity, an interaction formula giving higher values of R was introduced. The "deteriorated critical axial load" was then determined using design curves derived from Wood's own treatment of the stability of columns in continuous frames (37).

CONCLUDING REMARKS

In this review, it has been impossible to deal adequately with many topics which are of importance in achieving any complete understanding of the stability problems affecting the ultimate loads of steel building frames. One major area is that of lateral torsional buckling. This is of importance, not only for columns but also in portal frames where the apparent advantages of using deep haunches in the eaves can bring severe problems in this respect. Even with all the work that has been done in these areas, it cannot be said that completely satisfying design procedures are yet available.

Ostensibly, as soon as a compression flange has become fully plastic, virtually all lateral torsional stability disappears, but deeper theoretical considerations show that, because any subsequent plastic deformation involves interaction between minor axis flexural deformation, major axis flexural deformation and torsional warping deformation, no conclusions about the severity or otherwise of the stability problem can be reached without a complete three-dimensional study of the elasto-plastic deformations involved. In his studies leading to the formulation of his "new approach", Wood bypassed the apparently insuperably difficult problem of lateral torsional buckling in columns bent and restrained about both axes by excluding as not allowed designs in which lateral torsional buckling would have reduced estimated carrying capacity to any significant degree. Wood also bypassed the problem of how plastic bending about the major axis at the ends of a column length interferes with the effective continuity with beams framing into the minor axis, arguing that such continuity will only be compromised with increasing deformation (Shanley principle), whereas in estimating carrying capacity, one is concerned only with the earliest stages of collapse. A preliminary theoretical study (44) has explored the possibility of more advanced work on this topic, but has also shown that this represents an extremely difficult area in which to make progress.

A very great deal has been learned since the Second World War about stability as it affects the behaviour of structures beyond the elastic limit. It is a fascinating subject, and Wood was certainly one of those who found it so. The fact that he himself made most distinctive contributions to our understanding derived from the single-mindedness with which he pursued his goal of producing a rational, viable and economic design procedure for multi-storey frames. In the end, his creative thinking carried him into realms where practising engineers have been reluctant to follow - the new BS 5950, although owing much to his contributions, still fails to embody many of the concepts that he advocated. Despite the potential we now have for "sledge-hammering" our way into more and more deeper studies into elastic plastic stability problems using increasingly ambitious computer programs, the challenge of producing better, more completely convincing and yet still practically usable design procedures remains. Wood saw his "new approach" as a new start, are we to be too timid to follow?

REFERENCES

(1) Wood, R.H: "A new approach to column design", BRE Report, 1974

(2) Robertson, A: "The strength of struts", ICE Selected Engineering Paper No 28, 1925

(3) Engesser, F: Schweizerische Bauzetung, vol 26, 1895, p24

(4) Shanley, F.R: "Inelastic column theory", J Aer Sci, 1947, p261

(5) Tall, L. (ed): Structural Steel Design, Ronald Press, New York, 1974

(6) Johnston, B G (ed): Guide to Stability Design for Metal Structures, 3rd edition, John Wiley & Sons, New York, 1976

(7) Steel Structures Research Committee, First Interim Report 1931, Second Interim Report 1934, Final Report 1936, London HMSO

(8) Baker, J.F: The Steel Skeleton Volume One, Elastic Behaviour and Design, CUP 1954.

(9) Baker, J.F & Roderick, J.W: "The behaviour of stanchions bent in single curvature", Trans Inst Weldg, vol 5, 1942.

(10) Baker, J.F & Roderick, J.W: "Further tests on stanchions", Weld Res vol 2, 1948.

(11) Baker, J.F & Roderick, J.W: "The behaviour of stanchions bent in double curvature", Weldg Res vol 2, 1948.

(12) Baker, J.F, Horne, M.R & Heyman, J: The Steel Skeleton Volume Two, Plastic Behaviour and Design, CUP 1956.

An historical review

(13) Baker, J.F, Horne, M.R & Roderick, J.W: "The behaviour of continuous stanchions", Proc Royal Soc A, vol 198, 1949, p 493.

(14) Roderick, J.W: "The behaviour of stanchions bent in single curvature", BWRA Report No FE1-5/24, 1945.

(15) Roderick, J.W & Heyman, J: "Approximate methods of calculating collapse loads of stanchions bent in double curvature" Weld Res vol 2, 1948.

(16) Horne, M.R: "The flexural-torsional buckling of members of symmetrical I-section under combined thrust and unequal terminal moments", Quart J Mech App Math, vol 7, 1954, p410.

(17) Salvadori, M.G: "Lateral buckling of I-beams", Trans Amer Soc Civ Engrs, vol 120, 1955, p1165.

(18) Horne, M.R: "The stanchion problem in frame structures designed according to ultimate carrying capacity", Proc Instn Civ Engrs, vol 5, part 3, 1956, p105.

(19) Horne, M.R: "Safe loads on I section columns in structures designed by plastic theory", Proc Instn Civ Engrs, vol 29, 1964, p137.

(20) Horne M R, "The plastic design of columns", Publn No 23, British Constructional Steelwork Association, 1964.

(21) Institution of Structural Engineers and Welding Institute - Joint Committee on "Fully-rigid multi-storey welded steel frames", First Report 1964, Second Report 1971 (Instn Struct Engrs).

(22) Wood, R.H, Needham, F.H & Smith, R.F: "Tests of a multi storey rigid steel frame designed to the Joint Committee Report", The Structural Engineer, 1968, p107.

(23) Smith, R.F & Roberts, E.R: "Test of a multi-storey rigid steel frame of high yield steel", The Structural Engineer, vol 49, 1971, p451.

(24) Merchant, W: "Frame stability in the plastic range" Symposium on the Plastic Theory of Structures, Cambridge 1956 (British Welding Journal, vol 3, 1956, p366).

(25) Wood, R H: Discussion of reference 24, ibid vol 4, 1957, p26

(26) Wood, R H: "The stability of tall buildings", Proc Instn Civ Engrs, vol 11, 1958, p69.

(27) Cranston, W.B: "A computer method for inelastic analysis of plane frames", Cement and Concrete Association, Tra/386, 1965.

(28) Jennings, A & Majid, K.I: "An elastic-plastic analysis for framed structures loaded up to collapse", The Structural Engineer, vol 43, 1965.

(29) Majid, K.I & Anderson, D: "The computer analysis of large multi-storey framed structures", The Structural Engineer, vol 46, 1968.

(30) Merchant, W: "The failure load of rigid jointed framework as influenced by stability", The Structural Engineer, vol 32, 1954, p185.

(31) Salem, A: "Structural frameworks" PhD thesis, University of Manchester, 1958.

(32) Merchant, W, Rashid, C.A, Bolton, A & Salem, A: "The behaviour of unclad frames" Proc Fiftieth Anniversary Conference, Instn Struct Engrs, 1958.

(33) Low, M.W: "Some model tests on multi-storey rigid steel frames", Proc Instn Civ Engrs, vol 13, 1959, p287.

(34) Anderson, D, Lok, T.S & Wood, R.H: "Studies on the Merchant-Rankine formula", University of Warwick Dept Engineering, Research Report CH12, 1962.

(35) Anderson, D & Lok, T.S: "A limitation on the use of the Rankine-Merchant Formula" University of Warwick Dept Engineering Research Report CE13, 1982.

(36) Horne, M.R: "Elastic-plastic loads of plane frames", Proc Roy Soc A, vol 274, 1968, p343.

(37) Wood, R.H: "Effective lengths of columns in multi-storey buildings" The Structural Engineer, vol 52, 1974, pp235, 295, 341.

(38) British Standard 5950 "Structural use of steelwork in buildings" part 1, British Standards Institution 1985.

(39) Heyman, J: "The plastic design of tall steel buildings", Proc Univ Hong Kong Golden Jubilee Congress, 1961.

(40) Heyman, J: "An approach to the design of tall buildings", Proc Instn Civ Engrs, vol 17, 1960, p431.

(41) Holmes, M & Gandhi, S.N: "Ultimate load design of tall steel building frames allowing for instability" Proc Instn Civ Engrs, vol 30, 1965, p147.

(42) Horne, M.R & Majid, K.I: "The design of sway frames in Britain" Proc Conf on Plastic Design of Multi-Storey Frames. Lohigh Pa, 1965.

(43) Horne, M.R & Chin, M.W: "Plastic design of portal frames in steel to BS 968", British Constructional Steelwork Association, Pub No 20, 1966.

(44) Horne, M.R: "Failure of biaxially loaded I-section columns restrained about the minor axis" Engineering Plasticity (ed J Heyman & F A Leckie), CUP 1966.

2

THE STABILITY OF FRAMED STRUCTURES

U Vogel
Karlsruhe University, Federal Republic of Germany

SYNOPSIS

In the first section of this paper, general aspects of in-plane stability behaviour of mild steel frames in the elastic-plastic range are discussed. Recent developments in Europe for the ultimate limit state calculation of unbraced frames are covered in the second section. Finally, practical hints are given for the application for the second-order plastic hinge theory, using a dialogue procedure with desk-computers.

INTRODUCTION

This paper discusses the in-plane behaviour of unbraced frames with rigid joints. Braced frames with rigid joints are not included, because the stability check for this type of structure is usually done by the effective length method, so that the columns are regarded as beam-columns which are checked with the corresponding ultimate strength interaction formulae. Unbraced frames with sidesway, however, exhibit more complex behaviour and therefore they must be treated carefully in stability or ultimate strength calculations.

Since extensive research on frames with semi-rigid joints is still underway, this type of construction is also not included in the following, because it seems too early for the discussion of the final results. It should be mentioned, however, that a special task group of the ECCS Committee 8 (Structural Stability) will publish a report soon on this topic.

Attention is drawn to the fact, that only the in-plane overall stability behaviour of frames is discussed. Out-of-plane behaviour (e.g. lateral torsional buckling or biaxial bending), local buckling and serviceability (e.g. drift limitations) must be considered separately in actual design cases.

GENERAL ASPECTS OF FRAME STABILITY

It is well known that in the case of a statically indeterminate steel structure, which is not subject to instability or fatigue problems, the classical allowable stress design method leads neither to a knowledge of the actual stresses, the safety-factor nor of the load carrying capacity of that structure. The actual stress pattern cannot be calculated for the following reasons:

(i) residual stresses due to rolling, welding, straightening and erection,

(ii) stress concentrations in the vicinity of holes, notches and concentrated loads,

(iii) uncontrolled settlement of supports,

(iv) unavoidable imperfections or deviations from the mathematical model used in the structural analysis.

Such a structure is only safe because of the ductility of the material. Even if the actual stress pattern is known, the factor of safety cannot be calculated, because a statically indeterminate structure does not reach its full strength at the elastic limit load, i.e. the load at which the yield stress is reached in the most stressed fibre.

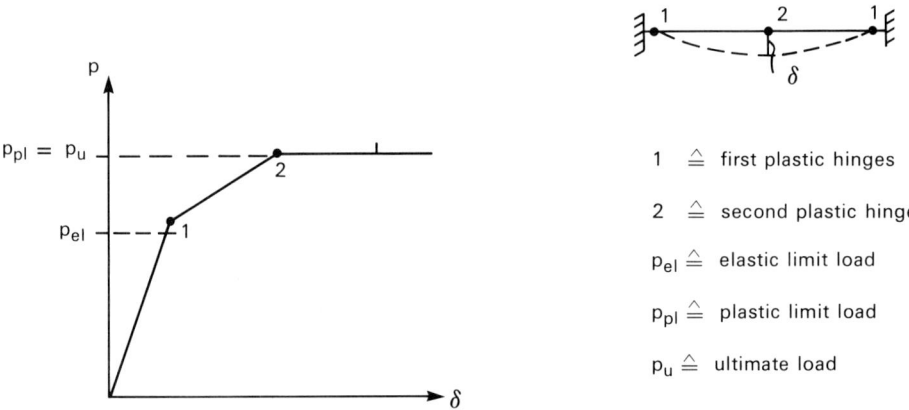

Fig. 1. Load-deformation-behaviour of a statically indeterminate beam of mild steel (lateral torsional and local buckling prevented)

To study the real behaviour of a steel structure plastic analysis is a much better approach than elastic analysis. This approach takes advantage of the ductility of mild steel not only qualitatively but also quantitatively. It shows that in the case of bending without compressive forces, a sufficient number of plastic hinges must be formed to produce a mechanism with at least one degree of freedom. This mechanism forms under constant load, called the plastic limit load. If no instability problems are involved, this limit load is independent of all kinds of geometric and material imperfections and can be regarded as the ultimate strength of the structure (Fig. 1.)

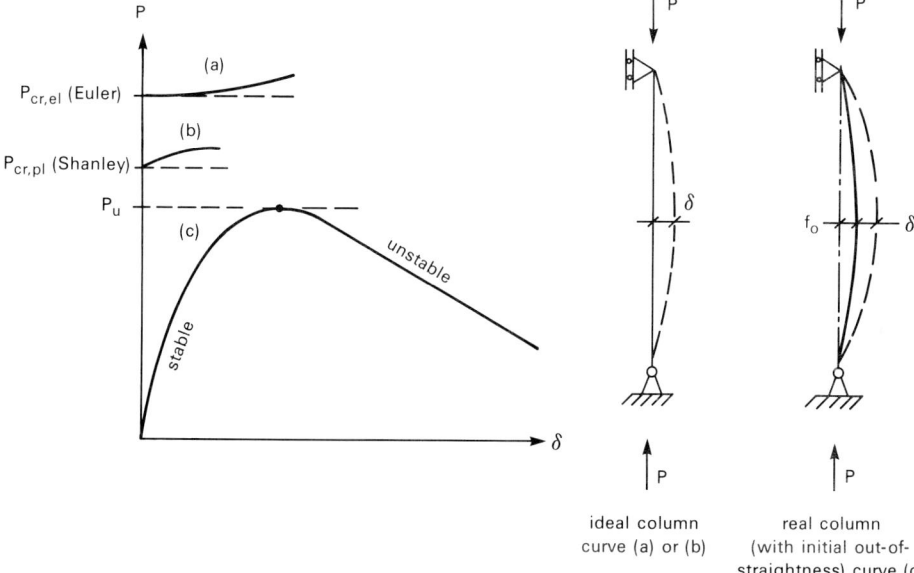

ideal column
curve (a) or (b)

real column
(with initial out-of-straightness) curve (c)

Fig. 2. Load-deformation-behaviour of mild steel columns

The behaviour of a steel structure, whose members are subject to axial compressive forces is somewhat different. The most simple case is the centrally loaded straight column without initial bending (Fig. 2). The classical approach to determine the failure of such a column is the well known equilibrium-bifurcation theory. Depending on whether the column is stressed in the elastic range or in the plastic range, when buckling with bifurcation of equilibrium occurs, curve (a) or (b), (Fig. 2) respectively is valid. A real column will never behave in such a way. Since geometric imperfections and residual stresses as well as material inhomogeneities are always present, the column will deflect from the beginning of loading (curve (c) in Fig. 2) and will reach its ultimate strength by divergence from equilibrium, i.e. at the point where equilibrium between external and internal forces is no longer possible because of decreasing internal strength due to plastic deformations. To calculate the nonlinear curve (c), it is necessary to take into account the influence of deformations, i.e to use a second-order (elastic-plastic) analysis.

The form of curve (c) in Fig. 2 is also characteristic for beam-columns with primary bending and for framed structures with members in bending and compression, the bending moments of which are caused by transverse and/or unavoidable geometric imperfections and load eccentricities. This behaviour can be approximated very closely by a second-order elastic-plastic analysis using the concept of plastic hinges (curve (a) in Fig. 3) instead of taking into account the actual spread of plastic zones (plastic zone theory, curve (b) in Fig. 3).

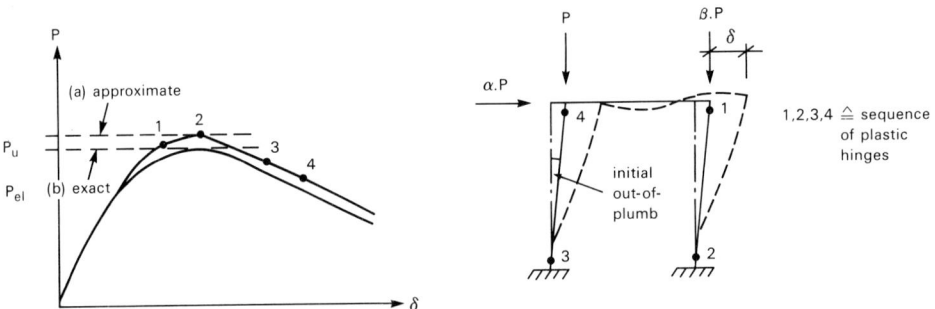

Fig. 3.   Load-deformation-behaviour of a mild steel frame

The difference between the curves of Fig. 1 and Fig. 3 is as follows:

> For a beam with no compressive forces the structure always forms a mechanism at the ultimate load P_u, and the equilibrium is not important beyond this point.

> In the case of a frame (or a beam-column), however, the structure might have failed before a complete mechanism has formed, and the equilibrium is always unstable beyond the load level P_u even if the last plastic hinge forms at the load level P_u- under which the structure collapses.

The stability of framed structures 33

But in both cases the ultimate strength can be determined by the same basic approach, which - for economic reasons - takes advantage qualitatively of the ductility of the material and shifts the design load from P_{el} up to P_u, (at least in all cases where problems of fatigue or brittle fracture are not involved).

As shown above in addition to plasticity effects, it is necessary to take into account the influence of deformations (second-order effects) and unavoidable imperfections in all cases where the danger of instability exists, i.e. for columns, beam-columns, frames, plates and shell-type structures, when compressive forces are involved. In all these cases, instability under ultimate load is accompanied by divergence of equilibrium - not by bifurcation.

These are the basic principles of ultimate strength design accepted widely by engineers all over the World. It is the only correct way to determine P_u as accurately as possible in accordance with the real behaviour of a structure. This basic concept is not a new one, because it has been used in several national design codes, e.g. in the DIN 4114 where for 40 years, the ω-method has been based on the ultimate strength of a mild steel column with unavoidable load-eccentricity and an ideal elastic-plastic stress-strain-relationship with a reduced yield stress. If this concept is to be reasonably logical, it is necessary - or at least desirable - to know the value of P_u, regardless of whether one uses a classical deterministic or a modern probabilistic concept of structural safety analysis. It must be admitted, however, that for the present state-of-the-art, it is sometimes very difficult to follow this concept. In such cases, approximations are introduced, e.g. the Merchant-Rankine formula for frames, as an alternative solution, within certain limits to elastic-plastic second-order analysis.

Though the concept of ultimate strength design is the basis of modern recommendations or codes (e.g. DIN 18 800 or EC 3), it is not, of course, always necessary, for the engineer to perform a second-order analysis taking into account all the effects of plasticity as well as all kinds of imperfections or favourable conditions, such as cladding. For the sake of simplicity some convenient simple formulae, usually of an interaction type, tables, and approximate solutions which can be used in practice are offered, since they lead to the same results, or at least very close to the results, of a more accurate ultimate strength analysis.

Finally it must be said, that the bifurcation theory has not become obsolete, because its results serve in many cases as a characteristic parameter or an upper limit in an ultimate design procedure. Similarily, the linear or nonlinear elastic theory cannot be abandoned, because it is still valuable and necessary for the analysis of a structure under service conditions or in fatigue problems. But it must be clear, that in general, allowable stresses do not make any sense in stability problems of frames.

RECENT EUROPEAN DEVELOPMENTS IN FRAME STABILITY ANALYSIS

General Remarks

Nowadays, with the help of computers, engineers are able to analyse almost every complex structure with difficult boundary conditions, geometric imperfections, residual stresses, arbitrary material behaviour and various types of loading. Almost everywhere in the world very sophisticated computer programs exist to predict the ultimate strength behaviour of such structures by the yield-zone theory. But it is still somewhat difficult to compare the results of different computer programs, because of different and frequently not clearly defined assumptions used in the development of these programs. On the other hand, there is still a need for simplified hand methods which may be used for a preliminary design or for the check of computer results.

Consequently, the Technical Working Group 8.2 of ECCS prepared and published a report two years ago on "Ultimate Limit State Calculation of Sway Frames with Rigid Joints" (1), (2), which takes care of these aspects. This report does not deal with a basically new approach to the ultimate limit state calculation of sway frames. But it gives an overview and some clarification of the better known methods. It was the intention of TWG 8.2 of ECCS that this publication should meet two objectives:

- to make the results of different computer programs following the plastic-zone theory more comparable by the use of the same important basic input parameters, upon which the members of 11 European countries have agreed unanimously.

- to illustrate, for ultimate limit state calculations the different possibilities for simplification or approximation and the range of their application.

Special care was taken to establish simple criteria to distinguish between systems for which the application of the first-order theory (either elastic or plastic) is accurate enough, and systems for which, for the sake of safety, it is an absolute necessity to use a second-order (either elastic or plastic) method. Also approximate or simplified second-order methods are shown, which take into account all the important parameters (such as imperfections, real material behaviour, geometrical changes, etc.) but are simple enough to:

- be used as hand methods (at least for the preliminary design of complex structures), and
- to be taught in the undergraduate level in the civil engineering education system.

Some of these simplifications are already included in the last draft of Eurocode 3 as well as in the new German Stability Code DIN 18 800 Teil 2 (3).

Some aspects of the ECCS developments for the limit state calculation of frames with rigid joints.

In the ECCS publication No. 33 (1), TWG 8.2 of ECCS recommends one of the following methods for the design of unbraced frames or, to check their stability, under static loading conditions:

1. Elastic-plastic analysis

    1.1 Ultimate strength theory (plastic-zone theory)
    1.2 Second-order plastic hinge theory
    1.3 First-order plastic hinge theory

2. Elastic analysis

    2.1 Second-order elastic theory
    2.2 First-order elastic theory

3. Merchant-Rankine-approach

None of these methods uses the effective length concept, because they are in the line of a full system design. This is valid at least in the UK also for the well known Merchant-Rankine approach, which is, however, empirical in nature and will not be discussed here. A few explanations of the other methods are given below:

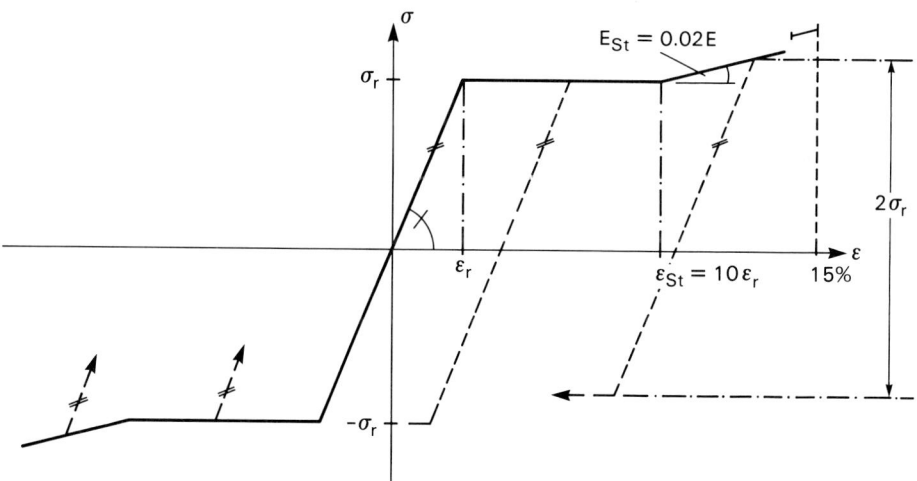

Fig. 4. Simplified constitutive law for uniaxial strain (for Fe 360 and Fe 510)

1.1 Ultimate strength theory (plastic-zone theory)

Today many computer programs based on the ultimate strength theory are available. Since they have been developed in different continents and countries of the world by different engineers or research groups, they use somewhat different assumptions and basic input parameters. This is true especially for stress-strain relationships, geometric imperfections and residual stresses. In order to get more comparable results, TWG 8.2 agreed upon certain proposals for these basic input data to be used in computer programs (see Fig. 4, Fig. 5 and Fig. 6).

It does not make very much sense to discuss such computer programs in detail, but it should be mentioned, that all of them take into account the spreading of plastic zones along the members and over the cross-sections. Such programs describe very realistically the actual load-deformation behaviour of steel framed structures. Therefore, even if not used in the daily engineering practice, they are very useful tools for:

- research projects
- checking the accuracy of simplified methods
- checking experimental results
- developing design charts

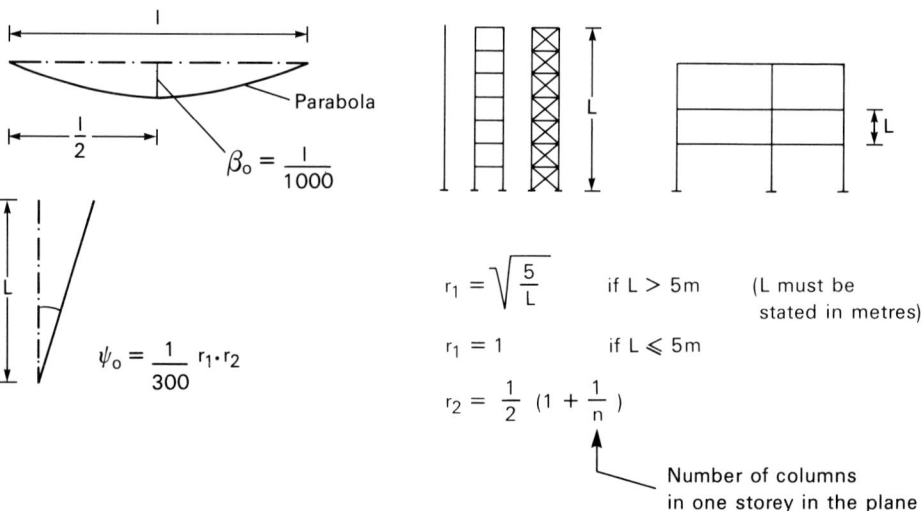

Fig. 5.   Geometric imperfections

The stability of framed structures

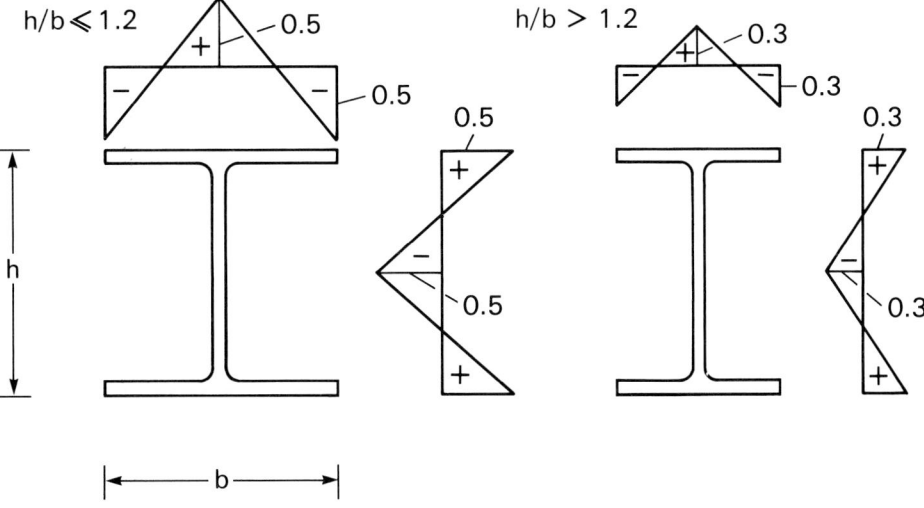

Fig. 6. Residual stress pattern $\bar{\sigma}_{res} = \sigma_{res}/235$ N/mm² for rolled I-sections

Range of application:

1. $\varepsilon = \ell\sqrt{N/EI} \leqslant 1.6$ in all columns, and
2. no plastic hinges between column end-points

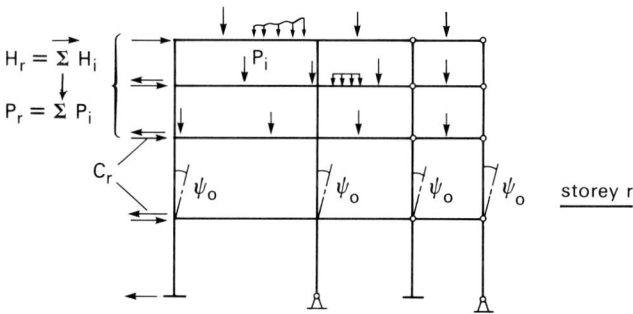

First-order analysis, but with the storey-shear:

$$Q_r = H_r + \psi_o P_r + 1.2\, \psi_r P_r$$

Fig. 7. Simplified second-order plastic hinge theory

1.2 Second-order plastic hinge theory

The only difference between the plastic zone theory and the second-order plastic hinge theory is that in the latter, strain-hardening is neglected and the spread of plastic zones is localised to plastic hinges. This method is much simpler than the plastic-zone theory, but almost always leads to the same results. However, a direct consideration of residual stresses is not possible. Therefore, the geometric imperfections must be enlarged to equivalent geometric imperfections, in order to include an approximation of the influence of residual stresses.

Usually (for $\varepsilon = \ell \sqrt{N/EI} \leq 1.6$), in sway frames, it is sufficient to use an initial out-of-plumb of $\psi_0 = 1/200$ for the columns, neglecting any initial crookedness, and applying the reduction factors r_1 and r_2 shown in Fig. 5.

Simplified second-order plastic hinge theory

In the formula given in the box in Fig. 7

$H_r =$ total sum of factored external horizontal working loads above storey r
$P_r =$ total sum of factored vertical working loads above storey r

$\psi_r =$ column slope of storey r, calculated iteratively by first-order plastic hinge theory (or estimated at the beginning and checked at the end of the calculation).

In order to avoid local instability of highly compressed slender columns the restrictions $\varepsilon \leq 1.6$ and no plastic hinges between column end-points are necessary.

In a correct second-order plastic hinge analysis one would have to use a storey shear without the factor 1.2 in the last term, in order to take care of the "P-Δ effect" (reducing the storey stiffness), and the well-known stability functions in order to take care of the "ε-effect" (reducing the member stiffness). The latter effect can be taken into account approximately by using the factor 1.2 in the elastic plastic part of the P-Δ term. This simplifies the calculation considerably.

If ψ has been estimated very well, the amount of calculation work does not differ from the application of a simple first-order plastic hinge analysis, but the deterioration of the load capacity of the structure, caused by the dominating P-Δ effect, has been properly taken into account and is physically correct.

1.3 First-order plastic hinge analysis

The next step of simplification is a first-order plastic hinge analysis but with a storey shear which takes into account only the drift force $\psi_0 P_r$, caused by the initial out-of-plumb.

The stability of framed structures

Range of application:

1. $\varepsilon = \ell\sqrt{N/EI} \leqslant 1.6$ in all columns,

2. no plastic hinges beween column end points, and

3. $$\left| \psi^I_r \cdot P_r \right| \leqslant \frac{1}{10} \cdot Q_r$$

Storey-shear:

$$Q_r = H_r + \psi_o P_r$$

Fig. 8. First-order plastic hinge theory for multi-storey sway frames

This method may be used, as shown in Fig. 8, if the conditions 1 and 2 of the simplified second-order plastic hinge theory and also criterion 3 in figure 8 are fulfilled. This criterion means approximately that the additional storey shear produced by the axial forces in the columns due to sway is smaller than 1/10 of the storey shear Q_r, and therefore may be considered as negligible.

Common simple single-storey building frames may be analysed using first-order plastic hinge theory neglecting any imperfections, if no plastic hinges develop between column endpoints, and the criterion 2 of Fig. 9 is fulfilled. This criterion follows the demand that the safety factor η against the Euler buckling load is at least 10 in a state just before the development of the last plastic hinge for the most unfavourable mechanism.

A more economic criterion has been established by a special task group (Anderson/Rubin) of ECCS-TC8 in the meantime and will be published soon as an addendum to (1).

2.1 Second-order elastic theory

This well-known method is already used widely in practice. Computer programs suitable for small desk computers are available. In simple cases, even calculations "by hand" are possible except the solution of the set of linear equations for the unknowns. Instead of an exact second-order analysis a simplified second-order elastic analysis using the P-Δ

Range of application:

1. no plastic hinges between column endpoints, and

2.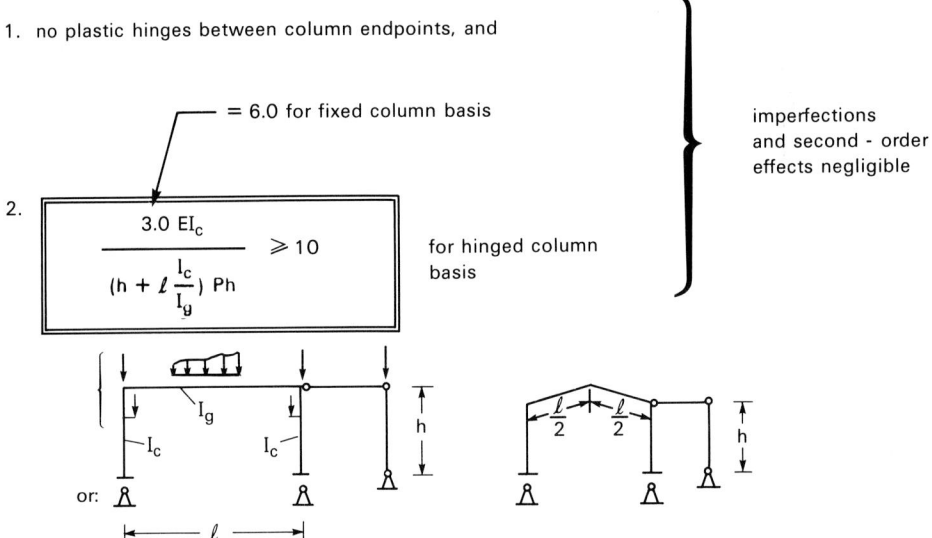

   $\dfrac{= 6.0 \text{ for fixed column basis}}{}$

   $\dfrac{3.0\, EI_c}{(h + \ell \dfrac{I_c}{I_g})\, Ph} \geqslant 10$ for hinged column basis

   } imperfections and second-order effects negligible

Fig. 9.   First-order plastic hinge theory for single-storey building frames

1	Range of application		$\varepsilon = \ell \sqrt{N/EI} \leqslant 1.6$
	Storey shear	:	$Q_r = H_r + \psi_o\, P_r + 1.2\, \psi_t\, P_r$
2	Range of application	:	$4 \leqslant \eta_r \leqslant 10$
	(and approximate affinity between second-order and first-order bending moments)		
	Storey shear	:	$Q_r = (H_r + \psi_o\, P_r) \cdot K_r$
	Magnification factor	:	$K_r = \dfrac{\eta_r}{\eta_r - 1}$
	(simplified formulae for the critical buckling factor η_r are given)		
3	if $\eta_r > 10$: first-order theory without imperfections		

Fig. 10.   Simplified second-order elastic analysis

The stability of framed structures 41

effect only, and shown under no.1 of Fig. 10, may be used. An elastic analysis is allowable with the same conditions as those required for the simplified second-order plastic hinge theory and with the same storey shear. In this case, only the dominating P-Δ effect is taken into account as second-order effect. One should be aware of the fact that a direct computation of forces and deformations without any iterative procedure is possible, if the displacement method is used (4). Only the values of the storey-stiffness (for each unbraced storey only 1 term!) show a direct available reduction against the values of the first-order theory. All other coefficients of the system matrix and also of the load-vector are the same as in a first-order analysis. As an alternative to this method another simplified second-order elastic analysis, using magnification factors - as shown in Fig. 10 under no.2 is applicable. Using these magnification factors a first-order analysis is performed, but it must be assured, that the distribution of the second-order bending moments shows approximately affinity to the distribution of the first-order bending moments.

Finally, as can be seen under no.3 of Fig. 10, the usual first-order elastic theory without any imperfections may be used if the critical buckling factor η_r of the system is at least 10.

Calibrating Frames

The author likes to draw the attention to a further publication on "Calibrating Frames" (5). As a result of discussion within TWG 8.2 ECCS, three different sway frames are defined in this paper, namely:

- a 6 storey 2 bay frame
- a rectangular portal frame
- a pitched roof frame

These frames represent certain common types of building frame and they may be used as a "calibrating system" in order to check the reliability of different computer programs, or some simplified methods discussed earlier for the ultimate limit state analysis of such frames. The results of the different methods published in the ECCS publication no.33 (1), and applied to these calibrating frames, are shown (e.g. in Fig. 11), as well as the use of certain criteria for the applicability of some approximate methods.

Finally, a comparison between test results and theoretical results based on the plastic-zone theory is given (see Fig. 12).

The three defined "calibrating frames" seem to be suitable for checking the reliability of different computer programs and simplified methods for the stability check or design approach, based on the ultimate load theory. It is also shown, that a more or less exact plastic-zone theory leads to an excellent agreement with test results. Therefore it can be concluded that the plastic-zone theory may be regarded as "calibrated" by these results.

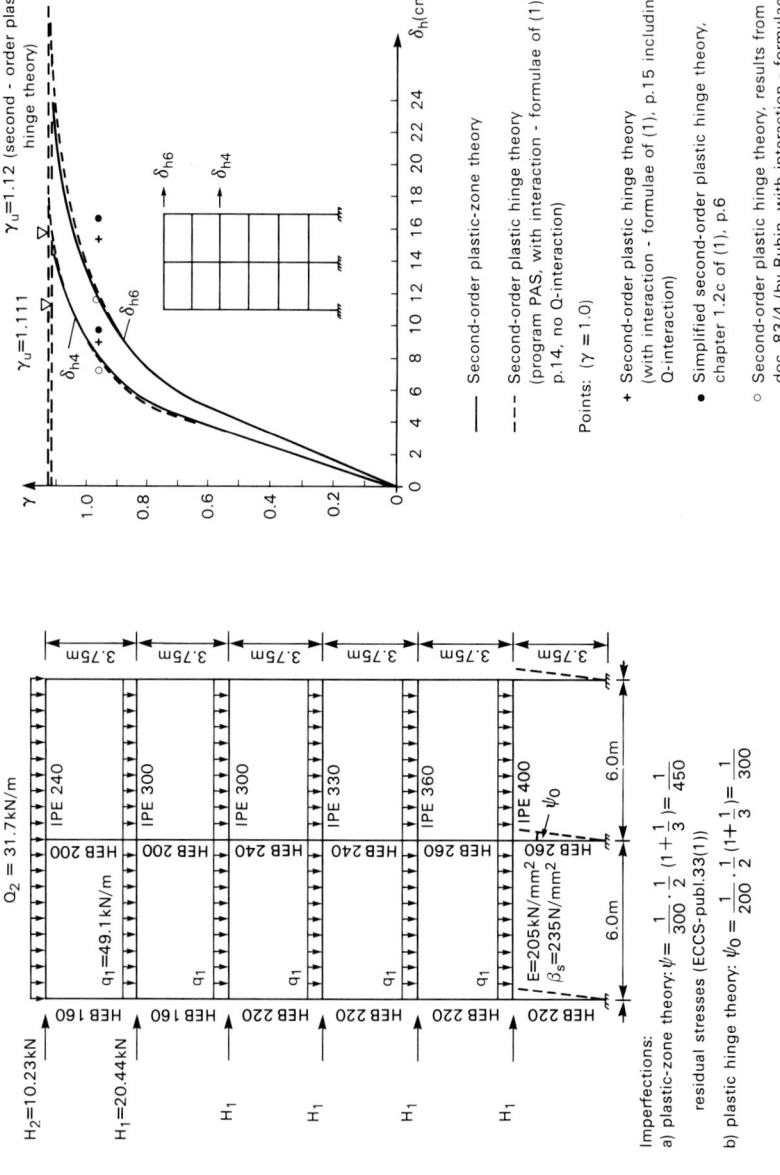

Fig. 11. "Calibrating Frame" 1, design load distribution ($\gamma = 1.0$) and load-deflection-diagram

The stability of framed structures

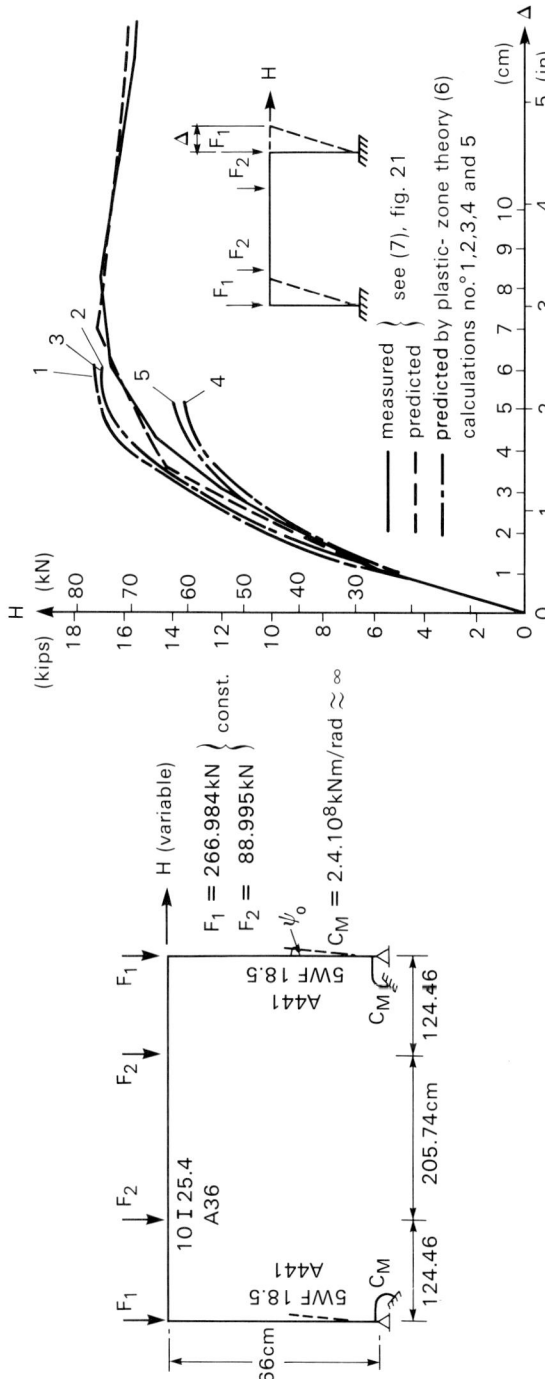

Fig. 12. Test frame, system, load distribution and load-deflection relationship

whereas the simplified methods described in (1) are "calibrated" by the plastic-zone theory.

Final Remarks

In this brief extract of the work of the ECCS on the analysis and design of unbraced frames, several simplified methods have been illustrated and the criteria are given for the range of their application. If one considers the fact, that very seldom are multistorey frames unbraced, because internal shear-walls, staircases or cores for elevators, etc usually perform this function, these elements will exist in many practical cases of low-rise construction. Therefore it is the opinion of the author, that these simplified methods will be used in practice, if they are taught in the engineering schools and if practical engineers become aware of the fact, that they are simple, safe and at least in the case of the simplified second-order plastic hinge method also economic when compared with other methods.

PRACTICAL APPLICATION OF THE SECOND-ORDER PLASTIC HINGE THEORY

General Remarks

In the following section practical hints are given for the application of a second-order plastic hinge analysis for the in-plane stability check of steel frames. More sophisticated methods for such an approach have been published, but the application in practice is very rare. The reason for this seems to be that many engineers hesitate to learn new theories, and like to stick to familiar methods. Therefore, the author proposes to use a method which follows exclusively the line of a well-known second-order elastic analysis. Hand calculations or preferably desk-computer programs and a display that allows a dialogue procedure, should be used for a step-by-step approach. This enables the engineer to use his experience and intuition to obtain quickly safe results for his design. It also gives him a much better insight into the real behaviour of a structure than the use of a completely automatic "black-box" program. A numerical example illustrates the procedure and shows how simple such a calculation can be.

Basic facts and different possibilities for a second-order plastic hinge analysis

Many engineers hesitate to apply the plastic hinge theory, because they believe that this would lead to the necessity of tracing the successive development of plastic hinges in numerous steps up to the ultimate limit load P_u, in order to show that $P_u \geq P_d$, with P_d being the design load. Indeed, this method would be very tedious, because:

- the static system changes with each newly developed plastic hinge.

- the "law of linear superposition" cannot be applied, since the normal

The stability of framed structures

  forces in the members also increase with increasing loads.

- the bending-moment capacity of plastic hinges vary with varying normal and shear forces.

For these reasons, the development of computer programs using the plastic hinge theory seems to be more difficult at least for skeletal structures than using the plastic zone theory.

Therefore the following general procedure is recommended:

1. The structure should be designed using the experience and intuition of the engineer.

2. It should be shown that the structure is in a stable state of equilibrium under the governing design loads.

This means that only one particular loading condition will be considered. The corresponding distribution of the normal forces in the members can be estimated very accurately by means of equilibrium conditions (see numerical example on page 50). With these constant member forces the "law of linear superposition" is fully valid in a second-order analysis, so that all the well-known linear static methods can be applied.

Different approaches following this general method are possible and have been published recently, eg the "M-θ method" by Oxfort (9) or Vogel (10) and a "Displacement method with introduction of plastic hinges" by Rubin (11). To apply these very sophisticated methods, however, special knowledge of the second-order plastic hinge analysis is necessary, which engineers in practice usually do not have. The advantages and disadvantages and other special aspects of these two methods are discussed in more detail in (8).

A third method is proposed in the following. For this method, the common knowledge of second-order elastic analysis is sufficient. If one wants to use the displacement method for hand calculations practical hints and design aids are given in (4), (12) and (13). Nowadays however, many computer programs are available in design offices or can be purchased very cheaply for small desk computers which are suitable for the second-order elastic analysis of skeletal structures. Even if these programs are used as "black boxes", their application in the method explained in the following can be recommended, if the user has basic knowledge of statics and a "sound engineering judgement".

The following special procedure can be used step-by-step in order to perform step 2 of the general procedure mentioned above (see also flow-chart on page 47):

Second-order elastic analysis

Analyse the given structure under the governing design load.

Check of the determinant D of the system matrix A.

If A is positive definite, the structure is in a stable state of equilibrium and one proceeds to the next step. If this condition is not fulfilled the structure fails in the elastic range and it has to be redesigned.

Check the yield condition.

If at each location $M_i \leq M_{pl.i}$, the structure does not fail under design load, the overall stability analysis is finished. If at some locations $M_i > M_{pl.i}$ continue to the next step.

Introduction of plastic hinges.

At all locations, where $M_i > M_{pl.i}$ (or close to $M_{pl.i}$) a plastic hinge is introduced by means of:

- implementation of a real (frictionless) hinge and

- application of an external couple of bending moments with the magnitude of the full plastic moment M_{pl} of the cross-section
  (if necessary reduced by the normal force N and/or the shear force Q).

Second-order elastic analysis of the new system (with one or more hinges) under design load and additional external couples of bending moments

Final checks:

- Check the determinant D and the system matrix A.
- Check the yield condition. If necessary, introduce further plastic hinges as explained above.
- Check the sign of the angle of rotation θ_i in the plastic hinges (see Fig. 13).

If all three checks are satisfied the structure is in a stable state of equilibrium under the design load. This design load P_d is then located on the ascending branch of the load-deformation curve. The calculation is finished since $P_d \leq P_u$ (see Fig. 14).

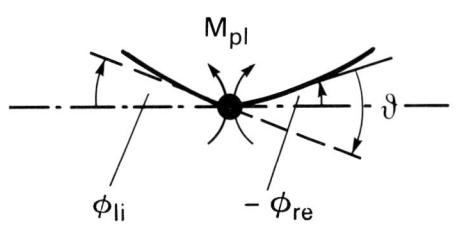 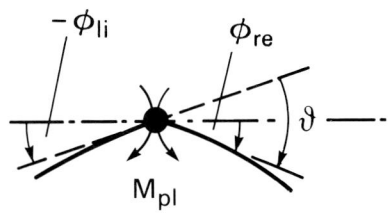

a) tension in lower fibre

b) tension in upper fibre

$$\vartheta = \phi_{li} - \phi_{re} \geqslant 0$$

$$\vartheta = \phi_{re} - \phi_{li} \geqslant 0$$

Fig. 13. Correct sign of rotation in plastic hinges

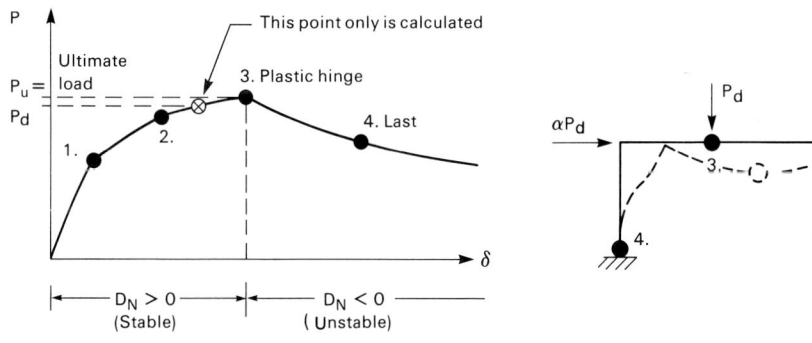

Fig. 14. Example for a load-deformation curve of a frame (P_u may correspond also to the 1, 2 or 4 plastic hinge)

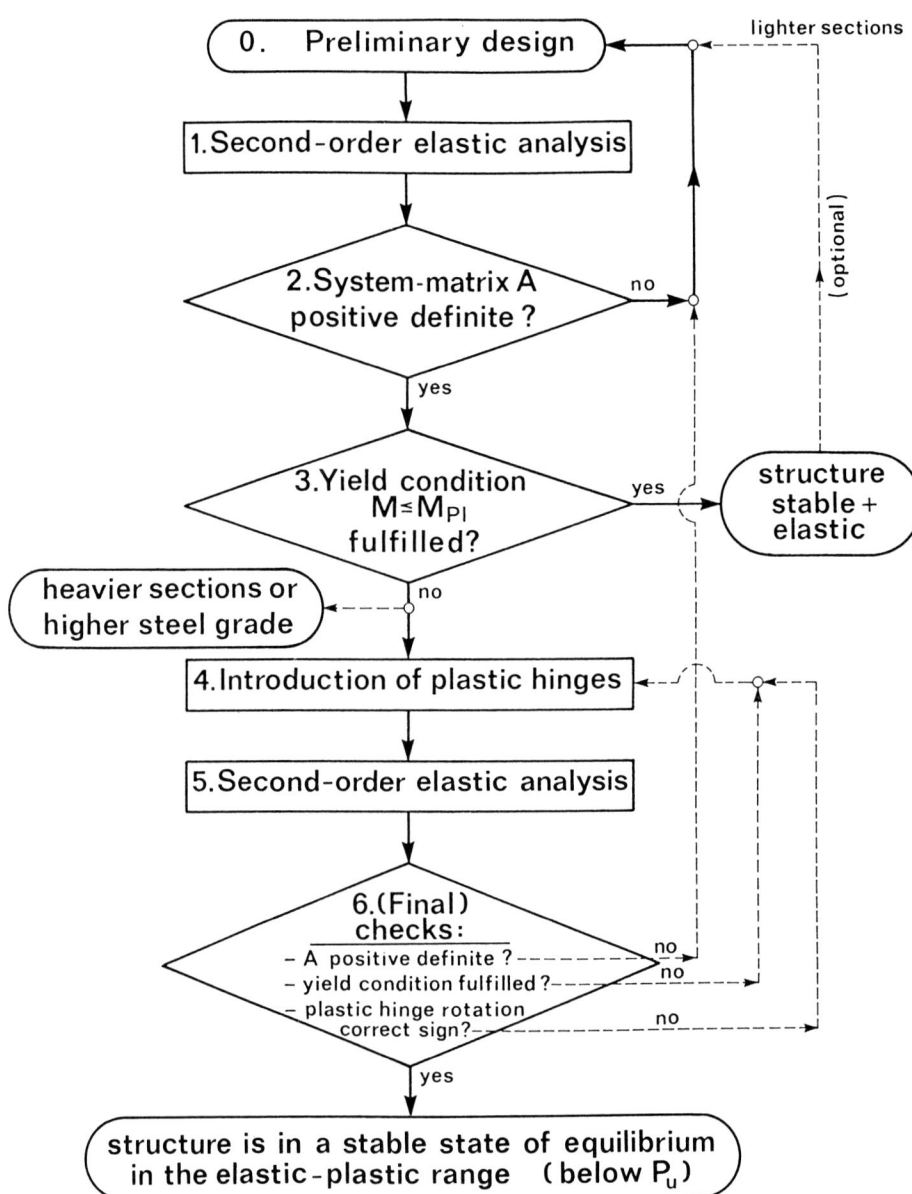

Flow diagram for design process

The stability of framed structures 49

This method is strongly recommended by the author, for the following reasons:

- The engineer may use familiar design tools, no new sophisticated special second-order plastic hinge theories have to be studied.

- If necessary the engineer can use his common engineering judgement to redesign the structure locally depending on the results of the relevant calculations.

- The engineer works in a dialogue procedure with the computers with the result of gaining more and more experience and insight into the real behaviour of the structure up to the failure load.

Numerical Example

General remarks:

The following example illustrates the method explained above. The complete calculation applying the second-order displacement method (4), (12) done "by hand" is documented in (14).

In the following, an interactive procedure with a computer, preferably with screen-display is simulated. Therefore only input data, results and checks are given and discussed. All calculations which are trivial or not visible if one uses a "black-box" program are excluded. For those readers who want to compare the results with their own possibly more accurate computer program, it is mentioned that some practical approximations have been used which do not affect the results significantly. These approximations are:

- longitudinal and shear strain of the members are neglected.

- the compression forces N in the columns have been estimated by equilibrium conditions at the very beginning and kept as constant values in all calculation steps in order to determine the stiffness-parameters $\varepsilon = \ell \sqrt{(N/EI)}$.

- the approximations on the extreme right hand side of the following equations have been used for the "stability functions" of the columns.

$$A' = \frac{\varepsilon(\sin \varepsilon - \varepsilon \cos \varepsilon)}{2(1-\cos \varepsilon) - \varepsilon \sin \varepsilon} = 4 - \frac{4}{30}\varepsilon^2$$

$$B' = \frac{\varepsilon(e - \sin \varepsilon)}{2(1-\cos \varepsilon) - \sin \varepsilon} = 2 + \frac{1}{30}\varepsilon^2$$

$$C' = \frac{\varepsilon^2 \sin \varepsilon)}{\sin \varepsilon - \varepsilon \cos \varepsilon} = 3 - \frac{6}{30}\varepsilon^2$$

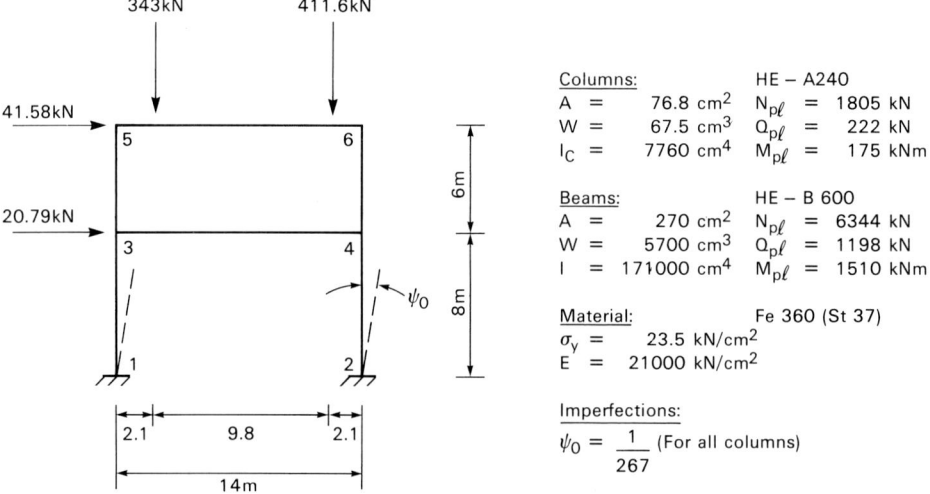

Fig. 15. System, design loads and imperfections

These approximations are good for A' and B' if $\varepsilon \leq 2.6$ and for C' if $\varepsilon \leq 1.7$. For the beams (with $\varepsilon \ll 1$), A' = 4, B' = 2 and C' = 3 (first-order stiffness) is accurate enough.

- For the full plastic moment capacity of the columns, the linearised interaction formulae

$$M_{pl,N} = 1.1 \, M_{pl} \left(1 - \frac{N}{N_{pl}}\right) \text{ was used.}$$

The stability of framed structures

Stability check of a 2 storey frame

INPUT DATA : (see Fig. 15)

To estimate the compression forces in the columns, the beams are treated as simply supported and hinges introduced at the mid-height of the columns. From equilibrium conditions one gets:

N_{35} = 344.39 kN $\qquad$ N_{13} = 317.65 kN
N_{46} = 410.21 kN $\qquad$ N_{24} = 436.95 kN

Σ = 754.60 kN $\qquad$ = 754.60 kN = 343 kN + 411.6 kN

CALCULATION:

1. Second-order elastic analysis of the given structure

The results for the bending moments are given in Fig. 16.

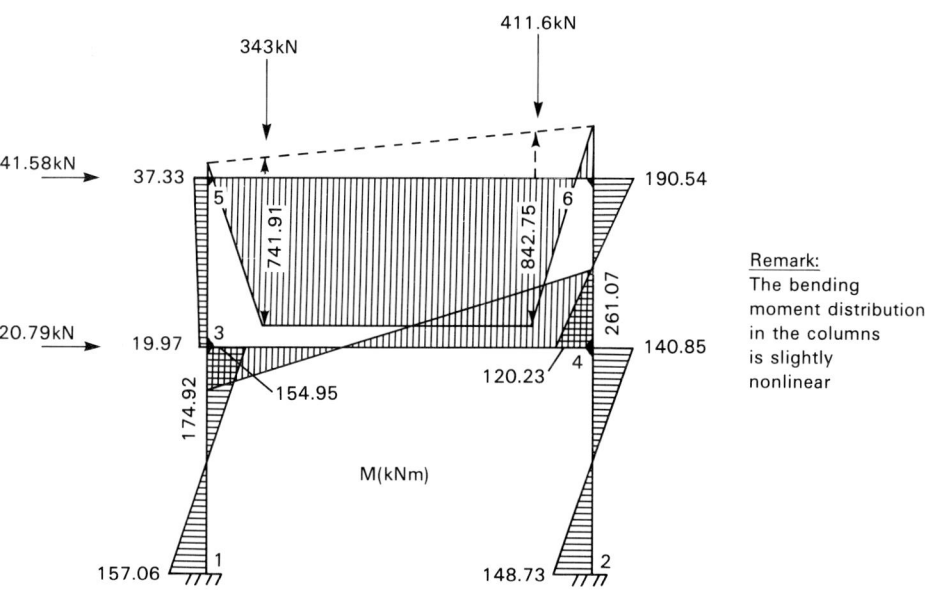

Remark:
The bending moment distribution in the columns is slightly nonlinear

Fig. 16. Bending moments due to the first calculation step

2. Check the determinant D

$D = +3.697 \cdot 10^9 > 0$ (and A positive definite), i.e. the elastic system is in a stable state of equilibrium.

3. Check the yield condition. In the columns at the nodes 1, 2 and 6:

$$M_{pl,N_{13}} = 1.1 \cdot 175 \cdot (1 - \frac{317.65}{1805}) = 158.62 \text{ kNm} > 157.06 \text{ kN}$$

$$M_{pl,N_{24}} = 1.1 \cdot 175 \cdot (1 - \frac{436.95}{1805}) = 145.90 \text{ kNm} > 148.73 \text{ kN}$$

$$M_{pl,N_{64}} = 1.1 \cdot 175 \cdot (1 - \frac{410.21}{1805}) = 148.75 \text{ kNm} > 190.54 \text{ kN}$$

It can be seen that the yield condition is not fulfilled at 2 and 6. This can be expected in the next step at 1 also, because of the loss of stiffness due to the development of plastic hinges.

The designer now has two choices:

- if he wants the structure to behave elastically under the design load he has to select heavier cross sections or a higher steel grade for the columns.

- check the stability in the elastic-plastic range.

The latter procedure is shown below:

4. Introduction of plastic hinges

The system given in Fig. 17 should be considered now (note that at point 1, a hinge is introduced also in order to avoid too many calculation steps). All other information about cross section properties, imperfections etc can be taken from Fig. 15.

5. Second-order elastic analysis of the new system.

Results are given in Fig. 18.

The stability of framed structures 53

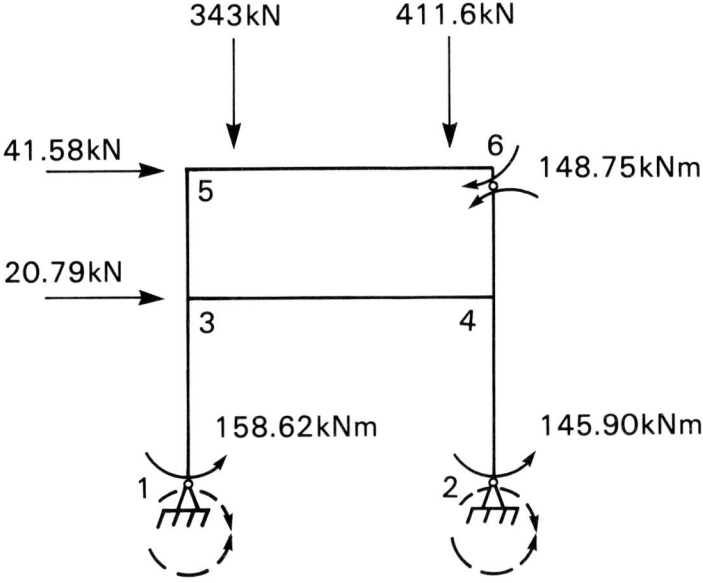

Fig. 17. New system with plastic hinges

6. Final checks

- $D = + 3.967.10^6 > 0$ (and A positive definite)(the new value of D is ~ 0.1% of the first value of D, this indicates that the structure is close to the ultimate limit load)

- from Fig. 18 it can be seen that the yield condition is satisfied in the complete structure (at 3 and 4 one can expect additional plastic hinges after a very slight increase of the design load; this also indicates that the structure is close to the ultimate limit load and therefore designed economically).

- a "user-friendly" computer program (or a hand calculation based on the results of Fig. 18) shows that the plastic hinge rotations, are:

$\overset{\frown}{\vartheta_1} = + 0.00035 > 0!$ (correct sign)

$\overset{\frown}{\vartheta_2} = + 0.00163 > 0!$ (correct sign)

$\overset{\triangleleft}{\vartheta_6} = \phi_{64} - \phi_{65} = - 0.00369 - (-0.01124)$
$= + 0.00755 > 0!$ (correct sign)

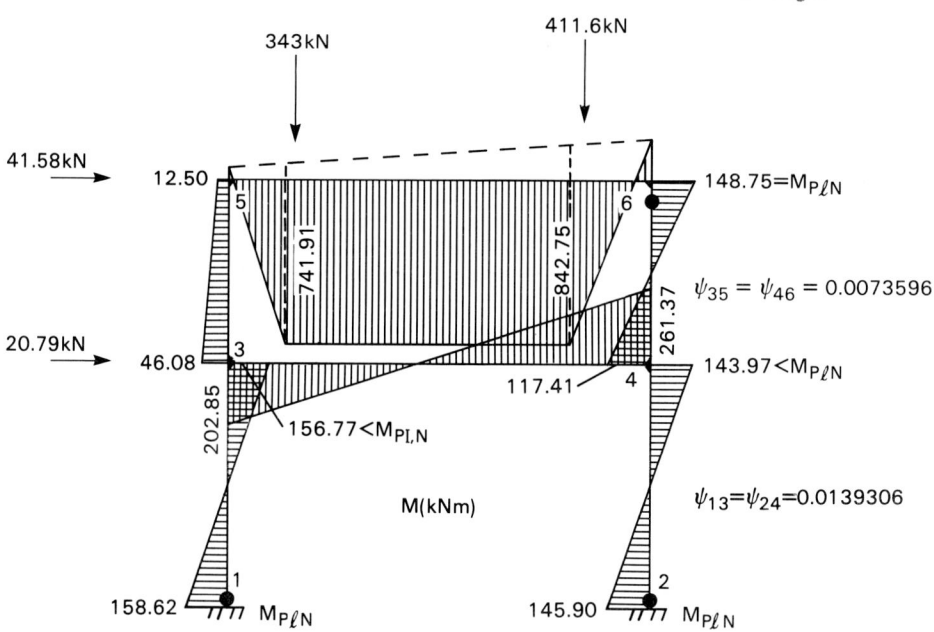

Fig. 18. Bending moments and column slopes of the new system with plastic hinges

All of the three necessary control checks are fulfilled. i.e. the structure is in a stable state of equilibrium in the elastic-plastic range under design load. It is mentioned that an additional elastic analysis under service load (eg $P_d/1.5$) shows that the structure is stressed below the elastic limit load. so that no danger of low cycle fatigue due to alternating plasticity under alternating wind directions exists.

Finally the compressive forces in the columns may be checked with the results of Fig. 18:

N[kN]:		Calculated	Estimated	Error [%]
N_{35} = 291.55 + 61.74 + (12.50-148.75)/14	= 343.56	344.39	-0.24
N_{46} = 51.45 + 349.86 - (12.50-148.75)/14	= 411.04	410.21	+0.20
N_{13} = 291.55 + 61.74 - (202.85+261.37)/14	= 320.15	317.65	+0.79
N_{24} = 51.45 + 349.86 + (202.85+261.37)/14	= 434.45	436.95	-0.57

It can be seen, that no N-iteration is necessary for the member stiffness parameters or for $M_{pl.N}$, so that the calculation using the linear super-position law is justified.

CONCLUSIONS

After a general discussion of frame stability, a practical way to check the in-plane overall stability of a framed structure on the basis of the second-order plastic hinge theory is shown using common methods (or computer programs) following the well known second order elastic theory. Usually only two or three different elastic systems have to be analysed if instead of the tedious (or expensive) calculation of the ultimate load P_u (in numerous calculation steps sometimes with severe convergence problems) the design load level P_d only is considered. Such an interactive dialogue-procedure with the computer enables the skilled engineer to use his experience, "sound engineering judgement" and intuition to accelerate the calculation or to redesign the structure in a suitable and economic manner. So dangerous errors which may result from the uncritical use of a complete "black-box" program based on an unfamiliar theoretical background can be avoided.

REFERENCES

(1) Vogel U. et al: "Ultimate Limit State Calculation of Sway Frames with Rigid Joints". ECCS-Pub.No. 33, First Edition, Rotterdam (1984).

(2) Vogel U: Some Comments on the ECCS Publication no.33 "Ultimate Limit State Calculation of Sway Frames with Rigid Joints". Construzioni Metalliche H.1 anno XXXVII (1985) pp 35-39.

(3) DIN 18 800 Teil 2 (draft), March 1988, Steel structures, stability, buckling of bars and skeletal structures, Beuth Verlag GmbH, Berlin

(4) Vogel U: "Practical Application of the Displacement Method for a Direct Second-Order Analysis of Regular Multistorey Sway Frames". In VERBA VOLANT - SCRIPTA MANENT en hommage á Charles Massonnet, Liége 1984, pp. 365-371

(5) Vogel U: "Calibrating Frames - Vergleichsrechnungen an verschieblichen Rahmen". STAHLBAU 54 (1985) S. 295-301

(6) Ackermann Th: "Traglastberechnung räumlicher Rahmen aus Stahl - oder Leichtmetallprofilen mit dünnwandigen offenen Querschnitten" (Ultimate limit state calculation of spatial frames of steel or aluminum profiles with thin-walled open cross-sections). Diss. University Karlsruhe (1981)

(7) Arnold P., Adams P.F. and Le-Wu: "Strength and Behaviour of an Inelastic Hybrid Frame".Journal of the Structural Divison, ASCE, ST 1, January 1968, pp. 243-266.

(8) Vogel U: "Praktische Hinweise zur Anwendung der Fließgelenktheorie II. Ordnung". In Der Metallbau im Konstruktiven Ingenieurbau (Festschrift Prof. Baehre) Karlsruhe Februar 1988, S. 359-376

(9) Oxfort J: "Anwendung des gemischten Kraft - und Weggrößenverfahrens (M- -Verfahren) der Theorie II. Ordnung zur vollständigen Berechnung beliebiger, biegesteifer Stahlstabwerke bis zur Traglast und plastischen Grenzlast". STAHLBAU 47 (1978), S. 139-145.

(10) Vogel U: "New German Rules on Plastic Design and Stability of Steel Structures". Lecture Notes (Guest Lectures, University of Liége 1978).

(11) Rubin H: "Das Drehwinkelverfahren zur Berechnung biegesteifer Stabwerke nach der Elastizitäts -oder Fließgelenktheorie I. und II. Ordnung unter Berücksichtigung von Vorverformungen". BAUINGENIEUR 55 (1980), S 81-92, Beispiele Hierzu S. 147-155

(12) Vogel U: "Nichtlineare Probleme der Baustatik I". Vorlesungsumdruck (Lecture Notes) WS 1988/89 (since WS 1975/76)

(13) Osterrieder P. and Ramm E: "Berechnung von ebenen Stabtragwerken nach der Fließgelenktheorie I. under II. Ordnung unter Verwendung des Weggrößenverfahrens mit Systemveränderung". STAHLBAU 50 (1981) S.97-104.

(14) Vogel U: "Limit State Design of Steel Structures". Lecture Notes (Guest Lectures, Tohoku University, Sendai, Japan 1987 pp. 68-74 and pp. 93-99).

3

FRAME ANALYSIS AND THE LINK BETWEEN CONNECTION BEHAVIOUR AND FRAME PERFORMANCE

D A Nethercot
University of Sheffield, UK

SYNOPSIS

The effects of connection behaviour on the structural performance of steel frames is described, with an emphasis on the role of the rotational stiffness and moment capacity of the connections. Historical aspects of the subject are traced and some of the major findings of the numerous studies of the past 10 years reviewed in more detail. It is suggested that, whilst research into both the analytical and experimental techniques necessary to study semi-rigid joint action and the production of results leading to an understanding of the subject have progressed well, translation of this information into the sort of design procedures likely to find favour with the industry still requires considerable attention.

INTRODUCTION

Steel frame structures comprise an assembly of connected members supporting the floors, cladding, partitions, etc.. The simplest approaches to their analysis and design concentrate on the behaviour of the steel skeleton - neglecting any structural interaction with the floors, etc., - with an emphasis on modelling the response of the individual frame members. Thus connections are assumed to behave in one of two idealised ways:

(i) as pins which allow free rotation between the ends of members and thus transmit no bending moments.

(ii) as if they were rigid and could maintain unchanged the initial angle between members, transferring whatever moment this requires.

Frame analysis therefore usually operates in terms of "pin-jointed" frames or "rigid-jointed" frames, using rather different techniques for the two classes, although the widespread acceptance of the matrix stiffness method and its implementation in computer programs has, of course, brought about some degree of unification. Even here though, it is normally necessary to make an initial selection of frame type when supplying the input data. For design purposes, variants of "simple construction", in which connections are assumed to be incapable of transmitting significant moments and the frame members are designed more or less in isolation, and "continuous construction", in which full interaction is assumed, are the norm in most countries.

The research literature, however, contains numerous attempts at a more complete treatment of the response of steel frame structures. For example, when computerised frame analysis was being developed, several attempts were made to extend frame analysis to include an interaction with walls, cores and cladding (1). Much work has been done on the behaviour of infilled frames (2) and the whole subject of stressed skin and diaphragm action provided by the metal cladding in low-rise structures has been explored (3) to the extent that its use is now quite widespread and a special part of BS 5950 is now in preparation. Similarly, attempts to incorporate something like the real behaviour of connections, especially beam-to-column connections, into frame analysis and design have appeared on many occasions during the past 80 years.

The earliest known tests on steel beam-column connections were those conducted by Wilson and Moore in 1917 (4). These used rivets as the fasteners but the findings are not at all unrepresentative of the behaviour of more modern forms of connection.

Two essential characteristics were identified from these and other early studies:

(i) the connection possessed a certain degree of rotational stiffness that tended to reduce as both load and deformation increased.

(ii) the connection was able to sustain a maximum moment greater than zero but less than the full moment capacity of the beam.

Such information is most conveniently displayed on a moment-rotation (M -ϕ) diagram in which the moment transmitted by the connection is plotted against the relative rotation of the connection.

These early investigators discussed the implications of their findings for frame analysis but, in the absence of both modern analytical techniques and computers, were unable to fully quantify the implications of semi-rigid joint action. Part of the work of the Steel Structures Research Committee in the UK in the 1930's (5) also highlighted through laboratory testing the true behaviour of connections but went a stage further with measurements on the performance of actual buildings under construction. These showed that significant moments were transmitted from the beams by the connections into the columns and that such moments bore little relationship to such notions as beam reactions acting at eccentricities equal to the centre of bearing of the beam-column connection. This clear recognition of semi-rigid joint action was followed up in the Committee's

Frame analysis 59

reports with suggested ways of incorporating it into design practice but the perceived complexity was such that traditional methods survived.

The fact that traditional assumptions of pin joints or rigid joints in steel frames do not really accord with the real behaviour of the connections has thus been known for more than 50 years and numerous research studies in the area may be located in the technical literature. At certain times there appears to have been bursts of activity e.g. the work of the Steel Structures Steel Committee in the 1930's and activity in a number of universities and research institutes in the past 10 years. In the case of the SSRC, measurements of real structures are perhaps the key element in recognising the phenomenon, whilst the later work appears to have been motivated by a combination of the transition to limit states design seeking to obtain better understanding of the performance of structures at all stages up to collapse and the availability and acceptability of computing as a means of conducting structural calculations. A particular reason for the author's involvement arose from work on the calibration of new codes of practice - BS 5950 and EC3 - where careful examination of the rules for the design of components by detailed comparison against test data contrasted with an inability to assess safety levels for complete frames due to the virtual absence of data on the true strength of such assemblies.

Whilst this paper is an attempt to provide a survey of the current link between connection behaviour and frame performance, it can only cover certain aspects of the subject. Progress in recent years has been considerable with more than a dozen centres - BHP Melbourne, BRE, TNO Delft laboratories and the Universities of Boulder (Colorado), Cachan, Innsbruck, Liege, Milan, Minnesota, Monash, Naples, Purdue, Queen's (Kingston), Sheffield, Warwick, etc. - being involved in quite substantial programs of study. In reviewing the subject the author has thus used illustrations and examples from those studies with which he is best acquainted; much worthwhile work has not for reasons of space been referred to explicitly.

HISTORICAL BACKGROUND - PRE 1970

Early work on linking joint behaviour to frame performance was largely limited to two topics:

(i) collection of M-ϕ data by means of laboratory testing.

(ii) inclusion of semi-rigid joint action in methods of frame analysis.

The first of these has, of course, been a continuous process with testing continuing up to the present day. Tests prior to 1940 dealt almost exclusively with riveted connections with some use of welding. By the 1950's bolts - either torque controlled or "black" -had replaced rivets. Only two-dimensional bare steel specimens under monotonically increasing load were considered. Several reviews and data collections extending through to the most recent sets of joint tests are available and serve as valuable sources for investigators

Moments in an end restrained beam

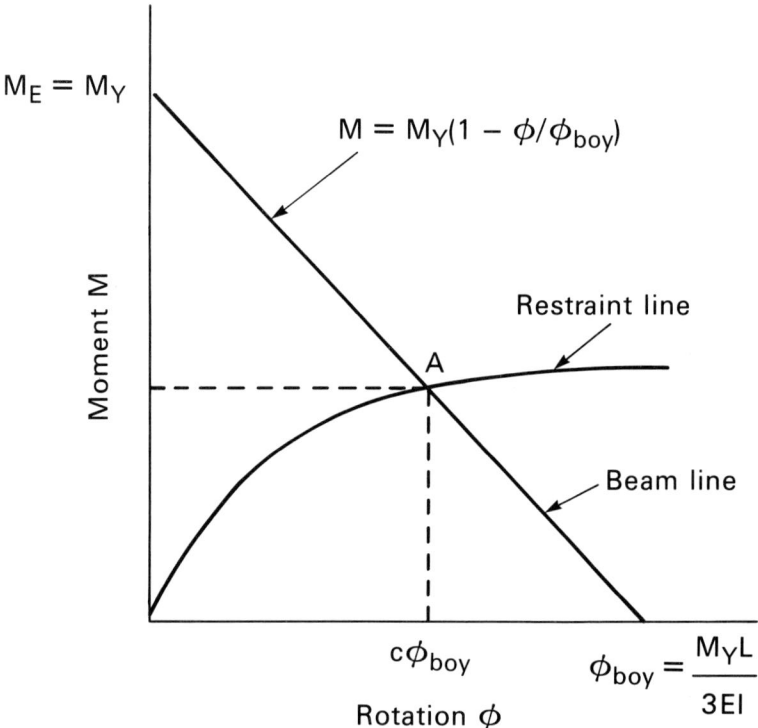

Equilibrium of beam and connection

Fig. 1. Use of beam line technique

wishing to study the effects of semi-rigid joint action.

Most of the conventional methods of indeterminate frame analysis e.g. slope-deflection, moment-distribution, etc. have been developed to include semi-rigid joint action (6). However, only linear elastic connection response - using the initial slope of the M-ϕ characteristic - was used. Once the matrix stiffness method had become established, it too was extended in this way, with bifurcation analysis for elastic critical loads being added. None of these studies could allow for the variation of connection stiffness with increased rotation and were thus likely to produce results erring on the unconservative side.

One significant early development that recognised the nonlinear nature of connection M-ϕ behaviour was the beam line method of Batho (7). This uses moment-area principles to relate beam end restraining moment to end rotation, superimposing the result on the connection M-ϕ curve as shown in Fig. 1. Their intersection corresponds to a matching combination of beam response and connection restraint.

More recent developments of the beam line technique (8) have identified two discrete parts for the beam response, corresponding to maximum span or end moment, and have used the construction as a means of quantifying rotation capacity requirements for connections.

CONNECTION BEHAVIOUR

Connection M-ϕ curves are a fundamental requirement for any study of the interaction of joint and frame behaviour. Many hundreds of M-ϕ tests, embracing more than a dozen different connection types (9), have now been completed; several compendia of results are available (10,11). Modern incremental ultimate strength frame analysis techniques (12) normally require that the M-ϕ curve be available in a convenient mathematical form and several schemes have been successfully used (13). Some progress has also been made with semi-analytical schemes in which a simplified behavioural model is coupled with some experimentally based correction factors (12). However, the generation of M-ϕ data by purely analytical means e.g. finite element method, whilst the subject of much study, cannot, at present, reliably account for all the subtleties present in connection response.

Fig. 2 presents typical M-ϕ curves for 3 different connection types, showing how stiffness and moment capacity vary with the type of detail used. Feedback from semi-rigid frame analysis (12) has identified two aspects of joint M-ϕ behaviour - not properly studied in most connection testing until very recently - as being particularly important:

(i)   accurate measurement of stiffness at low rotations

(ii) determination of unloading stiffness.

The first is required because even at ultimate load many joints in a frame will suffer only small rotations - unless failure is governed entirely by beam collapse through the use of

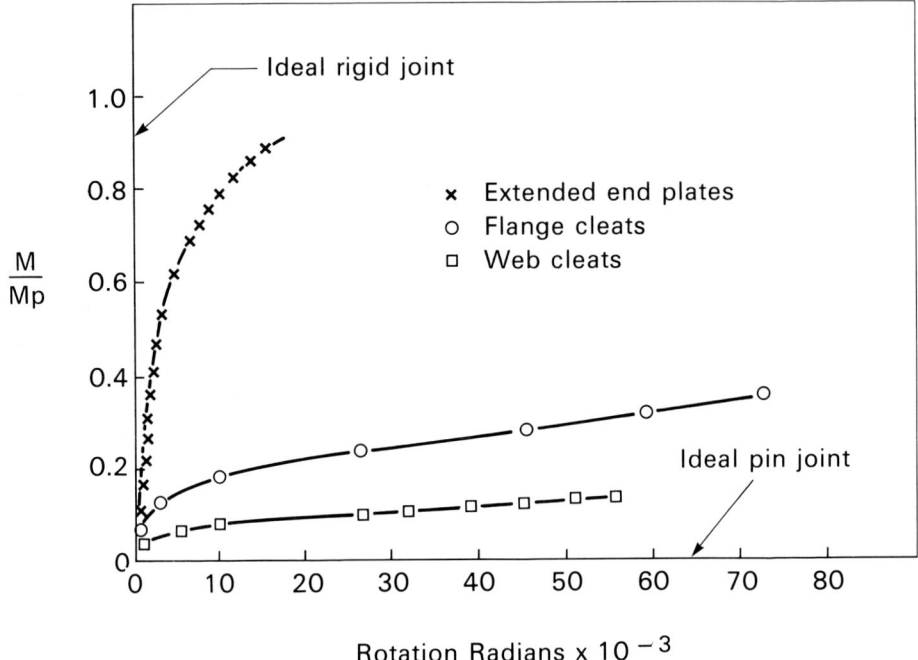

Fig. 2. Typical M-φ Curves

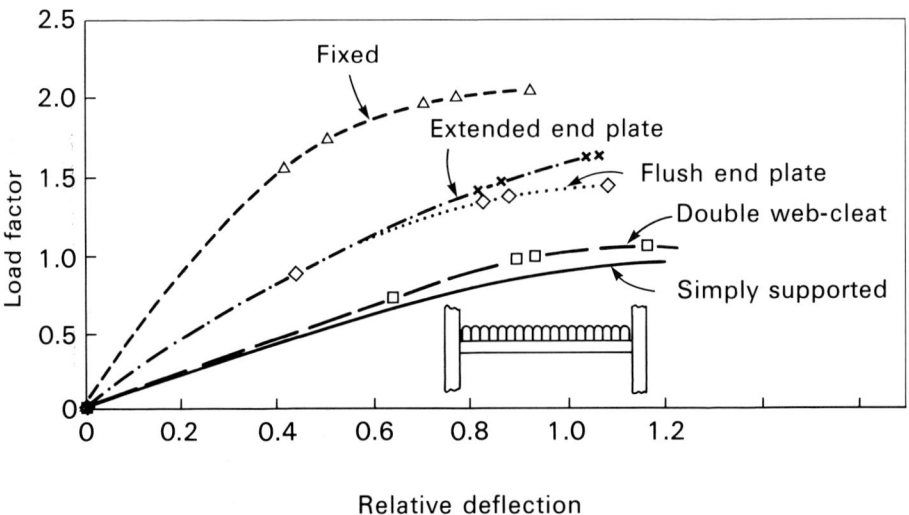

Fig. 3. Effect of end restraint on beam deformation (span/depth = 20)

Frame analysis 63

very stiff columns. Even under proportional loading reversal of connection rotation has been observed as the balance of beam-connection-column stiffness changes due to plasticity and instability effects, with the unloading connection at internal joints having the major influence as the ultimate condition is approached (14).

RESTRAINED MEMBERS AND SUBASSEMBLAGES

Beams

The in-plane response of an end restrained beam may readily be obtained by regarding the connections as nonlinear springs with a characteristic equal to their M-φ curve. Such an approach does, of course, neglect column flexibility and is thus only appropriate for beams in frames with comparatively stiff columns. Fig. 3 shows how the use of progressively more substantial connections reduces deflections leading to higher ultimate loads (15).

More recently (16) this approach has been extended to include lateral torsional buckling as a possible beam failure mode. Out of plane connection restraint i.e. lateral bending, twisting and warping, could not be included due to lack of the necessary M-φ data so the results essentially demonstrate the effect of more favourable in-plane moment patterns. Studies are, however, in progress (16) to generate the full set of connection data, although pilot tests have shown the three-dimensional problem to be considerably more complicated. For example decisions on the extent of the joint area, reference points for rotation measurement, etc. that could readily be resolved for the two dimensional case are open to several different interpretations (17). Fortunately progress with some of these issues is greatly facilitated by being able to assess the consequences of various assumptions from the results of restrained member analysis.

One simple finding has been that the torsional stiffness of even the most flexible of connections e.g. a double web cleat, is so much greater than the torsional stiffness of the beam that an assumption of end twisting being prevented will normally be quite reasonable.

Columns

The effect of connection restraint on the behaviour of axially loaded columns has been studied by several authors (18) and the main finding is as expected with progressively stiffer connections producing higher column curves as shown in Fig. 4. An effective length approach, including allowances for combined beam/connection restraint stiffness as a series system (19), may be used to represent the results in a simplified form. The influence of end moments, assuming these to remain directly proportional to the end load, may be included by using the normal type of interaction equations for beam-columns and basing the axial resistance term on the strength of the restrained column.

Of rather more practical interest is the behaviour of the type of subassemblage illustrated

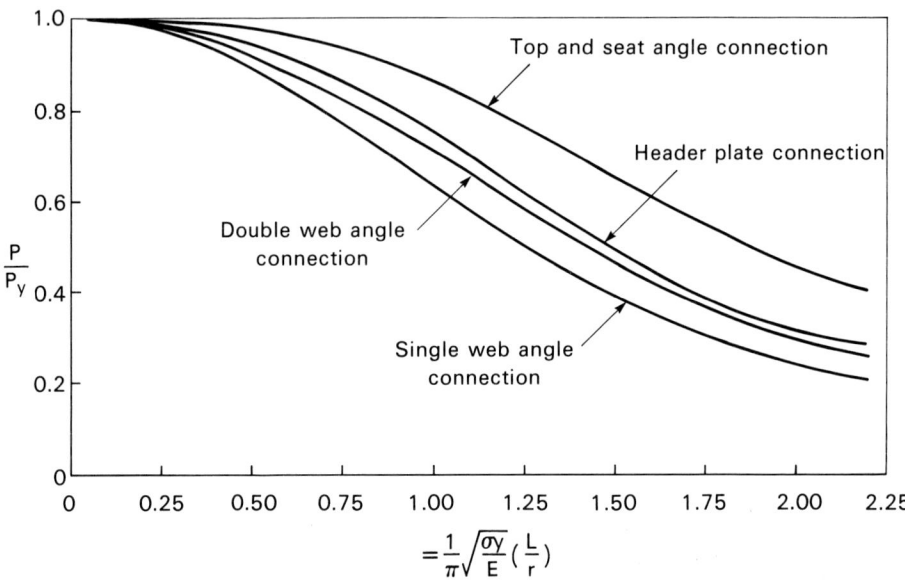

Fig. 4. Effect of connection type on axially loaded column strength

$$= \frac{1}{\pi}\sqrt{\frac{\sigma_y}{E}}\left(\frac{L}{r}\right)$$

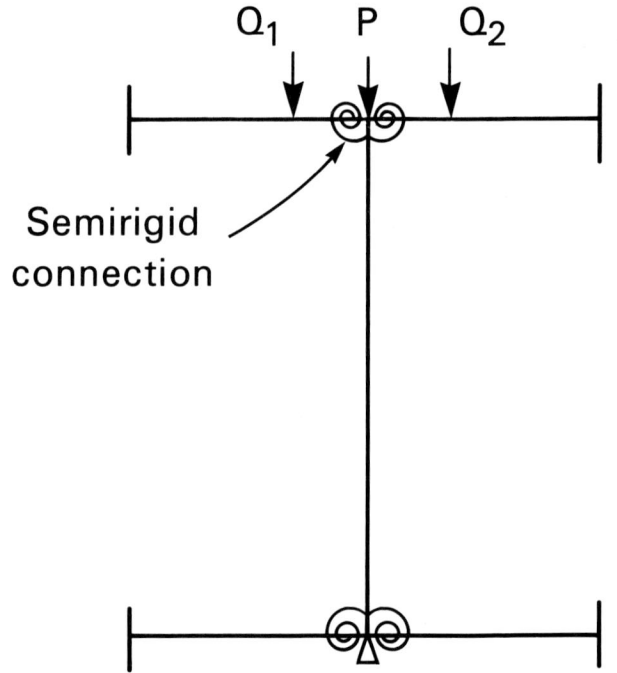

Fig. 5. Subassemblage for study of moment transfer

Frame analysis

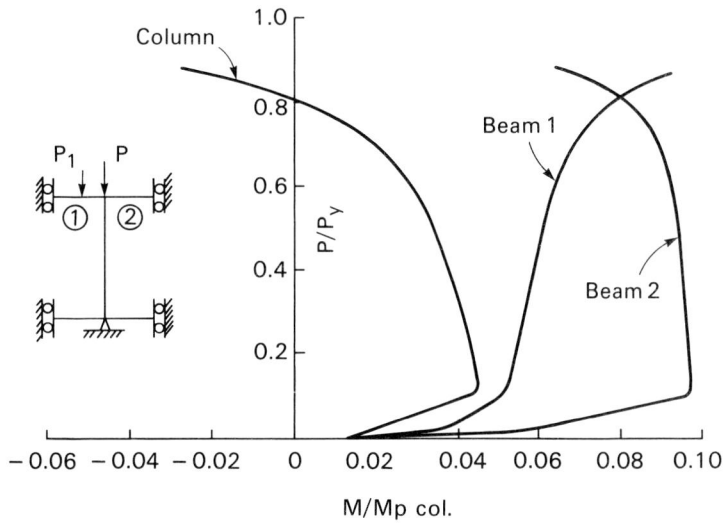

Fig. 6. Moment transfer for subassemblage of Fig. 5 (web cleat connections)

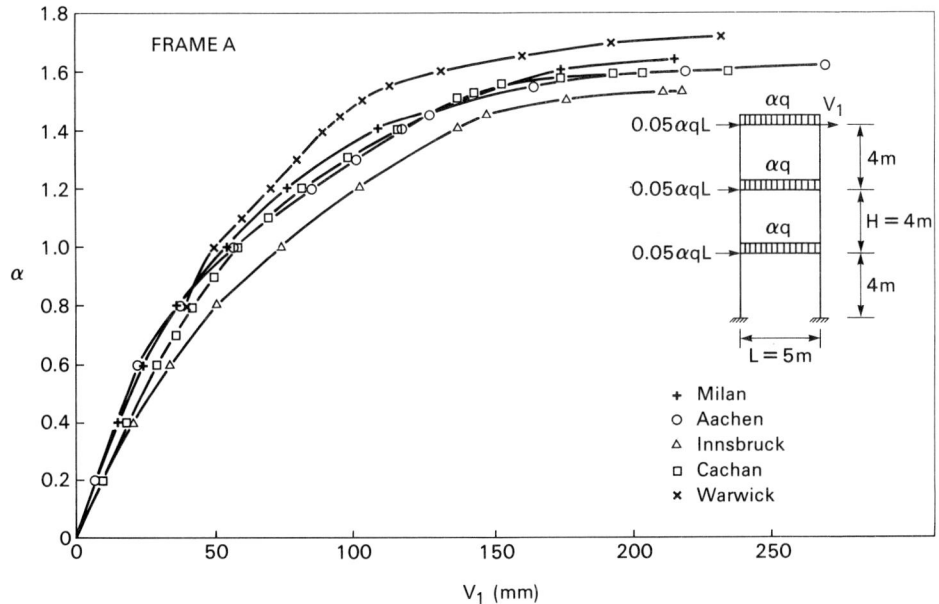

Fig. 7. Frame analysis from different programs

in Fig. 5. This enables the additional subject of the moment transfer from the beams through the connections into the columns to be studied (14,20,21). Fig. 6 shows the sort of result obtained with the moment at the column head reducing as failure is approached. Such behaviour does, of course, have important implications in terms of the correct set of member forces for which the column should be designed. This form of moment shedding, first observed in small scale tests on rigid jointed frames by Gent (22) and later seen in both series of BRE frame tests (23,24), was the key to Wood's intuitive "variable stiffness" column design approach (25). It was subsequently extended, again intuitively, to include semi-rigid action about the major axis (26). Sufficient numerical and experimental data are now available for both two and three dimensional response (21,27,28) to permit a proper calibration of the concept for semi-rigidly connected columns to be undertaken.

Frames

Whilst work on connections, members and subassemblages has provided much insight into the ways in which connection behaviour influences frame response, it is only by considering complete frames that the whole problem can be properly investigated. Numerical analysis techniques have now developed sufficiently that several research groups are able to conduct sophisticated ultimate strength analyses of flexibly connected frames allowing for such features as nonlinear loading/unloading connection characteristics, spread of plastic zones, residual stresses, initial deformations, etc. Fig. 7 presents a set of results for the same demonstration frame obtained from 5 such programs. Despite certain differences in approach all the results are quite similar.

Understandably few large-scale tests on semi-rigidly connected frames have been completed, with the recent BRE series being much the most substantial. Four three-storey by two-bay non-sway frames have so far been tested with one further test planned.

Fig. 8 shows one frame just prior to testing (29). Some 600 channels of data, covering member deflections, connection rotations and member strains (to permit the determination of member axial loads and moments) were recorded. The first two tests employed frames with heavy columns, fairly stiff endplate connections and used only beam loads, whilst the second pair used much lighter columns, cleated connections and included direct column loading to failure. Thus, the first pair essentially provided information on beam controlled failure whilst the second pair involved a more interactive column induced mode of collapse. Detailed comparisons of the predictions of the Milan program of Fig. 7 against the test results were made and Fig. 9 gives some idea of the level of agreement obtained.

Having validated the analyses both against one another and, more meaningfully, against realistic test data, it is now possible to utilise them in comprehensive behavioural

Frame analysis

Fig. 8. BRE test frame

studies. A recent Milan/Sheffield study (30) has looked at the frame of Fig. 10 in some detail for different load patterns and joint types.

Among the findings were :

(i) Subassemblage analysis can provide good estimates of force distribution but is less capable of accurately representing ultimate conditions.

(ii) Column continuity is surprisingly beneficial in enhancing column strengths; when acting in combination with quite flexible connections, sufficient restraint is available to critical column sections to produce failure loads almost equal to those obtained for rigid, full strength connections.

(iii) In some cases partial fixity gives a more balanced stress state in the columns, actually leading to a more favourable result than for an equivalent rigid jointed frame.

Parallel parametric studies on different aspects of semi-rigidly connected frames have been conducted in other centres, e.g. on sway frames by Ackroyd and his associates (31) and by Chen and his associates (32).

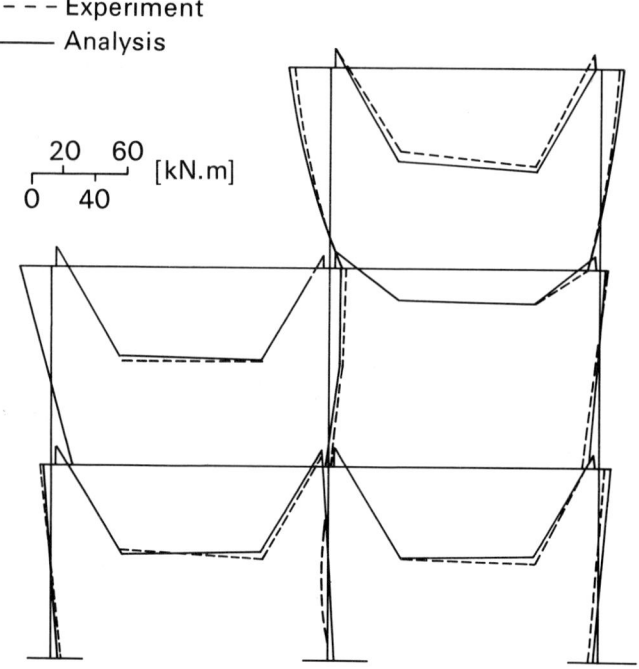

Fig. 9. Comparison of measured and predicted moments

Frame analysis

DESIGN

Early work on semi-rigid joint action was not taken up by the profession because of the real or perceived additional complexity of the resulting design procedures (33). Despite the volume of additional work conducted more recently, a danger clearly exists that this may be repeated. The conversion of the present set of research findings, augmented where necessary by supplementary work, is thus vital if the beneficial effects of utilising better representations of the true interactions between joint and frame behaviour are to be realised.

Several of the research studies referred to previously do contain suggestions, proposals and sometimes fully worked out but perhaps only partly validated design procedures. Groups such as the ECCS in Europe and the SSRC in North America have committees trying to synthesise the present body of knowledge into designer-oriented documents. However, the task, which is arguably more challenging than conducting the research in the first place, still needs considerable attention. Clearly the increasing use of computers for design work nowadays means that greater flexibility in terms of approach - especially in the level of calculation that can be managed - is available than was the case when the SSRC work was published. At one end of the level of design complexity it is certainly possible to include semirigid joint action in certain package programs (34). The challenge therefore lies at the simpler end where the need is to produce methods that clearly link the additional work with the benefits, allowing designers to judge for themselves those situations in which extensions to current methods are likely to prove cost-effective.

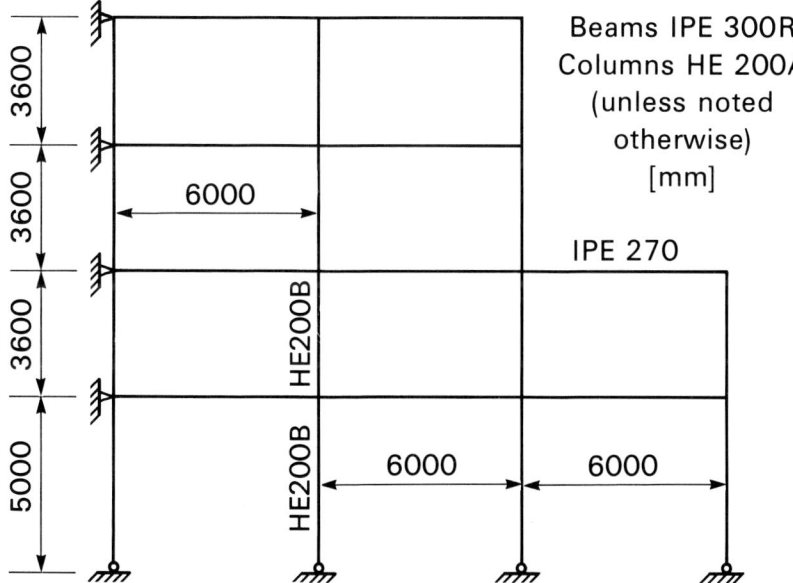

Fig. 10. Frame used for Milan/Sheffield study

THE FUTURE

Virtually all the foregoing review has been limited to the two-dimensional behaviour of bare steel frames. Whilst work in this area is continuing in several centres - some of it with very clearly defined objectives e.g. that at Warwick aiming to place the widely used "wind connection" method on a sounder scientific basis - three significant extensions are currently receiving considerable attention:

(i) three-dimensional behaviour

(ii) composite action

(iii) behaviour under dynamic loads.

The first of these has been briefly mentioned and is a logical removal of one of the limitations of the earlier studies.

Utilisation of composite action has the potential to provide very large increases in both stiffness and moment capacity as compared with bare steel connections with consequent additional benefits for frame performance.

Available test data on composite connections has been collated (35) and is now being supplemented by testing of arrangements representative of today's style of construction e.g. including the use of metal decking.

The aim is very much to understand the behaviour of typical details, to assess the benefits of limited modifications to these and to study the effects of the resulting connection behaviour on frame response. Work in the dynamic field is largely concerned with the response of semi-rigidly connected frames to seismic action(36) and thus requires test data for connection M-ϕ characteristics under reversed loading linked to frame analyses for cyclic loading.

CONCLUDING REMARKS

The development of links between an appreciation of the true behaviour of steel frame connections and their influence on the performance of frames has been reviewed. For the two-dimensional response of bare steel frames it has been suggested that the availability of techniques, results and understanding has progressed well; it remains, however, for this to be properly reflected in the availability of design procedures - clearly linking benefits to additional work - of the sort required by the industry. Current research seeks to remove some of the limitations of the earlier studies by applying the established methodology to topics such as composite action, three-dimensional behaviour and seismic response.

ACKNOWLEDGEMENTS

The subject matter for this review has drawn on the author's involvement both in research and in the work of several technical committees during the past 10 years. He gratefully acknowledges the contribution to the work at Sheffield of M Celikag, J B Davison, M A El-Khenfas, C Gibbons, S W Jones, P A Kirby, D Lam, D B Moore, C Poggi, A M Rifai, P A C Sims, J Surr, Y C Wang and R Zandonini, as well as those organisations who have assisted with funding.

REFERENCES

(1) Coull, A & Stafford Smith, B: "Structural analysis of tall concrete buildings", Proceedings Institution of Civil Engineers, Part 2, Vol 55, March 1973, pp 151-166.

(2) Wood, R. H: "Effective lengths of columns in multistorey buildings", The Structural Engineer, Vol 52, No. 7-9, July-September, 1976, pp 235-244, 295-302, 341-346.

(3) Davies, J. N & Bryan, E.R: "Manual of Stress Skin Design", Granada, 1982.

(4) Wilson, W. N & Moore, H. F: "Tests to determine the rigidity of riveted joints in steel structures", University of Illinois, Engineering Experiment Station, Bulletin No. 104, Urbana, USA, 1917.

(5) Steel Structures Research Committee, 1st, 2nd & 3rd Reports, Department of Scientific and Industrial Research, HMSO, London, 1931, 1934, 1936.

(6) Jones, S.W, Kirby, P.A & Nethercot, D.A: "The analysis of frames with semi-rigid connections - a state of the art report", Journal of Constructional Steel Research, Vol 3, No. 2, 1983, pp 2-13.

(7) Batho, C: "Investigation of Beam Column Connections", Steel Structures Research Committee, 2nd Report, London HMSO, 1934, pp 61-137.

(8) Kennedy, D.J.L: "Moment-rotation characteristics of shear connections", Engineering Journal, American Institute of Steel Construction, Vol 6, No.4, October 1969, pp 105-115.

(9) Nethercot, D. A: "Steel beam to column connections - a review of test data and its applicability to the evaluation of joint behaviour in the performance of steel frames", CIRIA Project Study 338, London, 1985.

(10) Goverdhan A. V: "A collection of experimental moment-rotation curves and

evaluation of predicting equations for semi-rigid connections", Doctoral Dissertation, Vanderbilt University, Nashville, Tenessee, 1985.

(11) Kishi, N & Chen, W. F: "Database on steel beam to column connections", CE-STR-86-26, School of Civil Engineering, Purdue University, 1986.

(12) Anderson, D, Bijlaard, F, Nethercot, D. A & Zandonini, R: "Analysis and design of steel frames with semi-rigid connections", IABSE Surveys, S-39/87, November 1987.

(13) Nethercot, D. A & Zandonini, R: "Methods of prediction of joint behaviour - beam to column connections", Stability and Strength of Connections, edited by R. Narayanan, Elsevier Applied Science Publishers, (in press).

(14) Nethercot, D. A, Kirby, P. A & Rifai, A. M: "Columns in partially restrained construction : analytical studies", Canadian Journal of Civil Engineering, Vol 14, 1987, pp 485-497.

(15) Nethercot, D. A, Davison, J. B & Kirby, P. A: "Connection flexibility and beam design in non-sway frames", Engineering Journal, American Institute of Steel Construction, 1988, (in press).

(16) Wang, Y. C, El-Khenfas, M & Nethercot, D. A: "Lateral torsional buckling of end-restrained beams", Journal of Constructional Steel Research, Vol 7, No.3, 1987.

(17) Celikag, M & Kirby, P. A: "Standardised method for measuring 3-dimensional response of semi-rigid joints", Connections in Steel Structures, edited by R. Bjorhovde, J. Brozzetti, and A. Colson, Elsevier Applied Science Publishers, 1987, pp 203-210.

(18) Nethercot, D. A & Chen, W. F: "Effects of connections on columns", Journal of Constructional Steel Research, 1988, (in press).

(19) Galambos, T. V: "Discussion of small end effects on strength of H-columns", Journal of Structural Division, ASCE, Vol 109, No. ST4, April 1982, pp 1067-1077.

(20) Rifai, A.M, Nethercot, D.A & Kirby, P.A: "Stability of column sub-assemblages with semi-rigid connections" Proceeding 2nd Regional Colloqium on Stability of Steel Structures, Tihani, Hungary, September 1986, pp 1/343-1/350.

(21) Davison, J.B, Kirby, P. A & Nethercot, D. A: "Column behaviour in PR Construction: experimental behaviour", Journal of Structural Engineering, ASCE, Vol 113, No. 9, September 1987, pp 2032-2050

Frame analysis

(22) Gent, A. R. & Milner, H. R: "The ultimate load capacity of elastically restrained H-columns under bi-axial bending", Proceedings, Institution of Civil Engineers, Vol 41, December 1968, pp 685-704.

(23) Wood, R. H, Needham, F. H & Smith, R. F: "Test of the multistorey rigid steel frame", The Structural Engineer, Vol 46, No.4, 1968, pp 107-119.

(24) Smith, R. F & Roberts, E. H: "Test of a fully continuous multistorey frame of high yield steel", The Structural Engineer, Vol 49, No. 10, October 1971, pp 451-466.

(25) Wood, R. H: "A new approach to column design", HMSO, 1974.

(26) Roberts, E. H: "Semi-rigid design using the variable stiffness method of column design", Joints in structural Steelwork, edited by Howlett, J. H, Jenkins, W. M & Stainsby, R., London, Pentech Press, 1981, pp 5.36-5.49.

(27) Rifai, A. M: "Behaviour of column sub-assemblages with semi-rigid connections", PhD Thesis, University of Sheffield, 1987.

(28) Wang, W. C & Nethercot, D. A: "Ultimate strength analysis of 3-dimensional column sub-assemblages with flexible connections", Journal of Constructional Steel Research, Vol 9, No.4, 1988, pp 235-264.

(29) Kirby, P. A, Davison, J. B & Nethercot, D. A: "Large scale tests on column sub-assemblages and frames", Connections in Steel Structures, edited by R. Bjorhovde, J. Brozzetti, A. Colson, Elsevier Applied Science Publishers, 1987, pp 291-299.

(30) Davison, J. B, Zandonini, R, Nethercot, D. A, Poggi, C: "Analytical and experimental studies of semi-rigidly connected brace steel frames", SSRC Annual Meeting & Technical Session, Minneapolis, April 1988.

(31) Ackroyd, M. H: "Non linear elastic analysis of flexibly connected frames", Connections in Steel Structures, edited by R. Bjorhovde, J. Brozzetti, A. Coulson, Elsevier Applied Science Publishers, 1987, pp 231-237.

(32) Lui, E. N & Chen, W. F: "Steel frame analysis with semi-rigid connections", School of Civil Engineering, Purdue University, CESTR-88-5.

(33) PD 3343 Supplement No.1 to BS 449 : Pt. 1, The use of structural steel in building, London, British Standards Institution, 1971.

(34) Edinger, J. A: "Non linear frame analysis using the AMSYS general purpose computer

programme", AISC research report, No. 83-1.

(35) Zandonini, R: "Semi-rigid composite joints, stability and strength of Connections", edited by R. Narayanan, Elsevier Applied Science Publishers (in press).

(36) Mazzolani, F. M: "Mathematical model for semi-rigid joints under cyclic loads", Connections in Steel Structures, edited by R. Bjorhovde, J. Brozzetti, & A. Colson, Elsevier Applied Science Publishers, 1987, pp 112-120.

4

THE DESIGN OF END-PLATE CONNECTIONS

D B Moore
Building Research Establishment, UK

SYNOPSIS

The historical development of end-plate design is traced and some of the major findings in the past 80 years are discussed in detail. The method of designing end-plate connections in the 1988 draft of EC3 is presented in full and the work on which it is based is identified and discussed. This method is compared with published experimental data on flush end-plate connections with three bolt rows and the accuracy with which it predicts a connection's moment-rotation curve is studied. Its application in the UK and the differences between current UK construction practice and that in the rest of Europe, on which the method is based, are also discussed.

1. INTRODUCTION

The integrity of a structure is dependent on the adequacy of both the members and the connections. Unfortunately, this is not generally reflected in the distribution of design effort and all too often, the design of structural connections is neglected. Current UK Codes of Practice (1,2) do little to redress this undesirable situation, as many of them give little guidance on connection design. With such a dearth of information on this subject, one might expect experienced engineers to, at the very least, check that the details chosen comply with the assumptions used to design the members. However, in many cases, the engineer is not involved in this part of the design/detailing process which is frequently left for the fabricator to complete.

Some of the types of beam-to-column connections used in multi-storey steel frame construction are shown in Figure 1. Choice of connection type is usually based on simplicity, duplication and ease of erection, all for economic reasons. Welded joints provide full moment continuity but are expensive due to the on-site welding involved for beam to column connections. In recent years bolted connections have increased in popularity. They have the advantages of requiring less supervision than welded joints and having a shorter assembly time, supporting load as soon as the bolts are tightened. They

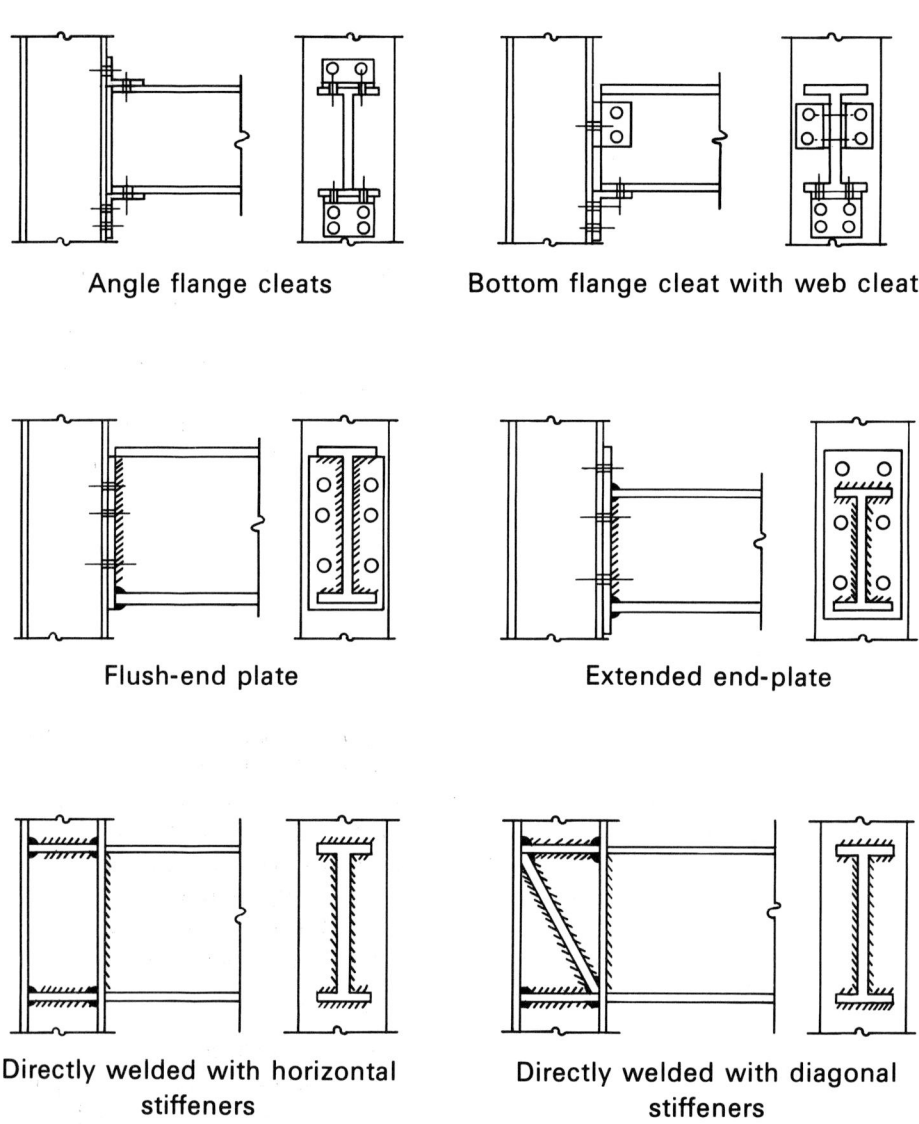

Fig. 1. Connection types

The design of end-plate connections 77

also have a geometry which is easy to comprehend and can accomodate minor discrepancies in the dimensions of beams and columns. However, when large forces are involved, bolted connections can be criticised for requiring extensive space which may conflict with the architechtural need for a "smooth line".

Of all the bolted connections shown in Figure 1, a recent survey (3) by the Building Research Establishment showed that flush and extended end-plates were the most popular. The behaviour of both these connection types has been the subject of many experimental and analytical studies and several methods for the design of the individual components (bolts, end-plate, column web and flanges) that constitute an end-plate connection are available. However, these methods have not been consolidated into a comprehensive design method for end-plate connections as such a development requires the determination of the moment-rotation characteristics of the connections.

This report reviews the major developments in connection design over the past 80 years with particular emphasis on flush and extended end-plates. A procedure developed in the Netherlands and adopted in the 1988 draft of Eurocode 3 - The Design of Steel Structures is presented in detail and compared with available experimental results for flush end-plate connections. Its accuracy at predicting the moment-rotation characteristics of a connection is examined and the differences between UK construction practice and that in the rest of Europe, on which the method is based, are also discussed.

2. HISTORICAL REVIEW OF END-PLATE DESIGN

Research into the behaviour of connections and their moment-rotation characteristics was first carried out in the early 1900's by Wilson and Moore (4) who conducted an experimental investigation to determine the rigidity of riveted joints in steel structures. It was not until the 1930's that the need to understand the behaviour of connections became more apparent and research in this area gathered momentum. With a view to providing a data base of experimental tests for semi-rigid design, separate investigations were conducted in the UK (5,6,7), the United States (8) and Canada (9). As a result of the work in the UK, PD 3343 (10) was published and later followed by PD 3857 (11). This later publication gives empirically derived recommendations for the design of four different classifications of top and bottom cleat connection and presents a number of tables and graphs to determine each connection's restraining moment. Figure 2 is a copy of its diagram No. 1 and shows the classification of connections. PD 3857 also includes a procedure for the design of stanchions. Although the recommendations given are very comprehensive, they are limited to the connections identified in Figure 2 and apply only to "the steel frameworks of building structures, formed with horizontal beams and vertical stanchions" and no-sway frames.

Probably the first attempt at producing a set of design procedures for beam-to-column connections was made by the British Constructional Steelwork Association in the mid 1950's and early 1960's in their series of "Black books" (12,13,14,15). These procedures

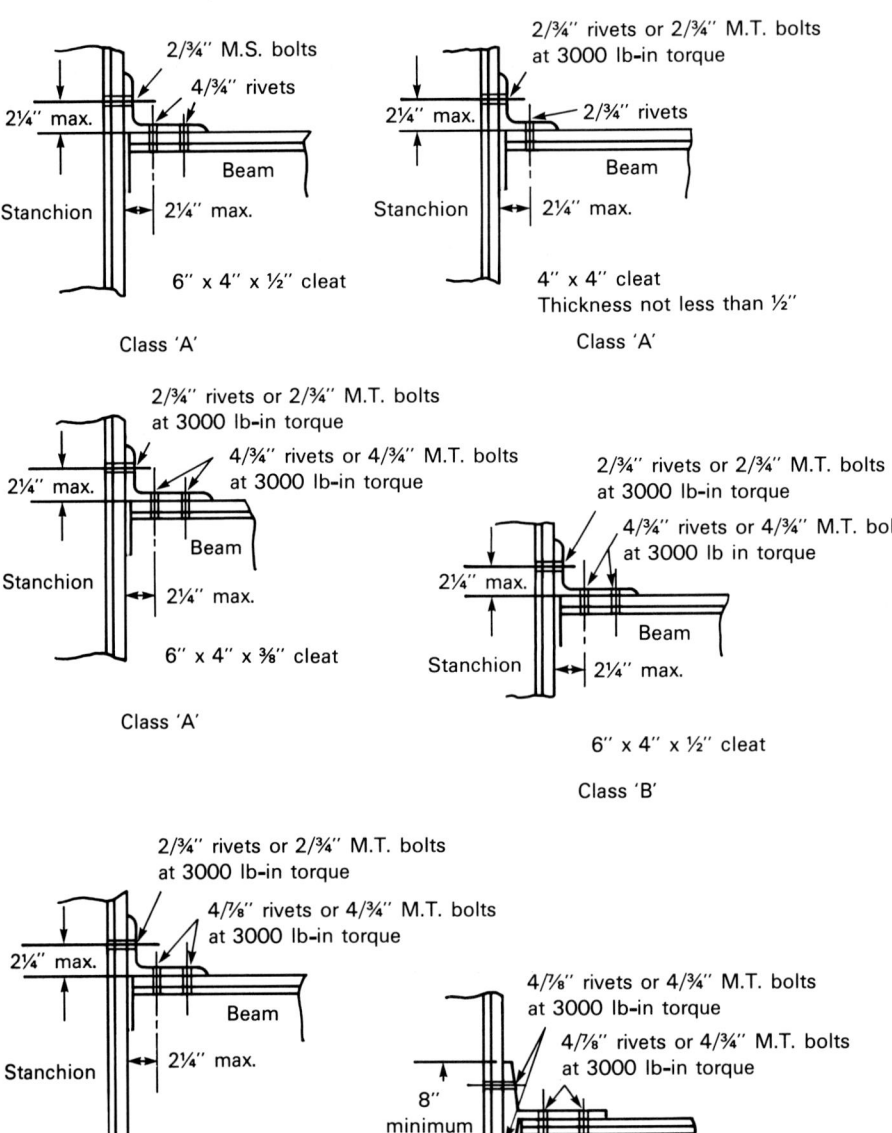

Fig. 2. Classification of connections in PD 3857 (11)

The design of end-plate connections   79

concentrate on the design of either the weld or the bolts and pay little attention to the bending of either the end-plate or the column flange. Indeed, the majority of the examples given for bolted connection assume the connection to be pinned and carry shear forces only.

It was not until 1959 that Schutz (16) used Douty's (17) work to propose a model for the analysis and design of end-plates in bending. This model is shown in Figure 3 and is based on a double cantilever. Plastic hinges are assumed to develop in the flange at the edge of the web and along the centreline of the bolts and prying forces are allowed to develop. With this model, Schutz predicted that the tension in each bolt is greater than that computed by dividing the total external tensile load equally among all the bolts in the joint, illustrating the influence of prying action. Schutz also suggested that the bolt force versus applied load curve can be divided into three separate regions, as shown in Figure 4. The first, represents the elastic response of the connection and in this region the bolt force increases linearly with the applied load. The second is the transition between elastic behaviour and the formation of a full plastic mechanism. In this region, Schutz speculated that the moment at the bolt holes reaches plasticity first, followed by the formation of a plastic hinge at the web. The third region of the curve represents the increase in bolt force due to strain hardening of the material in the plastic zones. This model is in good agreement with the bolt forces measured in T-stub connections and Schutz successfully used this model to design flush and extended end-plate connections.

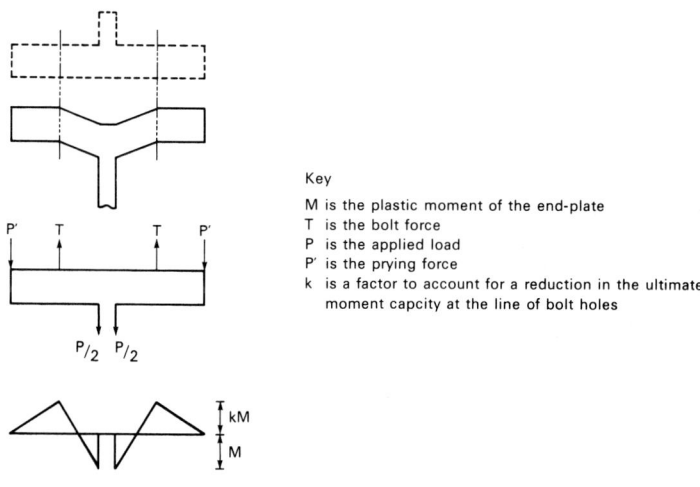

Key
M is the plastic moment of the end-plate
T is the bolt force
P is the applied load
P' is the prying force
k is a factor to account for a reduction in the ultimate moment capcity at the line of bolt holes

Fig. 3.   Failure model of an end-plate (after Schutz (16))

Using the concept of a double cantilever, Douty and McGuire (18), Nair et al (19), Sherbourne (20) and Packer and Morris (21) developed numerous models for predicting the behaviour of T-stub connections and their associated prying forces. The various formulas quoted for prying action differed widely for particular values of end-plate dimensions. Surtees and Mann (22), however, rejected these precise methods for determining prying action and chose to limit prying action by limiting the geometry of the connection.

In 1981 Horne and Morris (23) produced a design procedure for end-plate connections in which they assume that only the extremities of the connection behave plastically and that the rest of the connection remains elastic. As a result of this assumption, the bolt force distribution is limited to that shown in Figure 5a. This limitation was imposed because they believed that Grade 8.8 and H.S.F.G. bolts have insufficient ductility to allow plastic behaviour in the second and subsequent bolt rows i.e. it prohibits the bolt force distribution shown in Figure 5b. Horne and Morris also adopted Surtees and Mann's method for prying action which produced a design method easy to comprehend and without the complicated expressions for calculating prying forces. With the introduction of the BCSA publication 'Manual on Connections' (24), this method has become the UK standard for the design of end-plate connections. However, it gives little guidance on the connection's moment-rotation curve other than to suggest that adequate rotation capacity for plastic design can be achieved by either:-

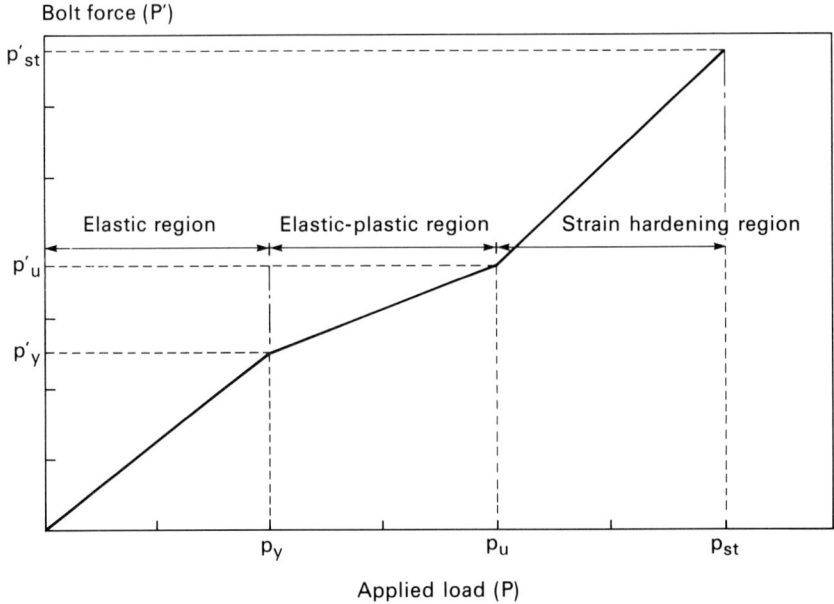

Fig. 4. Development of prying action

The design of end-plate connections

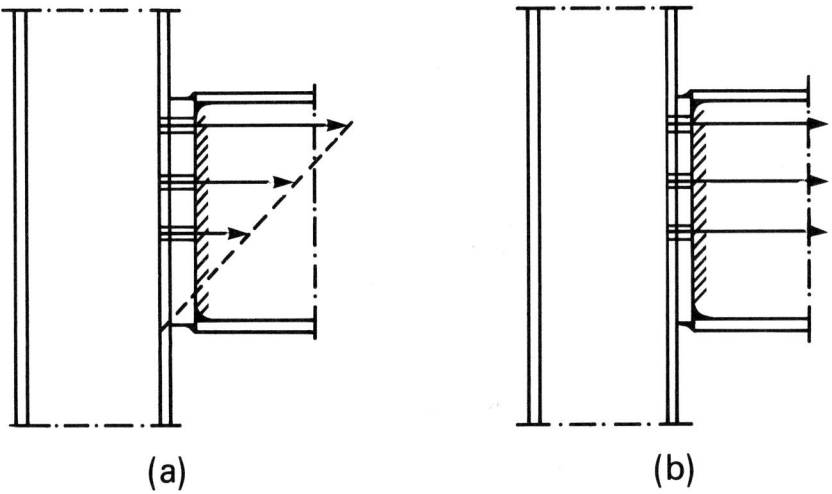

Fig. 5. Bolt force distribution models

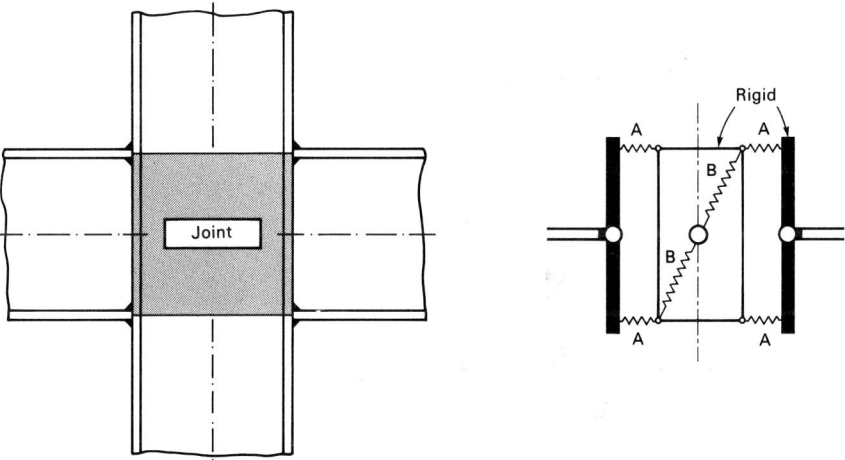

Fig. 6. Model of a fully welded connection (from Tschemmernegg (26))

a. allowing plastic deformations to occur within the connection
   OR
b. making the connection sufficiently strong that the plastic hinge forms in the beam adjacent to the connection.

Semi-rigid behaviour is an aspect of design which is becoming increasingly important and many national and international Codes and Standards permit semi-rigid joint action to be accounted for in design. A prerequisite to performing any frame analysis that seeks to include semi-rigid joint behaviour is the ability to predict a connection's moment-rotation curve. Numerous researchers (25) have studied this thorny problem and methods of predicting a connection's moment-rotation curve can be classified under the following headings:-

1. Empirical equations based on curve fitting of test data.

2. Methods which rely on a knowledge of the load-deformation curve of the key components of the connection to build up the moment-rotation curve.

3. Comprehensive finite element analysis.

Methods which adopt empirical equations can only be used in the range of connections for which the equations have been calibrated and a finite element analysis requires the ability to model bolt action and contact between plates at a level not yet attainable. Thus, at present, the ability to predict the moment-rotation curve with good accuracy is limited.

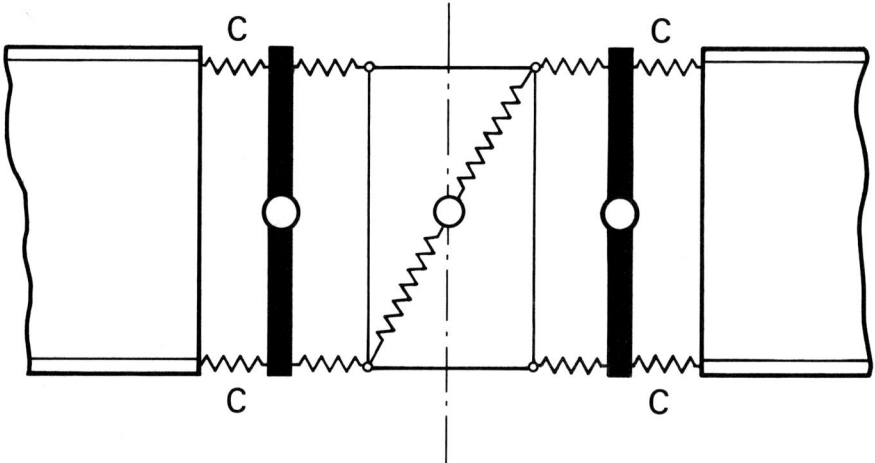

Fig. 7. Model of a bolted connection (from Tschemmernegg (26))

The design of end-plate connections

Recently a semi-analytical method of modelling a connection's moment-rotation curve has been proposed by Tschemmernegg (26). This method represents welded beam-to-column connections with a mechanical model which is composed of two sets of non-linear springs (illustrated in Figure 6). The first set of springs (A) account for the load introduction effect from the beam to the column, while the second set of springs (B) simulate the shear flexibility of the column web panel. This model has also been extended to bolted connections by the addition of a set of springs (C) to account for the deformations present in bolted connections. The model for bolted connections is shown in Figure 7. Both models have been calibrated against a large number of tests and the moment-rotation curves for fully welded and extended end-plate connections determined for all possible combinations of beam and columns made of European rolled I and H sections. Furthermore, a set of design tables (26) have also been prepared which give the design moment for the connection and the rotation associated with this moment. These tables are easy to use and for a fully welded connection all the designer needs to know is the size of the beam and the size of the column.

The draft European Code of Practice for The Design of Steel Structures (EC3,(27)) includes a design method for end-plate connections. The proposed method was developed in the Netherlands over a number of years and is based on a large experimental programme (28,29,30,31,32,33,34) on beam-to-column connections. This experimental work placed particular emphasis on the behaviour of unstiffened beam-to-column connections. These connections have the moment-rotation characteristic shown in Figure 8 and according to the

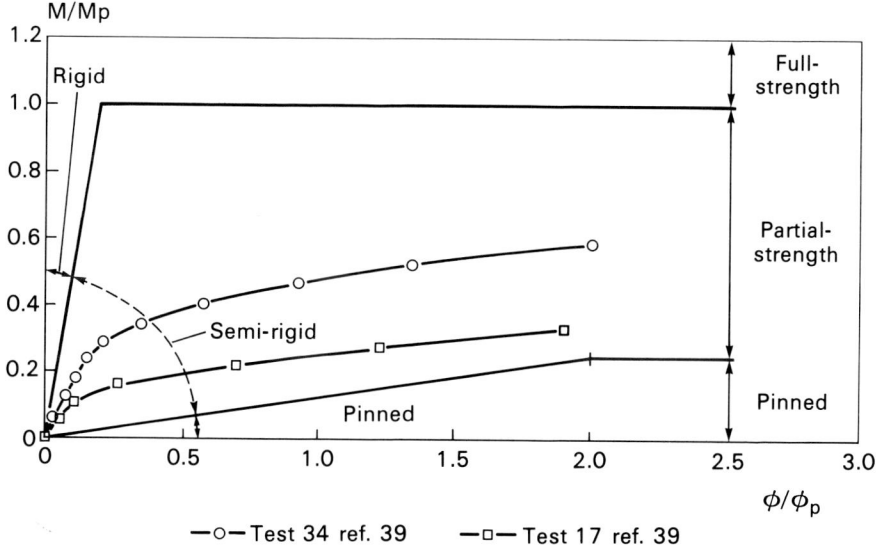

Fig. 8. Classification of beam-to-column connections (EC3 (27))

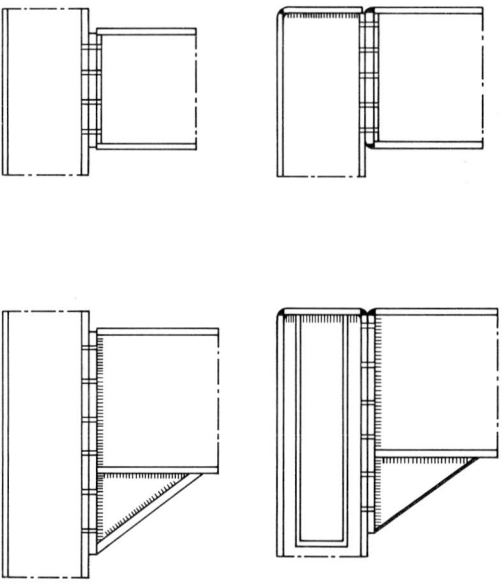

Fig. 9. Unstiffened beam-to-column connections

recommendations in EC3 are classified as semi-rigid, partial-strength connections. Designers may not be familiar with this classification of connection types and therefore a full explaination is given in the Appendix. Initially, the method was developed for the unstiffened flush end-plate connections (35) shown in Figure 9 but more recently has been extended to include extended end-plate connections (36). As this method is likely to become the European Standard for the design of end-plate connections, section 3 of this paper presents this design procedure in more detail.

3. EC3 DESIGN METHOD FOR END-PLATE CONNECTIONS

The design procedure outlined in this section enables the designer to determine the characteristics of semi-rigid, partial strength connections subject to the following restrictions (37):-

"a. The connections are formed between beams of welded or rolled I-sections and columns of European rolled I-section. Additional tests may be required for other sections.

b. Because the design philosophy is based on simple plastic theory in which geometrical non-linear effects are not included, the design rules only hold for braced frames.

c. The guaranteed minimum yield stress level σ_y of the sections is not allowed to exceed 260 N/mm^2.

d. The frame is subjected to predominantly static loading only."

3.1 Design Philosophy

The use of semi-rigid, partial-strength connections in multi-storey frames implies the formation of a plastic hinge within the connection at loads less than the factored design loads. The philosophy adopted in this design method is to allow each bolt row to attain its design strength - i.e. resulting in progressive collapse of the connection. This approach is best explained as an extension of existing design methods for end-plate connections.

Figure 10 illustrates two approaches for the design of a flush end-plate connection. In the approach shown in Figure 10a it is assumed that the centre of rotation is in line with the compression flange of the beam. With this assumption it is possible to compute the bolt forces at each bolt row. The connection attains its ultimate limit state when the uppermost bolt reaches its design strength. (This approach is, in many ways, analogous to the elastic design of beams, in which, when the extreme fibres reach yield the beam is assumed to have attained its design strength). One disadvantage of this method is that to acheive the assumed bolt force distribution, the designer often has to stiffen the column

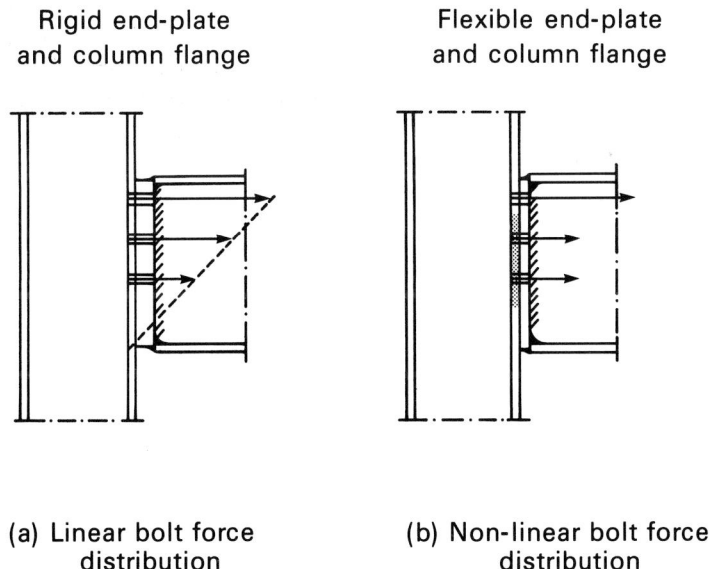

Fig. 10. Two approaches for the design of flush end-plate connections

flange, with the result that bolt fracture may occur before the column flange or end-plate deflect.

The approach shown in Figure 10b has been adopted by EC3. In this approach no assumption is made about the distribution of bolt forces. Instead, each bolt row is allowed to attain its full design strength (on the basis of the strength of the column flange or end-plate, whichever is the lowest) and the design moment capacity of the connection computed by adding together the products of the design strengths and corresponding lever arms. This model relies on adequate ductility of the connecting parts in the uppermost bolt rows to develop the design strength in the lower bolt rows. (An analogy can be drawn between this approach and the plastic design of beams). However, not all connections possess sufficient ductility to allow each bolt row to attain its design strength. Furthermore, this may also be prevented by other modes of failure (such buckling of the column web). Checks therefore have to be included which reduce the design strength of connections with limited ductility.

3.2 Design procedure

The main steps in the design process are illustrated in the flow chart in Figure 11. The designer first considers progressive collapse of the column flange and calculates the forces at each bolt row (as described in section 3.2 A). He then considers progressive collapse of the end-plate and calculates a further set of forces at each bolt row (as described in section 3.2 B). The two sets of bolt forces are then compared and the smaller of the two determined at each bolt row. The designer then proceeds by determining the strength of the column web in tension, compression and shear (equations for these are given in sections 3.3.3, 3.3.1 and 3.3.2 respectively). Next the sum of the forces at each bolt row is compared with the lowest mode of failure of the column web. If the column web fails before complete collapse of the column flange/end-plate then the forces at each bolt row starting with the lower most row are reduced until equilibrum is attained. The design moment of the connection is then calculated and the moment rotation curve determined with the method described in section 3.5.

The general procedure outlined above is described in more detail in the following sections:-

A. Consider progressive collapse of the column flange.

1. Determine the failure load of the top pair of bolts.
(In an extended end-plate, the top pair of bolts is the uppermost bolt row within the depth of the beam)

2. Determine the failure load of the column flange at the top bolt row.

3. Check for the lowest mode of failure.

The design of end-plate connections

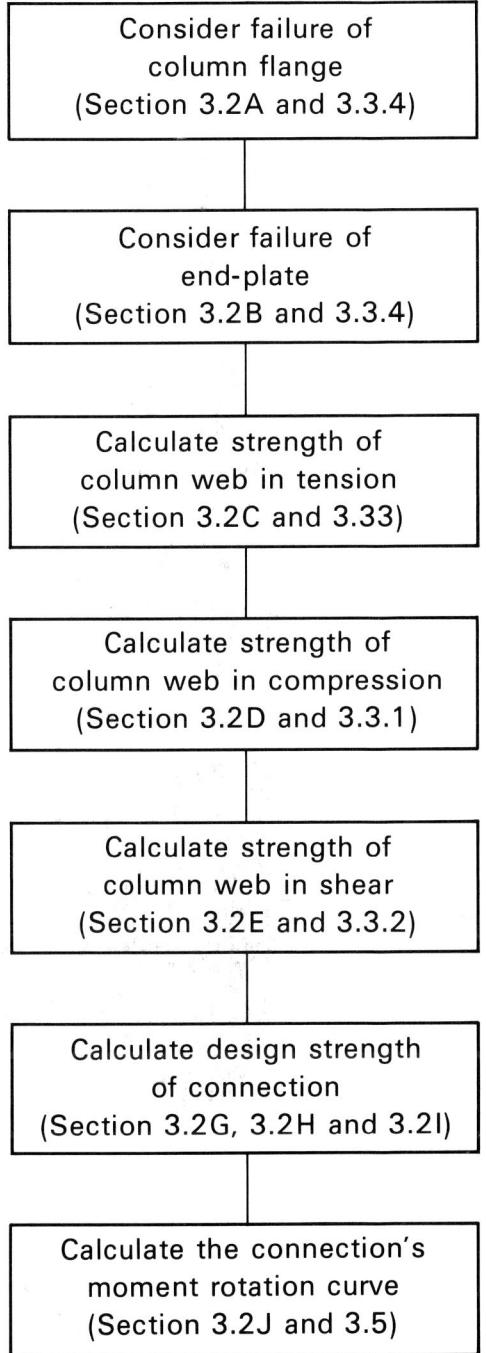

Fig. 11. Design procedure for end-plate connections

If bolt fracture is the lowest mode of failure

    4.1 Assume a linear distribution of bolt force and calculate the forces in each bolt row. This mode of failure is regarded as brittle and has limited or no rotational capacity.
The connection can only be used in plastic design if the hinge forms in the connected beam(ie. the connection is semi-rigid, full-strength).

    4.2 Determine the failure load of the column flange at the second and subsequent bolt rows.

    4.3 Check for the lowest mode of failure at the second and subsequent bolt rows.

    4.4 If yielding of the column flange is the lowest mode of failure then the failure load at the bolt row under consideration and all subsequent bolt rows should be reduced to zero.

If the column flange is the lowest mode of failure

    5.1 The failure load is that of the column flange.

    5.2 Determine the failure load of the column flange at the second and subsequent bolt rows. These are then the failure loads of the column flange at these positions.

B. Consider progressive collapse of the End-plate.

    1. Determine the failure load of the top pair of bolts.
(In an extended end-plate the top pair of bolts is the uppermost bolt row within the depth of the beam)

    2. Determine the failure load of the End-plate at the top bolt row.

    3. Check for the lowest mode of failure.

If bolt fracture is the lowest mode of failure

    4.1 Assume a linear distribution of bolt force and calculate the forces in each bolt row. This mode of failure is regarded as brittle and has limited or no rotational capacity.
The connection can only be used in plastic design if the hinge forms in the connected beam (ie. the connection is semi-rigid,

The design of end-plate connections

full-strength).

4.2 Determine the failure load of the End-plate at the second and subsequent bolt rows.

4.3 Check for the lowest mode of failure at the second and subsequent bolt rows.

4.4 If yielding of the End-plate is the lowest mode of failure then the failure load at the bolt row under consideration and all subsequent bolt rows should be reduced to zero.

<u>If the End-plate is the lowest mode of failure</u>

5.1 If the End-plate gives the lowest mode of failure then this is the failure load of the top bolt row.

5.2 Determine the failure load the End-plate at the second and subsequent bolt rows. These are then the failure loads of the End-plate at these positions.

C. Compare the failure loads of the column flange and end-plate at each bolt row and choose the lowest.

D. Determine the strength of the column web in tension.

E. Determine the strength of the column web in compression.

F. Determine the strength of the column web in shear.

G. If the loads given by D, E and F are greater than the sum of the bolt forces calculated in C then reduce the forces in the bolt rows starting with the lowest bolt row until equilibrum is attained.

H. Determine the lever arm for each bolt row by assuming that the connection rotates about the centreline of the connected beam's compression flange.

I. Calculate the design strength of the connection by adding together the product of the forces in each bolt row and the lever arm.

J. Calculate the moment-rotation characteristics of the connection.

The equations for calculating the design resistance of the critical components are given in section 3.3. In each case the equations have been developed using theoretical models (35) in accordance with the theory of plasticity and the failure mechanisms observed in tests.

3.3 Design Resistance of Critical Components

The equations given in each of the following sections are taken from a draft of Appendix 6A: Beam-to-column connections in EC3.

3.3.1 Column Web in Compression

The resistance of an unstiffened column web subject to compressive forces is given by the smaller of the following two expressions:-

$$F_{cd} = f_y \cdot t_w \cdot s(1.25 - 0.5|\sigma_n|/f_y) \tag{1}$$

$$F_{cd} = f_y \cdot t_w \cdot s \tag{2}$$

Where
- f_y is the yield strength of the column web
- σ_n is the maximum axial compression stress in the web of the column due to normal force and bending.
- t_w is the thickness of the column web
- s is the effective width of web in compression and is given by:- $t_{fl} + 2a\sqrt{(2)} + 2t_e + 5(t_f + r)$
- a is the throat thickness of a double fillet weld connecting the beam and the end-plate
- t_{fl} is the thickness of the beam flange
- t_e is the thickness of the end-plate
- t_f is the thickness of the column flange
- r is the root radius of the column

Equation 1 is the result of an extensive experimental programme of work on the influence of normal, bending and shear stresses on the buckling of European rolled sections subject to compressive forces. This research concluded that the collapse load was influenced by axial forces and bending moments in the column. This is accounted for by the term in brackets which becomes effective when the modulus of the bending and axial stresses in the column exceed half the value of the yield stress of the column web. Although no guidance is given in EC3 on how to calculate σ_n, Zoetemeijer (31) suggests the following equation:-

$$\sigma_n = N/A + M.e/I \tag{3}$$

Where
- N is the axial force in the section

M   is the bending moment in the section
I   is the moment of inertia of the section
A   is the area of the section
e   is the distance from the centre of the section to the transition from the web to flange.

If a web plate stiffener is used then the term t_w in equation 1 should be increased by half the thickness of the web plate.

If, alternatively a compression stiffener is used then the resistance of the column web should be taken as equal to that of the connected beam flange provided that the stiffener is the same thickness as the beam flange and that the welds are designed to transfer the compressive and shear forces.

3.3.2 Column Web in Shear

The resistance of an unstiffened column web subject to shear forces is given by:-

$$D_{sd} = 0.58 f_y . t_w (h-2t_f) \qquad (4)$$

Where
h   is the depth of the column

3.3.3 Column Web in Tension

The resistance of an unstiffened column web subject to tension forces is given by:-

$$F_{td} = f_y . t_w . b \qquad (5)$$

Where
b   is the effective width of the column web in tension which may be taken as the total effective length b_m of the column flange. This is defined in section 3.3.4.1.

If a web plate is used to reinforce the column web then the term t_w in equation 5 should be incresed by half the thickness of the web plate.

Alternatively if a tension stiffener is used then the resistance of the column web should be taken equal to that of the connected beam flange provided that the stiffener is the same thickness as the beam flange and that the welds are designed to transfer the tension and shear forces (see section 3.3.5 for the design of welds).

3.3.4 Design of Column Flange or End-plate

Research on end-plate connections (stiffened and unstiffened) suggests that the T-stub with an effective length b_m is a reasonable model of the tension zone of a column flange or an end-plate. In the T-stub, failure occurs by one of three mechanisms depending on the relative stiffnesses of the column flange, the end-plate and the bolts. These mechanisms (21) are bolt fracture (Mechanism A), yielding of either the column flange or end-plate and yielding of the bolts simultaneously (Mechanism B), and yielding of either the column flange or end-plate by itself (Mechanism C).

The resistance of either a column flange (stiffened or unstiffened) or an end-plate at any bolt row are given by:-

a. Mechanism A - Bolt fracture

$$F_{tfd} \text{ or } F_{ted} = 2B_t^* \tag{6}$$

b. Mechanism B - Simultaneous yielding of the bolts and column flange or end-plate

$$F_{tfd} \text{ or } F_{ted} = \frac{2b_m \cdot m_p + 2B_t^* \cdot n'}{m + n'} \tag{7}$$

c. Mechanism C - Yielding of the column flange or end-plate

$$F_{tfd} \text{ or } F_{ted} = \frac{4m_p \cdot b_m \, R}{m} \tag{8}$$

Where
- F_{tfd} and F_{ted} are the design resistances of the column flange and end-plate respectively
- m_p is the plastic moment of resistance of the column flange or end-plate
- b_m is the effective length of the column flange or end-plate at the bolt row under consideration.
- B_t^* is the design strength of an individual bolt in tension.
- m and n' are defined in Figure 12
- R is a reduction factor which is related to the stress level σ_n in the column flange.

The value of R is based on limited experimental work undertaken by Zoetemeijer et al (32) which concluded that a longitudinal stress less than 70% of the yield stress has negligible effect on failure of an unstiffened column flange and that stresses exceeding this value reduced the connection's stiffness at its ultimate design moment by a factor of 2. Figure 13 shows the way in which these results are interpreted in the design method.

The design of end-plate connections

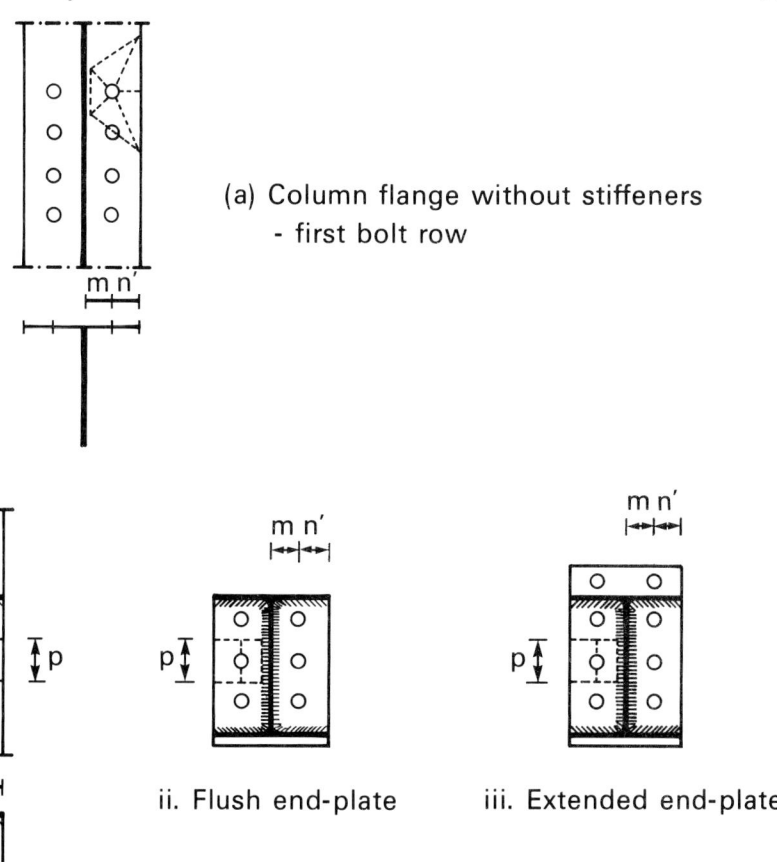

(a) Column flange without stiffeners - first bolt row

i. Stiffened column

ii. Flush end-plate

iii. Extended end-plate

(b) Column flange or end-plate - second and subsequent bolt rows

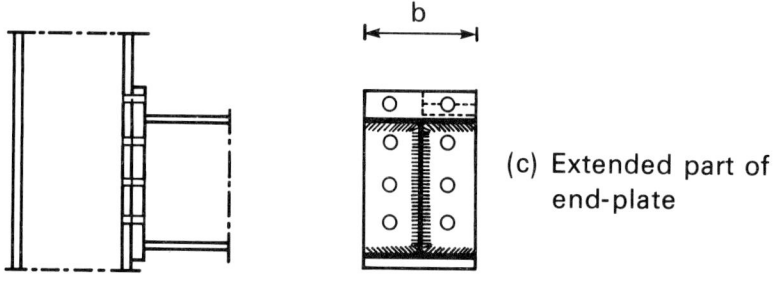

(c) Extended part of end-plate

Fig. 12. Yield line patterns

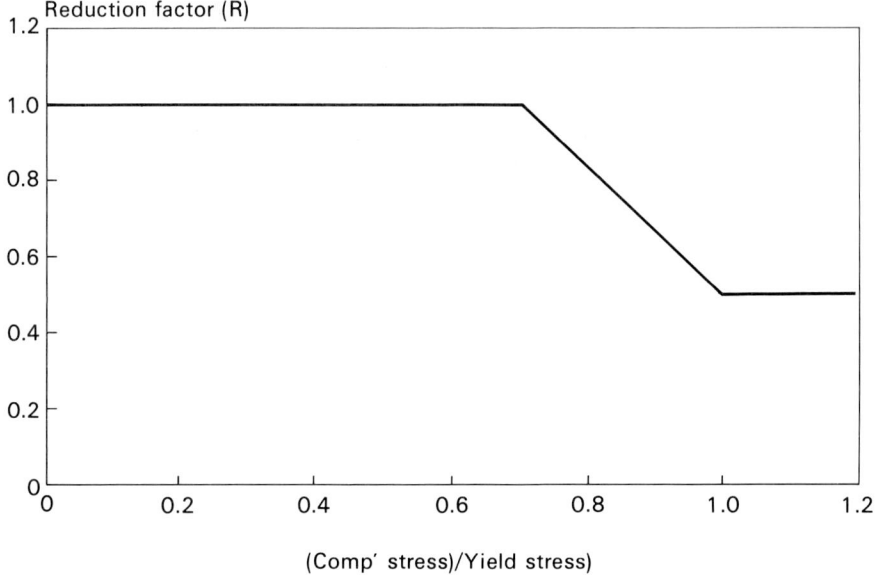

Fig. 13. Reduction factor for axial compression in an unstiffened column

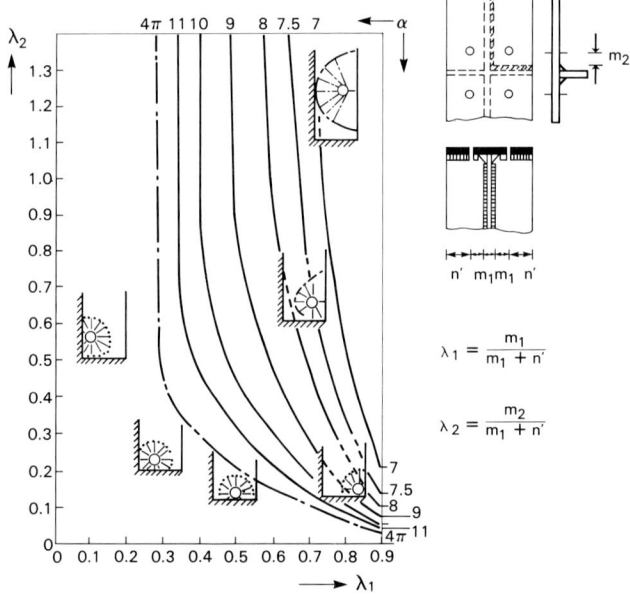

Fig. 14. Determination of α (after Zoetemeijer (35))

The design of end-plate connections

3.3.4.1 Evaluation of the effective length of Column flange/End-plate

a) Column flange without stiffeners : first bolt row

The yield line pattern developed by Zoetemijer (28) is used to determine the effective length b_m of the T-stub at the first bolt. The yield line patterns for different bolt hole positions are shown in Figure 12a.

$$b_m = 4m + 1.25n' \tag{9}$$

Where
m and n' are defined in Figure 12a

b) Stiffened column flange or end-plate : first bolt row

The resistance of a stiffened column flange or end-plate at the first bolt row position is given by:-

$$b_m = \alpha m/2 \tag{10}$$

where
α is a factor depending on the position of the bolt hole and can be found from Figure 14.

This expression and the chart given in Figure 14 were developed by Doornbos (38) from a series of tests on T-stub connections in which the column was stiffened with a tension stiffener.

c) Column flange or end-plate : second and subsequent bolt rows

Once again the expression for b_m at the second and subsequent bolt rows is taken from Zoetemeijer (28) and is given by:-

$$b_m = p \tag{11}$$

Where
p   is the vertical distance between bolt rows.

Figure 12b shows this effective length diagramatically (a similar diagram is given in Appendix 6A of EC3). This is a pictorial representation of the contribution that the second and subsequent bolt rows make to the resistance of the connection and is not a valid yield line pattern.

d) Extended part of end-plate

The resistance of the extended part of an extended end-plate is given by the smaller of the following expressions:-

$$b_m = 4m + 1.25n' \tag{12}$$

$$b_m = b/2 \tag{13}$$

Where
m, n' and b are defined in Figure 12c

Again Figure 12c is only a pictorial representation of the contribution that the extended part of the extended end-plate makes to the strength of the connection and is not a valid yield line pattern.

3.3.5 Weld Sizes

The rotational capacity of a connection is provided by bending of either the column flange or end-plate which induces bending in the weld between the column flange and tension stiffener and in the weld between the beam tension flange and the end-plate (shown in Figure 15).

To prevent premature failure as shown in Figure 16 the welds should satisfy the following criteria:-

$$a_s, a_d \text{ or } a_f \geq \frac{0.7 f_c . f_v . F_1}{4 m_2 f_y} \tag{14}$$

$$a_s, a_d \text{ or } a_f \geq \frac{0.7 f_c . f_v . F_1}{2(m_2 + n') f_y} \tag{15}$$

Where
a_s, a_d and a_f are weld throat thicknesses and are defined in Figure 16
f_c is a magnification factor which depends on the statical determinance of the structure.
$f_c = 1$ for statically determinate structures
$f_c = 1.4$ for braced frame
$f_c = 1.7$ for unbraced frames
f_v is a factor which depends on the bolt hole position
$f_v = 1$ if $\lambda_1 > 0.5$
$f_v = \dfrac{m_1}{m_1 + m_2}$ if $\lambda_1 \leq 0.5$

The design of end-plate connections 97

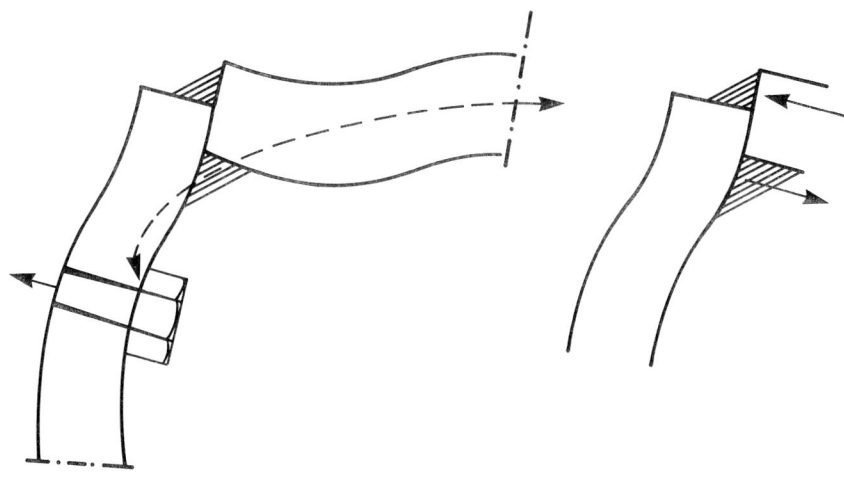

Fig. 15. Bending of the end-plate induces bending in the inner weld (after Zoetemeijer (35))

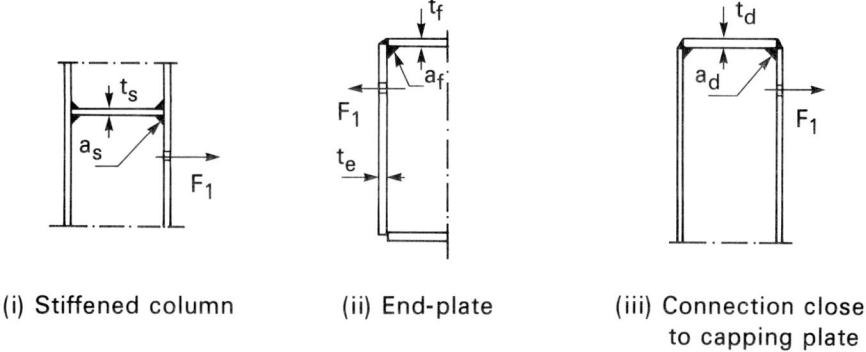

(i) Stiffened column   (ii) End-plate   (iii) Connection close to capping plate

Fig. 16. Welds used in three different positions (after Zoetemeijer (35))

λ_1 is used to calculate the strength of a stiffened column flange or end-plate at the first bolt row and is defined in Figure 14 (section 3.3.4.1 b).

F_1 is the force in in the first bolt row at the design moment of the connection

m_1, m_2 and n' are defined in Figure 16

Furthermore, if mechanism C is the governing failure mode of the column flange at the uppermost bolt row then the welds should also satisfy the following criteria:-

a_s, a_d or $a_f \geq 6mm$ (16)
and
$a_d \geq 0.4t_d$ (17)
$a_f \geq 0.4t_f$ (18)

Where
t_d is the thickness of the capping plate shown in Figure 18
t_f is the thickness of the beam flange shown in Figure 18

3.4 Determination of Moment Capacity

The moment capacity of the connection is determined from the product of the design strength of each bolt row and their corresponding lever arms, assuming the connection rotates about the centre line of the connected beam compression flange.

3.5 Determination of Rotational Stiffness

The rotational stiffness of an end-plate connection is given by the following expression:-

$$C = Eh^2 \left[\sum_{i=1}^{n} \left(\frac{F_i}{F_{id}} \right)^2 \frac{1}{C_i} \right]^{-1} \frac{M_{cd}}{2F_f h_1} \quad (19)$$

Where
C   is the secant stiffness of the connection at any moment M_c
E   is Young's modulus
h   is the distance between the first bolt row between the beam flanges and the centre of the beam's compression flange.
F_i is the actual force in part i of the connection at the moment M_c
F_{id} is the design resistance of part i of the connection
M_{cd} is the design moment of the connection
C_i is the stiffness factor of part i of the connection
F_f is the force in the beam tension flange due to the moment M_{cd}
h   is the centre to centre distance between flanges

The design of end-plate connections

The stiffness factors for each part of the connection are as follows:-

Shear stiffness of column web	$C_1 = 0.24 t_w$
Tension stiffness of column web	$C_2 = 0.8 t_w$
Compression stiffness of column web	$C_3 = 0.8 t_w$
Tension stiffness of column flange without stiffeners	$C_4 = \dfrac{t_f 3}{4 m^2}$
Tension stiffness of column flange with stiffeners	$C_5 = \dfrac{t_f 3}{12 \lambda_2 m_2}$
Stiffness of bolts	$C_6 = \dfrac{2 A_s}{l_b}$
Tension stiffness of end-plate	$C_7 = \dfrac{t_e 3}{12 \lambda_2 m_2}$

4. DISCUSSION

In this section the design method described in section 3 is compared with published experimental results on flush end-plate connections and the accuracy of the method at predicting the moment-rotation curve discussed. Attention is then focused on the restrictions described in section 3.0, the basis of these restrictions and the limtations they place on the design of end-plate connections within the UK.

4.1 Comparison with experimental results

In Figures 17a and 17b the design method is compared with published experimental data on flush end-plate connections with three bolt rows. The first of these figures compares the calculated design moment (Mc) of a connection with its experimentally determined collapse moment (Me). In this figure the ratios Me/Mc and Me/Mp are plotted on the vertical and horizontal axes respectively. Thus, the vertical axis is a measure of the method's accuracy at predicting a connection's collapse moment. Furthermore, values larger than 1.0 are low and safe estimate of the collapse moment while those below 1.0 are high and unsafe. Therefore, the method should give values above 1.0 for all connections. Figure 17a shows that for the majority of connections examined the ratio Me/Mc is above 1.0; the largest value is 3.04 and smallest 0.98. From these results it is concluded that the design method in EC3 gives conservative values of design moment for all flush end-plate connections with three bolt rows.

In Figure 17b the rotation (Rc) of a connection at the design moment (Mc) is compared with the measured rotation (Re) at the same moment. The ratio Rc/Re is plotted on the vertical axis and the ratio Me/Mp is plotted on the horizontal axis. Once again the vertical axis is a measure of the design method's accuracy but in this case values larger than 1.0 indicate an overestimate of the connection's rotational capacity (ie the connection is more ductile than it actually is) while values less than 1.0 indicate an underestimation (i.e. the connection is stiffer than it actually is). From Figure 17b it is observed that

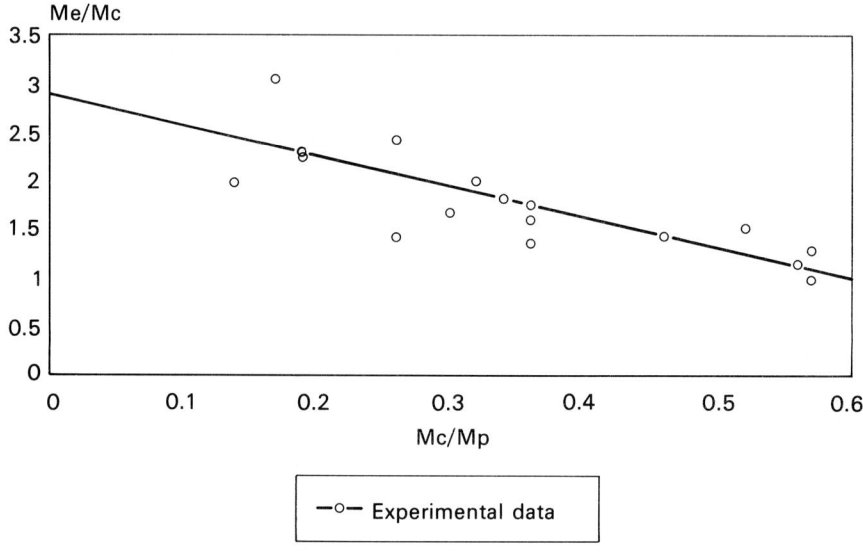

Fig. 17a. Comparison of moments for a Flush end-plate with 3 bolt rows

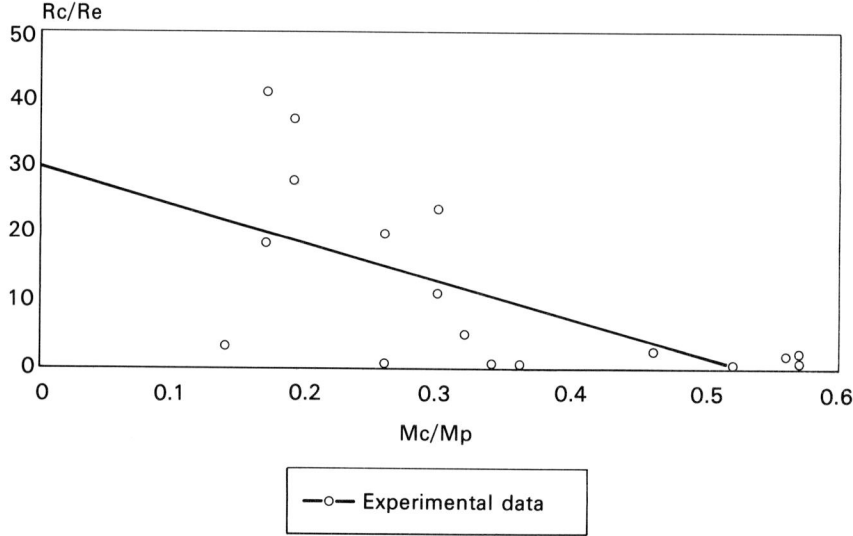

Fig. 17b. Comparison of rotations for a Flush end-plate with 3 bolt rows

The design of end-plate connections 101

the design method overestimates the rotational capacity of most of the connections examined and in some cases by as much as 4000%. Such large errors might suggest a re-think of the design method. However, the accuracy of the method should not be judged by its ability to predict connection behaviour alone but should include the sensitivity of a frame to changes in a connection's characteristics. In this regard Nethercot et al (42) suggest that as a structure becomes more extensive, the need for precise moment-rotation curves reduces. However, without further information and in view of the method's poor accuracy at predicting rotational capacity it is suggested that the rotations be used only to calculate beam and column deflections at serviceability.

Figures 17a and 17b also indicate that the method gives worse results for connections with low values of Me/Mp than for those with higher values. Examining the data in more detail it is observed that those connections with high values of Mc/Mp generally have Mechanism A or B modes of failure (i.e. bolt fracture) while those with low values fail by Mechanism C (yielding of the column flange/end-plate). The methods inability to predict accuractely the design moment of connections with Mechanism C modes of failure can be attributed to strain hardening of the column flange/end-plate material while the loss of accuracy on rotation can be deduced by studying the components of equation 19. When mechanism A is the mode of failure the connection derives most of its rotation from extension of the uppermost bolt row. Thus the expression for C6 dominates and predicts bolt extension accurately. When mechanism C is the mode of failure the connection derives its rotation from yielding of either the column flange or end-plate and in this case the expressions for C4 and C7 dominate. These expressions clearly, overestimate the flexibilty of the column flange. One explanation for this is that the they do not allow for preloading of the bolts.

4.2 Underlying assumptions

The connection tests described in references (28, 29, 30, 31, 32, 34, 35, 36, 39) were constructed from European rolled sections (40,41) with a guaranteed minimum yield stress of 235N/mm^2. These tests correctly reflect construction practice within Europe in the 1970's. Since then, construction practice within the UK has changed and rolled sections with a guaranteed minimum yield stress of 275N/mm^2 are now the norm. Thus the strict interpretation of restrictions (a) and (c) listed in section 3 disallow all Universal beams and columns and all British steels higher than grade 43. Such limitations would seriously disadvantage the UK steel construction industry. Research carried out by Zoetemeijer (36) on extended end-plate connections concluded that 'An actual yield stress of 1.5 times the guaranteed value of 240N/mm^2 has no disastrous effect on the rotational capacity' provided the extended end-plate is designed with the method described in section 3. If this effect can be shown to be valid for Universal sections of grade 43 steel then limitations (a) and (c) could be amended to include steels with a guaranteed minimum yield stress of 275N/mm^2.

Ductile failure of the connection depends on the relative strength of the bolts, end-plate and column flange. If the bolts are weaker than either the end-plate or the column flange a brittle mode of failure may occur. The majority of tests (on which the above design method was based) which failed in a ductile manner had ratios of bolt diameter to end-plate thickness between 1.5 and 2.0. In the UK it is common practice to have the thickness of the end-plate equal to the bolt diameter. The designer should therefore be aware that adopting this practice in conjunction with the design method of section 3 may at best produce connections in which only the uppermost bolts are fully loaded and at worst produce connections which fails in a brittle mode.

In section 3.3 no specific mention is made of prying action nor are any equation given to calculate its value. This is because prying action is allowed for in the equations for predicting the strength of either the column flange or end-plate. In his publication (28) "A design method for the tension side of statically loaded bolted beam-to-column connection" Zoetemijer addresses the problem of prying action and develops the following three expressions for the equivalent effective length of an unstiffened column flange:-

For prying force = 0.0 $\qquad b_m = (a + 5.5m + 4n')$

For maximum prying force $\qquad b_m = (a + 4m)$

For an intermediate value of prying force $\qquad b_m = (a + 4m + 1.25n')$

Zoetemeijer explains that the first equation has an inadequate safety factor against bolt fracture while the safety factor in the second equation is too high. He therefore suggests the third equation which allows for approximately 33% prying action. This approach is similar to the one proposed by Surtees and Mann (22) and simplifies the calculations by omitting complicated expressions for prying action.

5. CONCLUSIONS

The historical development of end-plate design is traced and some of the major findings in the past 80 years are discussed in detail. The method of designing end-plate connections in the 1988 draft of EC3 is presented in full and the work on which it is based is identified and discussed. This method is compared with pubished experimental data on flush end-plate connections with three bolt rows and the following conclusions are made:-

a. In the majority of cases the method gives safe values of design moment.

b. The method grossly overestimates the flexibilty of those connections which fail in a ductile manner.

c. In view of b. the calculated moment-rotation curve should only be used to predict the beam and column displacements at serviceability.

The design of end-plate connections

Its application in the UK and the differences between current UK construction practice and that in the rest of Europe, on which the method is based, are also discussed from which the following conclusions have been drawn:-

a. The method is limited to European beam and column section with a guarenteed minimum yield stress of $235N/mm^2$.

b. Construction practice in the UK is to use steels with a guaranteed minimum yield stress of $275N/mm^2$.

c. The UK practice of using bolt diameters equal to the thickness of the column flange when used in conjunction with the design method may results in connection which have brittle modes of failure.

6. ACKNOWLEDGEMENTS

The work described has been carried out as part of the research programme of the Building Research Establishment of the Department of the Environment and this paper is published by permission of the Director.

7. REFERENCES

(1) BS5950 "The structural use of steelwork in buildings. Part 1. Code of practice for design in simple and continuous construction: hot rolled sections". London, British Standards Institution, 1985.

(2) BS449 "The use of structural steel in building". London, British Standards Institution, 1948.

(3) Moore, D.B: "Survey of British beam-column steelwork connection practice". Building Research Establishment occasional paper (to be published).

(4) Wilson, W.M. and Moore, H.F: "Tests to determine the rigidity of rivited joints of steel structures". University of Illinois, Engineering Experiment Station, Bulletin 104, Urbana, USA, 1917.

(5) Steel Structures Research Committee, First Report, Department of Scientific and Industrial Research, London, HMSO, 1931.

(6) Steel Structures Research Committee, Second Report, Department of Scientific and Industrial Research, London, HMSO, 1934.

(7) Steel Structures Research Committee, Final Report, Department of Scientific

and Industrial Research, London, HMSO, 1936.

(8) Rathbun, J.C: "Elastic Properties of Riveted Connections". Transactions, ASCE Vol. 101, 1936.

(9) Young, C.R. and Jackson, K.B: "The Relative Rigidity of Welded and Riveted Connections". Canadian Journal of Research, Vol. 11, 1934.

(10) PD3343, "Recommendations for Design". Supplement No.1 to BS449, London British Standards Institution, 1959.

(11) PD3857, "Amendment No.1 to BS449". London, British Standards Institution, 1959.

(12) BCSA, "Welded Details for Single-storey Portal Frames". BCSA publication No.9, 1955.

(13) BCSA, "Steel Frames for Multi-storey Buildings. Some Design Examples To Conform With The requirements of BS449:1959". BCSA publication No.16, 1961.

(14) BCSA, "Examples of the Design of Steel Girder Bridges in Accordance with BS153:Parts 3A,3B & 4". (As at October, 1963). BCSA publication No.22, 1963.

(15) BCSA, "Single Bay Single Storey Elastically Designed Portal Frames". BCSA publication No.24, 1964.

(16) Schutz, F.W. "Strength of Moment Connections Using High Strength Bolts". Proc. of National Engineering Conference AISC, New York, pp98-110, 1959.

(17) Douty, R.T: "Characteristics of Flexible Flange Connections and Fasteners". A thesis presented as partial fulfillment of the requirements for the Degree Master of Science in Civil Engineering, Georgia Institute of Technology, June 1957.

(18) Douty, R.J and McGuire, W: "Research on Bolted Moment Connections - A progress Report". Proceedings of National Engineering Conference, AISC, New York, 1963, pp48-55.

(19) Nair, R.S. Birkemore, P.C and Munse, W.H: "Behaviour of bolts in Tee-Connections Subject to Prying Action". Structural Research Series, University of Illinois, Urbana, Illinois, 1969.

(20) Sherbourne, A.N., "Bolted beam-to-column connections". The Structural Engineer, Vol.39, 1961.

The design of end-plate connections

(21) Packer. J.A and Morris. L.J: "A limit State Design Method for the Tension Region of Bolted Beam-to-Column Connections". The Structural Engineer, Vol.55. 1977. pp446-458.

(22) Surtees. J.O and Mann. A.P: "End-plate Connections in Plastically Designed Structures". Conference on Joints in Structures. University of Sheffield. 1 (Paper 5). 1970.

(23) Horne. M.R and Morris. L.J: "Plastic Design of Low-Rise Frames". Constrado Monograph. Granada Publishing. London. 1981.

(24) Pask. J.W: "Manual on Connections for Beam and Column Construction Conforming with the requirements of BS449:Part 2:1969". BCSA. publication No.9/82. 1982.

(25) Nethercot. D.A and Zandonini. R: "Methods of Prediction of Joint Behaviour - Beam to Column Connections". Structural Connections - Stability and Strength. Edited by R Narayanan. published by Elsevier Applied Science (to be published in 1989).

(26) Tschemmernegg. F and Huber. K: "Rahmentragwerke in Stahl unter besonder Berucksichtigung der steifenlosen Bauweise". Swiss Institute of Steel Construction. Zurich.1988.(In German).

(27) Eurocode No.3 "Design of Steel Structures Part 1 - General Rules and Rules for Buildings". EC3:Part 1:19xx. Prepared for the Commission of the European Communities. August 1988.

(28) Zoetemeijer. P: "A Design Method for the Tension Side of Statically Loaded. Bolted Beam-to-Column Connections". Heron 20. No.1. Delft University. Delft. The Netherlands. 1974.

(29) Back de. J and Zoetemeijer. P: "High strength bolted beam to column connections. The computations of bolts. T-stub flanges and column flanges". Report No. 6-72-13. Stevin Laboratory. Delft University of Technology. March 1972.

(30) Zoetemeijer. P: "Geboute balk - Kolomverbindingen met korte kopplaat Uitvoeringsvormen en berekeningswijzen". Report No. 6-75-20. Stevin Laboratory. Delft University of Technology. 1975(in Dutch).

(31) Zoetemeijer. P: "The influence of normal-. bending- and shear stresses on the ultimate compression force exerted laterally to European rolled sections". Report No. 6-80-5. Stevin Laboratory. Delft University of Technology. 1980.

(32) Zoetemeijer, P and Munter, H: "Influence of an axial load in the column on the behaviour of an unstiffened beam-to-column end-plate connection. Tests". Report No. 6-84-1, Stevin laboratory, Delft University of Technology, January 1984.

(33) Witteveen, J, Stark, J.W.B, Bijlaard, F.S.K and Zoetemeijer, P: Design rules for welded and bolted beam-to-column connections in non-sway frames". ASCE National Spring Convention, Portland, USA, April 1980.

(34) Zoetemeijer, P: "Bolted beam to knee connections with haunched beams. Tests and computations". Report No. 6-81-23, Stevin Laboratory, Delft University of Technology, 1981.

(35) Zoetemeijer, P: "A design method for bolted beam to column connections with flush end plates and haunched beams". Report No. 6-82-7, Stevin laboratory, Delft University of Technology, June 1982.

(36) Zoetemeijer, P and Munter, H: "Proposal for the standardization of extended end plate connections based on test results. Tests and analysis". Report No. 6-83-23, Stevin laboratory, Delft University of Technology, December 1983.

(37) Bijlaard, F.S.K: "Requirements for welded and bolted beam-to-column connections in non-sway frames". Joints in Structural Steelwork. Edited by J.H.Howlett, W.M.Jenkins and R.Stainsby, Pentech Press, 1981, pp2.119-2.137.

(38) Doornbos, L.M: "Design method for the stiffened column flange, developed with yield line theory and checked with experimental results". Thesis, Delft University of Technology, 1979 (in Dutch).

(39) Zoetemeijer, P: "Semi-rigid bolted beam-to-beam column connections with stiffened column flanges and flush end-plates". Joints in Structural steelwork. Edited by J.H.Howlett, W.M.Jenkins and R.Stainsby, Pentech press, 1981. pp 2.99-2.118.

(40) Euronorm 19-57 Published by the Commission of the European Communities

(41) Euronorm 53-62 Published by the Commission of the European Communities

(42) Nethercot, D.A, Davison, J.B, and Kirby, P.A: "Connection flexibility and beam design in non-sway frames". American Society of Civil Engineers Covention, New Orleans, September 1986.

8. APPENDIX

Definition of a semi-rigid, partial strength connection

When calculating the force distribution in a steel frame in accordance with elastic theory, the connections can be assumed to behave as: pinned, rigid or semi-rigid. Pinned connections can undergo the necessary rotations without producing substantial moments, whereas, rigid connections have to transfer moments without producing substantial rotations. Semi-rigid connections transfer moments and their rotations may be substantial and influence the force distribution in the frame.

When using plastic design, the connections can be assumed to be either full-strength or partial strength. A connection is considered to be 'full-strength' when the calculated design strength is greater than that of the connected beam. Conversely, a connection may be classified as 'partial-strength' when the calculated design moment is less than that of the connecting beam.

Thus a connection can be semi-rigid or rigid and at the same time be either partial-strength or full-strength.

5

DEVELOPMENTS IN LIGHTWEIGHT DECKING AND CLADDING

E R Bryan
University of Salford, UK

SYNOPSIS

The paper reviews the functional requirements of cladding and the state of Codes of Practice. The different requirements of sheeting in roof sheeting, roof decking and wall cladding are considered, and the calculation procedures for sheeting are summarized. The advantages of testing are emphasized.

New developments in decking profiles for composite sheet steel/concrete floor decks are considered both with regard to the slabs alone and also with regard to composite action with the supporting beams. The advantages and disadvantages of dry composite floor decks are also considered.

Finally, further developments in cladding are reviewed and examples given of the structural benefits of cladding in practice.

INTRODUCTION

In recent years there has been a quiet revolution in the methods of cladding industrial and commercial buildings. The walls and sloping roofs of such buildings, which used to be clad with corrugated protected metal or asbestos cement sheeting, are now largely covered with profiled sheeting in coated steel or aluminium in a wide variety of colours and surface finishes. The metal sheeting may be used by itself or in combination with other materials, and thermal insulation may be effected separately using mineral wool and lining panels or integrally using sandwich panels. Plastic components, including translucent panels, are often an essential part of the system, and curved sheeting is frequently used at the eaves and apex to simplify the building details and improve the appearance. The method of fixing the sheets has also changed dramatically, so that not only are the fastenings much less obtrusive, they are also much more efficient and ensure that the cladding plays a full part in the total structural performance of the building.

Another development is that roof slopes have become much less steep over the past decade or two, so that decking profiles are used for such roofs as well as nominally flat roofs. The design of decking profiles has become much more sophisticated, with rolled-in stiffeners in the flanges and webs, so that deeper profiles can be used to span directly between the main frames of buildings.

Although indented profiled sheets have been used for some time for composite sheet steel/concrete floors, it is only in the last few years that this type of construction has experienced the present high rate of expansion. As a result there is currently a considerable programme of research work under way to improve the spans of such floors and to investigate new types of shear connectors in ensuring composite action between the floor and the supporting beams.

Other types of dry composite floors, using light gauge steel in conjunction with boards of other materials, have been largely neglected in this country. However, investigations have been carried out abroad and the results could be readily adapted for British use if required.

The above introduction to developments in lightweight decking and cladding is developed in the paper. However, one important aspect which is not covered in depth is that thin-walled metal panels have a load-bearing function as well as a space-covering function. Traditionally, buildings have had a structural steel skeleton and the cladding has been required mainly to exclude the weather. With the new developments which are taking place, it could well be that two-dimensional panels might have a much larger part to play in the structural behaviour of the building so that dependence on the frames is lessened. This could open up a new market for lightweight construction.

FUNCTIONAL REQUIREMENTS

The general functional requirements of cladding have been listed by Baehre (1) as follows: (i) load carrying capacity, (ii) stiffness, (iii) durability, (iv) fire protection, (v) sound insulation, (vi) climatic protection, (vii) room environment (viii) servicing facilities. Requirements (i)-(iii) are overriding and are independent of the type and use of the building, whereas the other requirements are building-dependent and are subject to quality demands. Often minimum requirements for (i)-(vi) are specified by the building authorities, but improved standards may be required by the owner. Requirements (iv)-(vii) can often be enhanced by the addition of mineral or wood-based board material to the cladding as shown in Table 1.

It is interesting to note the results of a survey of costs of four industrial buildings made by two of the country's largest steelwork contractors (2). In three of the buildings, the cost of the cold rolled steel members and sheeting in the walls and roof exceeded the cost of the hot rolled steelwork. Yet it is safe to assume that the latter accounted for most of the design effort.

Developments in lightweight decking and cladding

Building materials		I	II	III	IV	V	VI	VII
■ Suitable for high requirements ▫ Suitable for moderate requirements □ Suitable for low requirements — Unsuitable		Load bearing capacity	Stiffness	Durability	Fire protection	Sound protection	Climatical protection	Room environment
1	Unstiffened sheet panels (Galv)	□	□	□	—	—	—	—
2	As 1, Galv + coated	□	□	■	—	—	—	—
3	Stiffened sheet panels (Galv)	■	■	□	—	—	—	—
4	As 3, Galv + coated	■	■	■	—	—	—	—
5	20mm Fibre reinforced concrete	□	□	■		—		
6	13mm Gypsum board			□	□		—	■
7	26mm Gypsum board	□	□	□	■	□	—	■
8	13mm Plywood	□	□	□			—	□
9	Fibre board					—	—	■
10	Chipboard	—	—		—			□
11	Fibre board (mineral binding)	□	□	□	■		—	■
12	Mineralwool	—	—	□	■	■	■	—
13	Plastic foam			□		□	■	—

Table 1. Qualities of various building materials

Hence, overall efficient structural design can only be achieved when full consideration is given to effect of the cladding. The design of the frame may be more important from the point of view of safety, but less important from the points of view of serviceability, economics and overall aesthetics.

CODES OF PRACTICE

In Britain there are two Codes of Practice (3,4) for sheet roof and wall coverings, but these are orientated towards installation and performance rather than design. Part 6 of BS 5950 (5), now in preparation, will remedy this situation. The design of composite sheet steel/concrete floors is covered by Part 4 of BS 5950 (6), and the contribution of the cladding to the overall structural behaviour will be covered by Part 9 of BS 5950 (7).

In Europe, the European Convention for Constructional Steelwork (ECCS) has published a number of recommendations for the design, testing and use of profiled sheet in building (8-12). These documents have formed the basis of several National Standards on the subject, and further recommendations on the design and use of sandwich panels will no doubt be the definitive document for further Standards in this area. The ECCS recommendations seem likely also to have an important influence on the Eurocodes and ISO Standards in these fields.

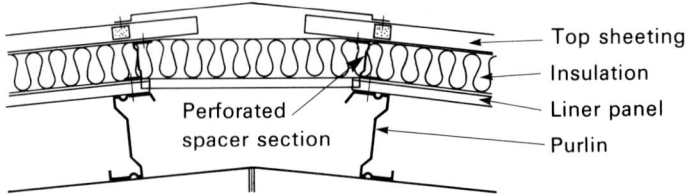

Fig. 1. Roof sheeting, insulation and liner panel

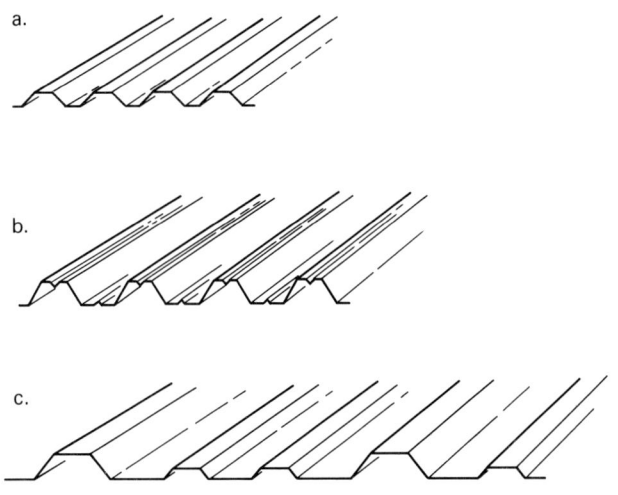

Fig. 2. Roof sheeting profiles

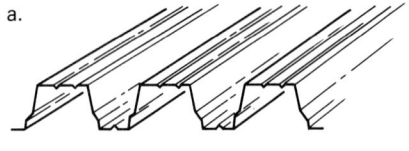

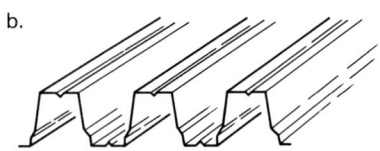

Fig. 3. Roof decking profiles

Developments in lightweight decking and cladding

SHEETING, DECKING AND CLADDING

Profiled metal sheeting may be used in quite different functions in the roofs and walls of a building as follows:

Roof sheeting

Sheeting used as the outer waterproof skin of a pitched roof, with the insulation inside, is known as "cold roof" construction. In this situation weather-tightness, durability and insulation are all-important. For maximum weather-tightness, the end laps should be adequate (sealed for low pitches) and the sidelaps should be under the crests of the sheeting. The troughs are generally wider than the crests in order to disperse rain water more quickly. It is no longer necessary for fasteners to be in the crests, since modern self-drilling screws with neoprene washers can effect a perfect seal. The durability and appearance of the sheeting depends largely on the choice of coating. In this country, the British Steel Corporation offers a range of Colourcoat and Stelvetite finishes. Insulation is usually effected by a glass fibre quilt held in place by means of a thin gauge steel liner panel fitted over the purlins (Fig. 1).

Because the roof sheeting is supported on purlins, longer spans mean not only deeper sheeting but also heavier purlins. It has been found that the most economical roof spans in practice are of the order of 2 to 3 metres, so scope for innovation in design is limited.

For many years, the simple trapezoidal profile (Fig. 2a) was used for sheeting, but this profile is now complemented by profiles with rolled-in stiffeners (Fig. 2b) which are more efficient in bending, and by profiles with a very wide bottom flange supported by intermediate stiffeners (Fig. 2c) which are efficient in shedding water in large span buildings with low roof pitches.

The steel quality of many sheeting profiles is about 280 N/mm^2 but some manufacturers are using material with a yield stress of 350 N/mm^2 and even up to 550 N/mm^2. Under these conditions, the minimum acceptable thickness of sheeting is no longer regarded to be 0.7 mm, and thicknesses of 0.5 mm are used quite widely in particular applications. As a result, the design criterion may well be the ability to withstand the effects of a concentrated load. Research is in progress to define the "walkability" of sheeting more objectively, since the current position depends far too much on subjective judgements.

Because roof sheeting is usually continuous over several spans, the maximum span may be determined by strength as well as by the usual deflection limitation of span/100 to span/200. Wind suction, particularly in local areas such as the eaves, gable or apex, is likely to be a more important design case than dead plus imposed load, and so proper attention must be given to the design of fixings.

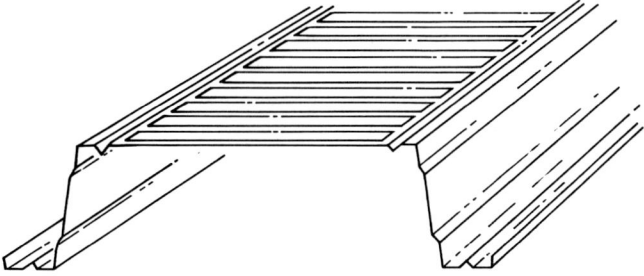

Fig. 4. Deep cassette decking profile

a.
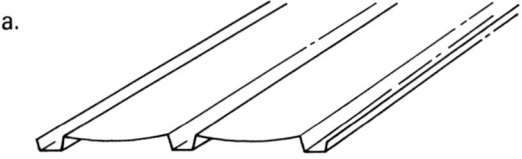

b.
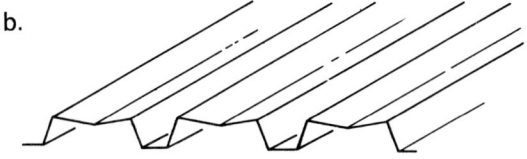

c.
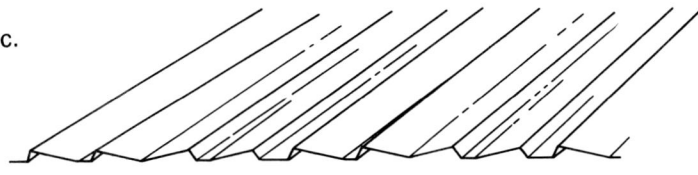

Fig. 5. Typical wall cladding profiles

Developments in lightweight decking and cladding 115

Roof decking

If the insulation and waterproof membrane is placed on top of the sheeting, then the sheeting is termed "decking" and the method is known as "warm roof" construction. The method is usually reserved for flat or nearly flat roofs, and demands considerable technical "know-how" and experience. A comprehensive guide to good practice is available (13).

In order to reduce the span of the insulation between crests, the crests are usually wider than in roof sheeting and the troughs are narrower. The sidelaps are made in the troughs. Because the decking spans directly between main or secondary beams without the need for purlins, it is usually advantageous for long spans to be used. The profiles shown in Fig. 3 have a depth of about 100 mm and can be used for spans between 6 and 8 metres for U.K. loading. Since the permissible deflection of decking is limited to span/250, the design criterion is usually that of stiffness rather than strength.

The most recent development in long span decking profiles is the 200 mm deep cassette profile, stiffened in both directions as shown in Fig. 4. It can span up to 10 metres and can thus span directly between main frames without the need for secondary members. An assessment of the design has been made at the University of Salford (14).

Wall cladding

For many applications, the profiles used for roof sheeting are also suitable for vertical cladding. In many respects the loading is similar except that the requirement for the profile to sustain a concentrated point load no longer applies.

Since the walls are often the most prominent feature of a building, the choice of wall cladding is often determined more by aesthetic considerations than structural performance. In fact, a number of profiles have been specially developed to give different impressions of scale and light and shade (Fig. 5). In addition, the sheeting may span vertically or horizontally or even diagonally. Curved sheeting around the apex or eaves of a building may be used to good effect. The choice of colour and texture of wall cladding is probably even more important than for roof sheeting.

CALCULATION PROCEDURES

The analytical model for a single trapezoidal profile in bending is shown in Fig. 6a. The effective width b_{ef} of the compression flange is determined from the Winter expression and the depth e_c to the approximate position of the neutral axis of the reduced section is calculated. The effective portions of the compression zone of the sloping web, s_{ef_1} and s_{ef_2}, measured from the top corner and approximate neutral axis respectively, are then determined from standard expressions. Using the doubly reduced cross section, the new section properties are calculated so that, for first yield in the compression flange as

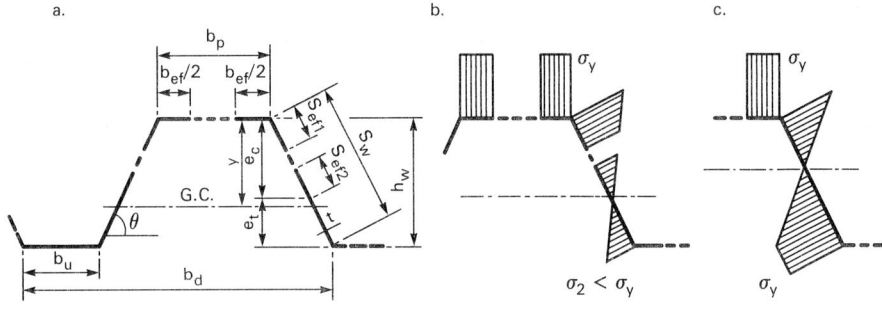

Fig. 6. Analytical model for trapezoidal profiles in bending

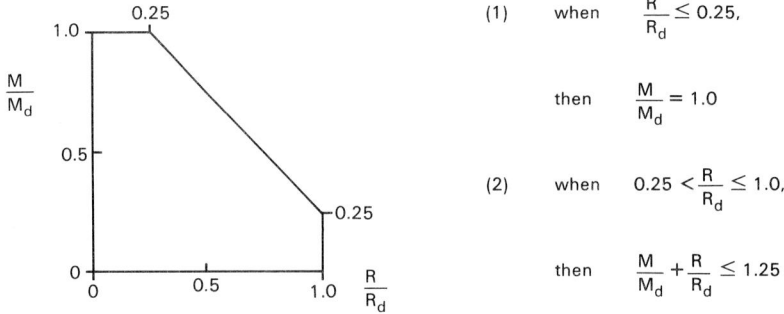

(1) when $\dfrac{R}{R_d} \leq 0.25$,

then $\dfrac{M}{M_d} = 1.0$

(2) when $0.25 < \dfrac{R}{R_d} \leq 1.0$,

then $\dfrac{M}{M_d} + \dfrac{R}{R_d} \leq 1.25$

Fig. 7. Interaction diagram for permissible values of ultimate moment M and support reaction R at an internal support

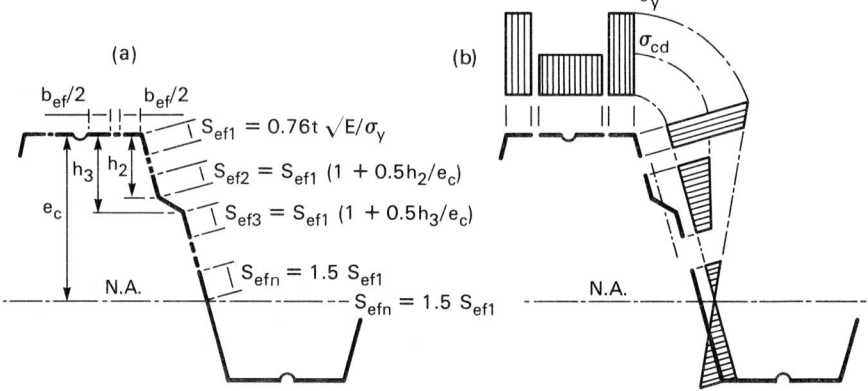

Fig. 8. Analytical model for stiffened profiles in bending

Developments in lightweight decking and cladding 117

shown in Fig. 6b, the ultimate moment of resistance of the section, M_d, may be found. If first yield occurs in the tension flange, as shown in Fig. 6c, then the tension flange is allowed to become plastic and an increased ultimate moment is calculated when the compression flange also reaches yield. For continuous sheeting, where the bending moment changes sign over the support, the above calculation has to be repeated in the case of unsymmetrical sheeting. The ultimate moment of resistance over the support will not generally be the same as that near mid-span.

Based on test results, a design expression is available for the support reaction capacity R_d of a trapezoidal profile. Then, at an intermediate support, the permissible support moment M and reaction R may be gauged from the interaction diagram and formulae of Fig. 7. This interaction diagram was obtained experimentally from a large number of tests.

For stiffened profiles (Fig. 8a), the rolled-in stiffener is treated as a strut restrained by an elastic medium representing the remainder of the profile. Hence, these struts - and their associated effective widths - are designed to a reduced stress as shown in Fig. 8b.

In calculating deflections of profiled sheeting it is strictly necessary to consider the effective width of the compression flange at the characteristic load, although conservative results will always be obtained if the effective width in the strength calculation is taken. It is noted that the second moment of area of the section varies with the magnitude of the bending moment along the member, but in practice a single value for an average bending moment will suffice.

TESTING OF PROFILED SHEETING

It is seen that the above calculation procedure, although rational, depends on data obtained from testing, particularly in the value of R_d and the interaction formula. As the data must represent the lower safe limits of many test results, they must usually be conservative for any particular sheeting profile.

In order to obtain the most economic profile it is therefore desirable to carry out tests. The savings which can accrue from such tests for mass-produced components such as sheeting, will normally well repay the effort. Four tests are specified in the European Recommendations (9):

Mid-span bending under distributed load

This test is to determine the bending moment capacity at sections where the shear force is negligible. The test load on a simply supported span may be applied by means of an air bag (Fig. 9), vacuum chamber or by two or four line loads applied by jacks or dead load.

Fig. 9. Airbag test on deep cassette profile

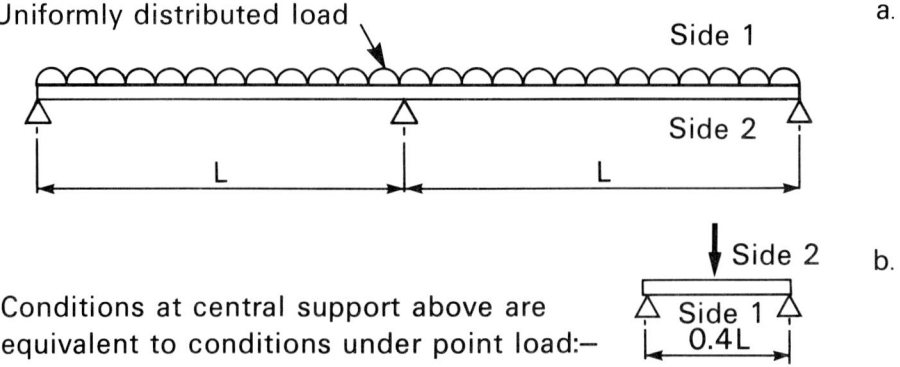

Fig. 10. Equivalent support test for double span sheeting

Fig. 11. Test for combined bending and reaction at support

Developments in lightweight decking and cladding

Bending capacity at an internal support

This test, to determine the capacity of the sheeting under a combination of bending moment and support reaction may be carried out on the full scale (Fig. 10a) or on a simply supported span of 0.4 times the full span (Fig. 10b). The latter test gives the same ratio of moment to reaction as the two span test, but is much cheaper to perform. Such a test in progress is seen in Fig. 11.

Capacity at end support

The purpose of this test is to determine the capacity of the sheeting under a concentrated reaction in the absence of bending. The tests show that the permissible reaction at the end of a sheet is approximately half that at an intermediate support.

Capacity to resist concentrated loading

It has been mentioned that this condition might well be the design criterion for thin or shallow sheeting. A concentrated load of 1kN on an area of 100 mm x 100 mm is specified in the European Recommendations and the percentage recovery required after removal of the load is stated.

Correlation of test results with theory

Since the design expressions in the European Recommendations (10) are based to some extent on test results, it is not surprising that there is generally fair correlation between the test results and theory. However there can be large individual variations as

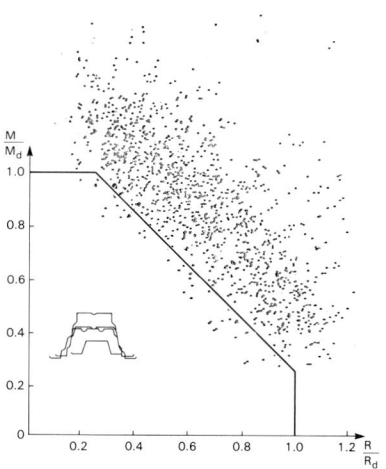

Fig. 12. Interaction between bending moment and support reaction for 1500 trapezoidal profiles

shown in Fig. 12 which gives the results of some 1500 tests of bending capacity at an internal support on 27 stiffened and unstiffened profiles (15). The amount of scatter relative to the interaction diagram is apparent. Hence, in the development of any particular profile it may well be worth obtaining specific test data for that profile. In fact, this is what has been done in practice.

COMPOSITE SHEET STEEL/CONCRETE FLOOR DECKS

In the last few years there has been a significant increase in the use of this type of floor. The profiled steel sheets have been developed specifically to ensure composite action with the concrete by means of embossments or indentations in the webs and/or flanges to develop horizontal shear capacity.

One of the limitations of this type of floor has been the ability of the steel decking to carry the weight of wet concrete without excessive deflection. In practice this has limited the length of unpropped spans to about 3.6 metres. A method of reducing the deflection is to use double-span sheets, but this imposes a high bending moment over the intermediate support. By designing for a certain amount of "plastic" rotation in the sheets over the support (16), longer spans and greater economy can be achieved while maintaining an adequate factor of safety against failure during concreting. Design expressions have been given for a number of sheeting types.

Advantage has also been taken of the use of lightweight aggregate concrete in composite floors. This allows the profiled sheets to be used on greater spans as formwork; it provides a greater fire resistance than normal weight concrete, and it enables economies to be effected in the structural framework from lower self weight.

Running parallel with the above developments has been the further use of the floor slab in composite action with the supporting beams. Traditionally the shear connection has been made by "through-the-sheet" welded shear studs, but recently a new type of connector has been developed which is shot-fired through the sheet into the top flange of the beam.

Research and Codes of Practice

The design of composite structures will be dealt with in Eurocode 4, which is presently under preparation. Parts of this Code, in particular the design treatment of composite slabs and beams, are the subject of research at Salford University. For slabs, there is a collaborative SPRINT study (Strategic Programme for Innovation and Technology) which is concerned with the harmonization of testing procedures across Europe. For beams, information is being gathered on push-off test behaviour of shear connectors and the performance of beams using shot-fired connectors. A joint SERC research programme between Salford University and University College, Cardiff will supply information for the forthcoming draft British Standard BS 5950: Part 3: "Code of practice for design in composite construction" and for Eurocode 4.

Developments in lightweight decking and cladding 121

Fig. 13. Bending test on composite slab using Cofradal sheeting

It is also envisaged that the ECCS recommendations for composite floors (8) will be redrafted to reflect the latest developments in theory and practice.

Composite slab performance

Under the SPRINT programme, 8 simply supported tests have been carried out on composite slabs using French sheeting (Fig. 13) and 2 tests on slabs using Dutch sheeting (17). The test procedure was according to Type 2 tests in BS 5950 Part 4, i.e. a dynamic loading of 10,000 cycles followed by a static test in which the load was progressively increased until failure occurred. From the failure loads, the reduction line values, m_r and k_r, were obtained from a plot of $V_E/B_s d_s \sqrt{(f_{cu})}$ against $A_p/B_s L_v \sqrt{(f_{cu})}$ (Fig. 14). In these expressions V_E is the maximum experimental shear force, B_s is the width of the test slab, d_s is the effective depth of the slab to the centroid of the profiled sheet, f_{cu} is the concrete cube strength, A_p is the cross sectional area of the profiled sheet and L_v is the design shear span. The reduction line is the experimental regression line reduced by 15%. Having determined m_r and k_r, the shear-bond capacity of the particular sheeting, for any depth of slab and for spans between the limits tested, can be obtained from a design expression in BS 5950: Part 4.

It is interesting to note that the French composite slabs failed by sudden end slip, i.e. inadequate horizontal shear connection, whereas the Dutch slabs failed without sudden and excessive end slip, i.e. adequate shear connection. The reports of the British, French

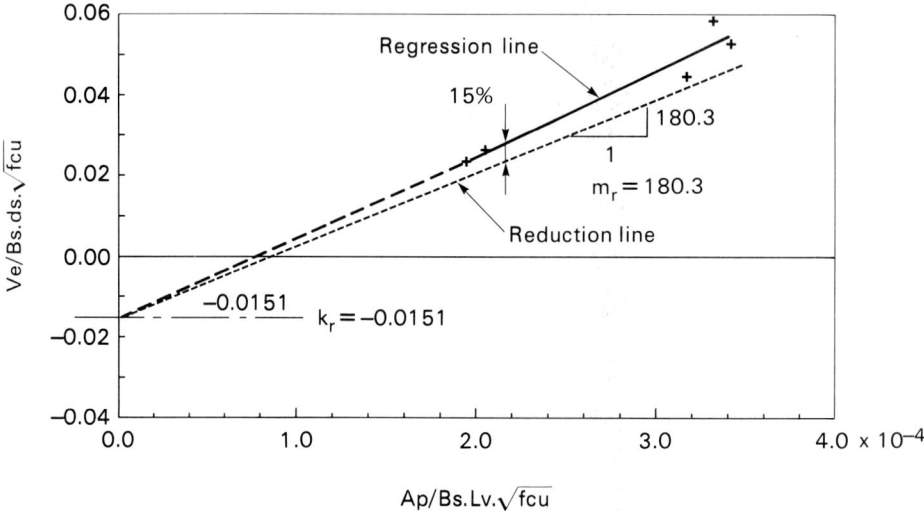

Fig. 14. Regression line and reduction line plot for composite slab using Cofradal sheeting

	Section	Effective section	Equivalent flange thickness t_{Si} (mm)	Relative moment of inertia I/I_{St}
1	(b, 60, t_F)	$b_e = b$	0,7	2,3
2		$1/2\, b_e$	0,7	1,0
3	12mm Plywood	$b_e = b$, 1,1 / 0,7	0,7 + 0,4 = 1,1	2,6
4	13mm Gypsum board	$b_e = b$, 0,82 / 0,7	0,7 + 0.12 = 0,82	2,4
5	12mm Plywood	$t_F = 2,7$, 0,7 / 0,7	0,7 + 2,0 = 2,7	4,7

Table 2. Comparative properties of dry composite sections

Fig. 15a. Composite beam prior to concreting; corrugations running perpendicular to span

Fig. 15b. Composite beam showing shot fired shear connectors; corrugations running parallel to span

and Dutch participants to the SPRINT programme are currently being compared and it seems likely that they will support the proposal in Eurocode 4 that the appropriate National Standards may be used for the determination of the regression line.

Composite beam research.

Eight 7.5m span composite beams, using shot-fired shear connectors and lightweight concrete, have recently been tested at Salford University (Figs. 15a and 15b) under static and cyclic 2 point loading (18,19) The degree of partial shear connection, i.e. the number of connectors used compared with the number required for full interaction, expressed as a percentage, was varied in the beams which were instrumented for deflection, end-slip, strain and slip between the steel sheet and beam at intermediate points. The results of six tests are shown in Fig. 16. The fully composite stiffness line is shown, together with the Johnson approach given in the draft BS 5950 Part 3.1. The actual failure loads at various degrees of design partial shear connection, $k_{(design)}$, are also compared with the loads at a deflection of span/200. Two approaches will be used in analysing the beam sections for strength. Both are given in the draft code. The first is a linear design rule proposed by Johnson and the second is an "exact" (stress block) method.

A preliminary analysis of the results of the composite beam tests shows that shot-fired shear connectors are very satisfactory for degrees of partial shear connection above 40% and are suitable for beams up to 10m span.

In addition to the above full scale beam tests, push-off tests on shear connectors (Fig. 17) are being carried out in an effort to correlate the results of the two types of test. If a definitive push-off test can be formulated, then this will reduce the need for expensive composite beam testing.

DRY COMPOSITE FLOOR DECKS

In contrast with composite floor decks using sheet steel and concrete, composite floor decks using light gauge steel in conjunction with other dry materials (20,21) have not been widely used in this country. Examples of such materials include gypsum board, plywood, fibre board, chipboard, mineral wool and plastic foam.

The effect of composite behaviour between the sheet steel and the other materials is due to (a) additional resistance of the sheet steel to buckling (b) two-way composite action for sheeting and panels, and (c) linear composite action for cold-formed sections and beams. In (a), not only does the bonding (or other fastening) of the building board inhibit buckling of the compression flange of the steel profile, it also increases the effective width of the flange. The effect of the building board gives an equivalent flange thickness of the section as shown in Table 2, and increases its second moment of area. This is especially noticeable if the building board can be used to close an open cold-formed member (compare the relative I values of sections 2 and 5 in Table 2).

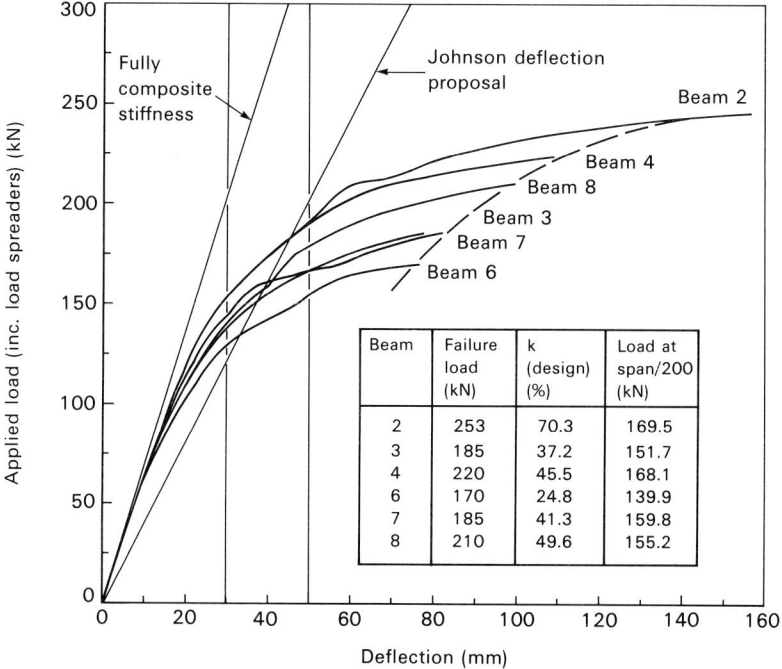

Fig. 16. Load/deflection curves for six composite beams

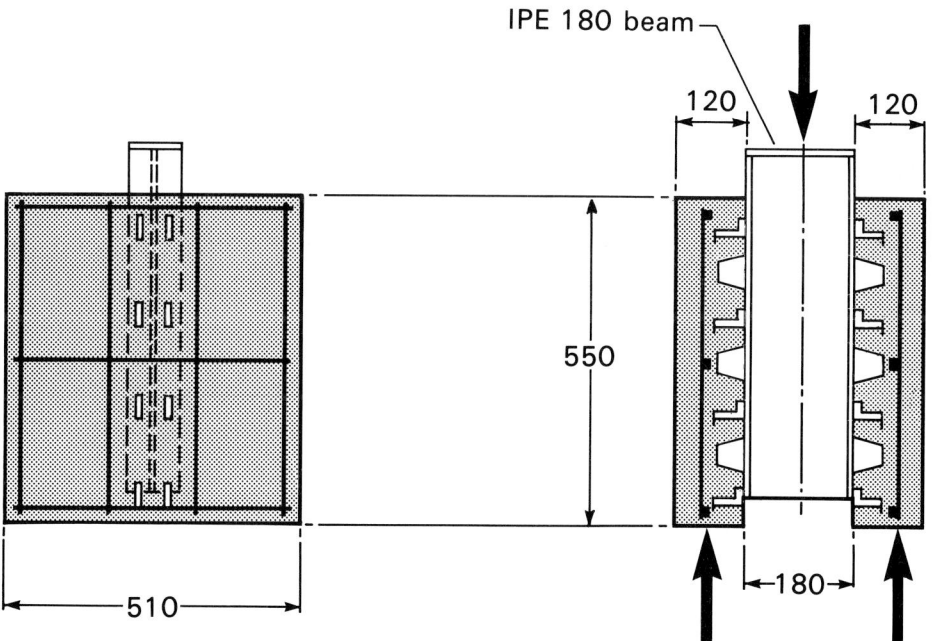

Fig. 17. Arrangement for push-off tests on shear connectors

Although the above advantages of dry composite floors appear attractive, it must be remembered that there are difficulties. The compatibility of the composite materials must be carefully considered; the question of who fabricates the composite beam section must be decided - in particular any bonding of materials should be carried out in factory conditions and not on site; the brittle nature of some insulating boards makes them susceptible to damage by impact, and the site control must be of high quality. Some of these problems may well account for the slow development of this type of flooring.

Fire protection and sound insulation

The required fire resistance of lightweight floors is generally achieved by mineral wool, glass fibre or gypsum boards. Suspended ceilings of gypsum, vermiculite, perlite or mineral fibre are often used. General rules for thicknesses and types cannot be given as the fire resistance depends on the general design of the floor.

A characteristic feature of dry lightweight floors with regard to sound insulation is the relatively small mass and the multi-layer construction. In order to attain sufficient insulation against airborne and footstep sound, the floor has to be constructed of adequate layers of flexible and rigid elements. If lightweight partition walls are used, special measures may be necessary to avoid sound bridges.

Floor stiffness and vibration

Lightweight floors are vibratory systems which can be excited by foot traffic and may be felt as unpleasantly springy in certain unfavourable combinations of vibration characteristics such as amplitude, natural frequency, acceleration and damping. At present, the detailed analysis of these parameters is replaced by criteria based on the stiffness of the floor.

Human sensitivity to low-frequency vibrations is very high and this makes it difficult for the engineer to protect people from environmental vibrations in different kinds of buildings. As a rule, the fundamental frequency of any floor should be above 6 Hz if at all possible.

In the course of a study of this subject (22), dynamic tests were carried out on a lightweight floor system consisting of the deep cassette profile shown in Fig. 4 with 19 mm thick flooring plywood screwed to the top flange, and 13 mm gypsum ceiling board attached to small Z sections running transversely across the bottom flanges. Structural modifications to improve the vibrational characteristics of this and other floors, are suggested.

The vibration of long span floors was the subject of a recent workshop meeting convened by the Steel Construction Institute (23). It dealt not only with lightweight floors, but with floors of all types and one of the outcomes of the meeting was a proposal to prepare a

Developments in lightweight decking and cladding

design guide on the subject.

FURTHER DEVELOPMENTS

Sandwich panels

This type of panel, illustrated in Fig. 18, has become very popular in recent years. The face material is usually thin gauge steel or aluminium and the core is usually a foamed plastic. Panels with flat or lightly profiled faces, Figs. 18a and 18b, are used mainly for walls, while panels which are more heavily profiled, Fig. 18c, are used mainly for roofs because the bending stiffness of the profile constrains the tendency of the foam core to creep under sustained load such as self-weight and snow.

Sandwich panels combine high load-carrying capacity with low self-weight; they have excellent and durable thermal insulation; they act as a reliable water and vapour barrier, and they are suitable for mass production. However, continuous foaming lines are very capital intensive, so that for low volume production batch manufacture is often used.

Because the core of a sandwich panel is so flexible, its shear deformation must be incorporated into the analysis together with other important criteria (24) In particular, temperature differences between the two faces - which arise because of the excellent thermal insulation of the core - can give rise to stresses which exceed those due to wind or snow.

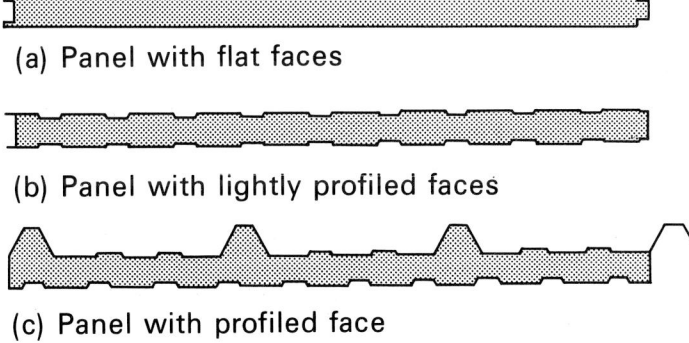

Fig. 18. Sections of typical sandwich panels

European Recommendations for the design of sandwich panels are at present being prepared by ECCS Working Group TWG 7.4, and it is expected that these will form the basis for national and international codes on the subject.

Frameless steel structures

Because steel sheeting has such high in-plane shear strength and shear stiffness, it may be used as a shear diaphragm acting either in conjunction with a conventional steel framework, or with very light edge members. The demonstration folded plate roof shown in Fig. 19 (25) had longitudinal apex and valley members only 5 mm thick, yet spanned 21.6m and carried an imposed load of 0.75 kN/m^2 with a central deflection of only 36mm. It collapsed at the calculated load of 2.3 kN/m^2. The economy and efficiency of this type of roof have not yet been taken up in this country, although there are many examples in the U.S.A.

Another type of frameless steel structure which has been investigated is one which consisted solely of curved profiled sheets. The strength of the structure depended not only on the bending strength of the profiled sheets, but also on diaphragm action arising from the non-deformability of the stiff gable ends. Although the span of such a structure depends very much on the depth of the profile and the process used to form the curved sheets, spans of 8m can be achieved, which may well be adequate for small standard buildings.

Microprofiling

Another recent development to improve the properties of sheet steel used for profiled steel sheet is "microprofiling". This consists of rolling very small corrugations in the sheet before forming it (Fig. 20). The effect of these small corrugations is to raise the buckling stress of the sheeting and also, due to extra cold forming, to raise the yield strength of the sheet.

STRUCTURAL BENEFITS OF CLADDING IN PRACTICE

It was mentioned in the introduction to this paper that metal sheeting can have an important effect on the structural behaviour of a building. This is illustrated by the following three examples from practice:-

Swagebeam frame

This building system, with frames of cold formed steel, is shown in Fig. 21. For a building of 12m span x 24m long x 4m high to eaves, with a roof pitch of 15°, the calculated stiffening effect of 0.6mm thick steel sheeting is as follows:

> Dead plus imposed load: reduction of 13% in the eaves moment and 16% in the apex deflection.

Fig. 19. Folded plate roof of light gauge steel

Fig. 20. "Microprofiling" of steel sheet

Fig. 21. Swagebeam light gauge steel frames

Fig. 22. Lee Valley ice-rink building

Fig. 23. Erection sequence for stressed arch building

Wind load: reduction of 34% in the eaves moment and 72% in the eaves deflection.

It is thus seen that the sheeting does not have a great effect in resisting vertical load (because of the low roof slope) but it does have a profound effect on the sidesway deflection. Moreover the sidesheeting eliminates the need for wall bracing to cater for wind on the gable end. If the building is lined with steel lining panels, the stiffening effect is even more pronounced.

Arched building for ice-rink

This building, shown in Fig. 22, has arches of 40m span x 10m high, spaced at 10m centres. The roof consists of 200mm deep x 1mm thick cassette decking profiles, spanning directly between the arches. The decking was required to act as a curved diaphragm to take the wind forces on the gable end of the building and to provide lateral support to the flanges of the arch ribs. The rigidity provided by this construction was enormous.

Stressed arch buildings

A novel development pioneered in Australia, (26) is shown in Fig. 23. Massive lattice girders fabricated from hollow sections, are erected flat on the ground. Prestressing cables are then threaded through the bottom chords, which have gaps in them. As the cables are tensioned, the gaps close and the lattice girders, complete with the sheeting and all services, are formed into curved arches. Buildings of 120m span and over have been erected and great savings are claimed over normal construction methods, due to the fact that all work is carried out at ground level. Again, as in the ice rink building, wind on the gable end is taken by diaphragm action in the sheeting.

CONCLUSIONS

The paper shows that substantial developments are now taking place in the design and use of lightweight decking and cladding, so that these components, which constitute a large part of the cost of a building, may be effectively used.

ACKNOWLEDGEMENTS

The author is grateful to his colleagues who have supplied information for this paper: Professor R. Baehre, Technical University, Karlsruhe; Professor J. M. Davies, Mr D. C. O'Leary and Mr D. A. B. Thomas, University of Salford.

REFERENCES

(1) Baehre, R: "Developments in cold-formed sections in Europe". Developments in thin-walled structures, Vol.3, edited by J. Rhodes and A. C. Walker. Elsevier Applied Science Publishers 1987.

(2) Bryan, E.R: "European Recommendations for cold-formed sheet steel in building". Proceedings fifth international specialty conference on cold formed steel structures. University of Missouri-Rolla, November, 1980.

(3) British Standards Institution: "Code of practice for sheet and wall coverings". CP 143: Part 10: 1973.

(4) British Standards Institution: "Code of practice for performance and loading criteria for profiled sheeting in building". BS 5427: 1976.

(5) British Standards Institution: "Code of practice for design of light gauge sheeting, decking and cladding". BS 5950: Part 6 (in preparation).

(6) British Standards Institution: "Code of practice for design of floors with profiled steel sheeting". BS 5950: Part 4: 1982.

(7) British Standards Institution: "Code of practice for stressed skin design". BS 5950: Part 9 (in preparation)

(8) ECCS: "European Recommendations for the design of composite floors with profiled steel sheet". 1975.

(9) ECCS: "European Recommendations for the testing of profiled metal sheets". 1977.

(10) ECCS: "European Recommendations for the design of profiled metal sheeting". 1983.

(11) ECCS: "European Recommendations for good practice in steel cladding and roofing". 1983.

(12) ECCS: "European Recommendations for the stressed skin design of steel structures". 1977.

(13) Tarmac Building Products Ltd.: "Flat roofing - a guide to good practice". RIBA Publications Ltd., 1982.

(14) Bryan E. R: "Long-span steel decking". The Structural Engineer, Vol.57A, No. 4, April, 1979.

(15) Baehre R. and Fick K: "Berechnung und bemessung von trapezprofilen - mit erlatrungen zur DIN 18807". Berichte der Versuchanstalt fur Stahl, Holz und Steine der Universitat Fridericiana, Karlsruhe, 1982.

(16) Bryan E. R. and Leach P: "Design of profiled sheeting as permanent formwork" CIRIA technical note 116, 1984.

(17) O'Leary D. C., Thomas D. A. B. and Duffy C. T: "Results of shear bond capacity tests on composite slabs with Prins/Cofradal". Report Nos. 88/215 to 218, Department of Civil Engineering, University of Salford, April-June 1988.

(18) Thomas D. A. B. and O'Leary D. C: "The behaviour of composite beams with profiled steel sheeting and Hilti HVB shot-fired shear connectors (Part 1)". Report No. 88/214, Department of Civil Engineering, University of Salford, April, 1988.

(19) Thomas D. A. B. and O'Leary D. C: "Composite beams with profiled-steel sheeting and non-welded shear connectors". Steel Construction Today, Vol.2, August, 1988.

(20) Baehre R. and Urschel H: "Lightweight steel based floor systems for multi-storey buildings" Proceedings seventh international specialty conference on cold-formed steel structures. University of Missouri-Rolla, November 1984.

(21) ECCS: Committee AC1: Lightweight steel based floor systems for multi-storey buildings, 1984.

(22) Ohlsson S: "Floor vibrations and human discomfort" Chalmers University of Technology, Goteborg, 1982.

(23) Steel Construction Institute: "Vibration of long-span floors". Steel Construction Today, Vol.2, No. 1, February, 1988.

(24) Davies J. M: "Design criteria for structural sandwich panels" The Structural Engineer, Vol.65A No.12, December, 1987.

(25) Davies J. M, Bryan E. R. and Lawson R. M: "Design and testing of a light gauge steel folded plate roof". Proc. Instn.Civ.Engrs, Part 2, June 1977.

(26) Gatzka B. et al: "Strarch building system" Australian Builder, VBP 1171, June 1986.

6

NEW DEVELOPMENTS IN COMPOSITE CONSTRUCTION

R P Johnson
University of Warwick, UK

SYNOPSIS

Current developments in design methods for continuous beams and columns are outlined. Examples are given of why little of the extensive literature on composite structures is of use to code drafting committees. The use of partial shear connection in beams of long span relies on connectors having greater slip capacity than has yet been established. It is suggested that paradoxes can be found in our attitudes to codes of practice.

INTRODUCTION

Dr Randal Wood will long be remembered for his stimulation of the research of others as well as for the quality and range of his own work. The writer first experienced this in connection with yield-line theory, soon after first using it in a design office in 1956. Later, there were many discussions at Cambridge with Dr E N Fox on Wood's puckish hypothesis (1) that exact solutions did not necessarily exist, even for quite simple problems. I hope Dr Morley will be referring to the interesting work that resulted.

Twenty years later, during one of his last visits to the University of Warwick, Randal was bubbling with excitement over some errors he had found (2) in the 1948 theory still used within the rubber industry for the design of elastomeric bearings. The subject was the influence of axial compression on buckling in shear. It was not some trivial matter, such as all existing bridge bearings being unsafe; it was that the conceptual model failed to explain some of the test data. He wanted us to drop all other work in favour of an "enthusiastic new attack" on this subject.

Wood's interest in composite action first developed with reinforced concrete in mind. His paper "Composite action in frame buildings" (3), on interaction between beams and floor slabs, includes both elastic and plastic solutions. Much of it is relevant to composite frames of structural steel and concrete, the subject of this paper.

Even when limited to buildings, this is now a wide field. The topics chosen below are of current interest to those now drafting Eurocode 4, Part 1 (4) and BS 5950: Part 3 (5). There has for twenty years been a swelling stream of reported research. For example, references (6) to (9), all published since 1985, total 2020 pages and include about 190 papers. These papers cite at least one thousand other papers as references.

No individual or committee can possibly assimilate all this material - and there may well be as much again published in languages other than English, German, or Russian.

Much of this work is ignored by those who design and build structures. In the U.K., composite structures have become widely used for buildings only within the last decade. Designers are wise to ignore much of the published research. They ask instead for good codes of practice; but the paradox is that there has never yet been a comprehensive British code for composite structures for buildings.

It is clear that the codes now being drafted will in places be more conservative than some methods now being used, and will require new checks to be made. Industry may well conclude that it can do better without codes than with them! For the leading practitioners, this is true.

This suggests that code drafters are at fault. But they too get little help from published research.

WHY ARE RESEARCH PAPERS SO RARELY USED?

The writing of a paper for publication is dominated by a restriction on length and often by shortage of time. It is not entirely the authors' fault, but the fact is that nearly all papers have several of the following characteristics, and consequently are of little or no value as a basis for design rules in codes of practice.

(1) No clear definition of the domain of applicability of the conclusions reached.

(2) No critical review of previous research relevant to that domain.

(3) Conclusions that are recommendations for more research.

(4) A design method that needs data which will not be available to the designer, or which itself depends on other variables (e.g., a load-slip curve for an anchored reinforcing bar).

(5) Test results that omit crucial data (e.g., for a stud shear connector, the size of its weld collar, the properties of the steel, and the load-slip curve at slips exceeding about 2 mm).

New developments in composite construction 137

(6) Test results that exceed proposed design resistances mainly because strengths of the materials far exceed the proposed design (i.e., factored) values.

(7) Theory based on unvalidated assumptions, or that fails to take account of imperfections likely to occur in practice.

(8) Conclusions applicable only within a particular environment of specifications and practice. This is the main barrier to making use of North American work in European codes. It applies, for example, to the Autostress method of allowing shakedown in composite bridge decks (10).

(9) An investigation based on literature in one language only, leading to a theory that is not checked against test data reported in another language. It is not sufficient that the theory predicts the author's test results!

TENSION STIFFENING AND CRACK-WIDTH CONTROL

An example of item (9) above can be found in the literature on cracking in tension flanges of reinforced concrete beams. It would be convenient if hogging moment regions of continuous composite beams could be designed by the rules for reinforced concrete. To check whether this is feasible, it is necessary to have a good mathematical model for the observed behaviour of composite beams, not necessarily simple enough to be used in design. Work on such a model is in progress at the University of Bochum. It makes extensive use of ideas on cracking in reinforced concrete developed in two papers published in W. Germany (11, 12). These papers refer to 93 other publications, of which 86 are in German and so are unlikely to be known to British engineers. Neither paper refers to any of the work of the Cement and Concrete Association on cracking that has been the basis of British practice for the last twenty years; yet the Bochum model predicts quite well the results of tests, including the British tests on composite beams (13).

COMPOSITE COLUMNS

There are other examples of the parallel development, in different countries, of design methods with little evident similarity, that fit most of the limited test data quite well. We will all find in the Eurocodes some methods new to us.

For composite columns, none of the three methods used in the U.K. have survived in Eurocode 4. The demise of the "cased strut" method of BS 449 is long overdue. The Basu and Sommerville method of BS 5400 is excellent as a standard by which to assess other methods, but it is rather complex for use in buildings. The Smith and Johnson method (14), given in the 1976 draft of BS 5950, was included in draft Eurocode 4 as a simple alternative. A much wider range of composite columns is used on the Continent than in the UK and the limited scope of this method did not appeal in several other Member States of the EEC. It will therefore be omitted, and only one method, unfamiliar in the UK, will be given.

ANALYSIS AND DESIGN OF CONTINUOUS BEAMS AND FRAMES

Methods for both global and local analysis have to be related to the slendernesses of steel components in compression. A system has become established with four classes: (1) plastic, (2) compact, (3) semi-compact, and (4) slender. Until now, the class of a cross-section has been that of the more slender of its compression flange and web, and that of a beam has been that of its most slender section. The slenderness limits for composite sections in classes 1 and 2 are lower than those for steel sections, to reflect differences in both the required and the available rotation capacities.

In composite beams, there is a great need to redistribute moments from supports to midspan regions. It has long been a problem that using slab reinforcement to strengthen a support region worsens the class of the section, as it rapidly increases the depth of web in compression.

Both plastic hinge analysis (for Class 1 only) and plastic analysis of sections (allowed for Classes 1 and 2) are of significant advantage, in comparison with elastic analysis. This is why it has recently been proposed (for BS 5950: Part 3.1 (5)) that the class of a section shall be that of its compression flange, even when the web is in a more slender class. There is also a tendency towards slenderness limits, which define the class boundaries, more generous than those given in draft Eurocodes 3 and 4. Relaxation is certainly justified in hogging moment regions, because most studies of instability in these regions have underestimated the beneficial effects of the steep moment gradient. Care will be needed, though, in extending these ideas to midspan regions of steel beams. Tests on composite beams have shown (15) that midspan regions reach their plastic moment of resistance even when the web depth/thickness ratio exceeds 150.

The applicability of plastic hinge analysis to composite beams is restricted by the low rotation capacity of some types of cross section. Research on this (16,17) has led to the inclusion in recent codes of rules that allow plastic hinge analysis only when adjacent spans of continuous beams are of similar length. There is a tendency in current practice to use Class 1 members with span ratios as high as 2.5:1. These have to be designed as if in Class 2, using elastic analysis and only about 20% redistribution of moments. It has been argued (18) that the redistribution ratio should be increased to about 70%. Even this would not often cause midspan hinges to form, so this is not plastic theory by another name. When this is done, yielding also occurs at internal supports at serviceability load levels. This too is thought to be acceptable, provided that the resulting increase in deflections is allowed for. This involves consideration of shakedown (18, discussion).

When using redistribution, it is dangerous to rely on the ductility of welded fabric of high-tensile steel wires, provided primarily to control cracking. It has been found in

New developments in composite construction

tests (19) that such fabric can fracture when the mean tensile strain at its level is as low as 0.3%, due to the concentration of elongation at cracks in the concrete slab. At the ultimate limit state, one should rely only on unwelded bars at least 10 mm in diameter.

It is one thing for experienced designers to use these ideas in particular structures, in which strict crack-width control at internal supports is not a requirement; but it is far more difficult to so specify them in a code of practice that they can be used for any structure, by an apprentice designer, and with due regard to the consequences elsewhere in the design process. There may be implications for the treatment of large holes in webs; and the bottom steel flange is more susceptible to lateral buckling if an assumed redistribution does not occur, than if it does. Also, the critical case for lateral instability is when only one of an adjacent pair of spans is fully loaded, and it is the unloaded span which is at risk.

Draft Eurocode 4 was criticised for giving little guidance on inelastic behaviour in composite frames, as distinct from continuous beams. This was because of the lack of data on rotation capacity of columns. Work now in progress is showing that in composite columns with high axial load, rotation capacity can be low. As so often in this subject, the wide range of feasible geometrical arrangements impedes the development of simple design rules.

DUCTILITY OF CONCRETE AND OF SHEAR CONNECTORS

In seeking to use more redistribution in beams, designers are thought to be on firm ground; but in other areas there is need for vigilance. We are familiar with concretes that can be assumed not to crush until the extreme-fibre strain exceeds about 0.0035; but concretes with cylinder strengths exceeding 45 to 50 N/mm^2 are increasingly more brittle, so most inelastic design procedures should be associated with an upper limit to the strength of the concrete. This is rarely done at present.

A different and more urgent problem of ductility arises with the use of partial shear connection in simply-supported beams. It is the problem of sudden and progressive failure in longitudinal shear. Calculations for full shear connection are done without consideration of the effects of slip, even though the connection is not effective until some slip occurs. This process is well validated for beams in buildings; and in bridges the presence of travelling and repeated loading ensures a surplus of connectors for static loading, which limits maximum slips.

The introduction of profiled steel sheeting has led to pressure for the use of partial shear connection; and spans in buildings now sometimes exceed 15 m. On the basis of tests, mostly on beams of span less than 7 m, and virtually none on spans longer than 10 m, it is now accepted practice in the UK to use partial shear connection down to 50% of full connection. In the USA, the 1986 Manual of Steel Construction (20) specifies no minimum level of shear connection, but states in the Commentary that 25% is a "practical minimum value".

Stud shear connectors in fact have a limited slip capacity, after which resistance declines steeply. The slip capacity depends on their degree of containment by the concrete slab as well as on the ductility of their shanks and welds. Data are very sparse, because nearly all reported push tests were terminated at slips between 2 mm and 4 mm, in the belief that maximum load had been reached.

Parametric studies by full-range numerical analysis of simply-supported and continuous beams (21, 22) show that for a given degree of shear connection, the maximum slip before a beam reaches its design load (determined as a function of the degree of shear connection) increases with span. It is simply a scale effect; but the required slip capacity at a given span depends in a complex way on the proportions of the cross section and the relative strengths of the steel and concrete. Some results (22) for degrees of shear connection (N/N_f) between 0.5 and 1.0 are shown in Fig. 1, in which L is the span.

The lines are tentative design rules. If, for example, a designer is using connectors which have an established slip capacity of 5 mm, but then fail, and if the span is 12.5 m, then point C on line ABCD shows that N/Nf should be not less than 0.8. Four of the computed results for a 5-mm slip limit lie on this line, eleven lie below safe) and two lie above (unsafe). Similar work for plate girders with unequal flanges gives slightly more restrictive results, due to the more severe residual stresses.

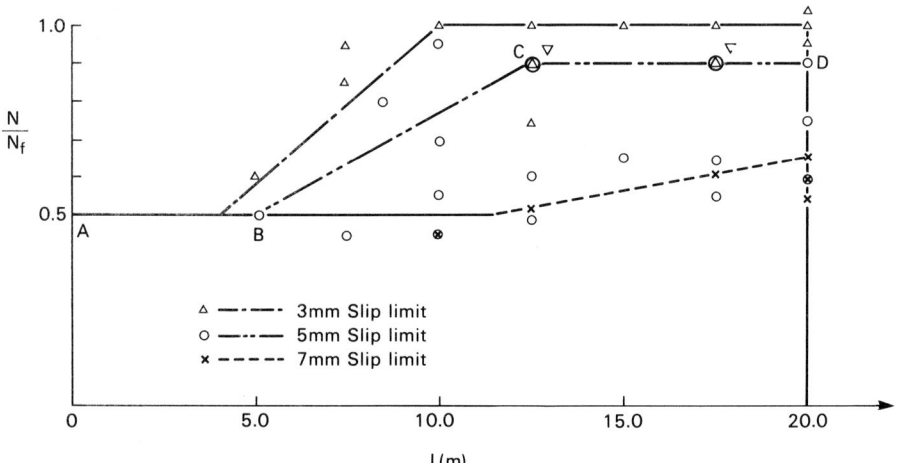

Fig. 1. Minimum shear connection required, for given slip limit and span, for members with rolled steel sections

Three laboratory tests have been done (23) on beams 12 m long each with one shear span of 9.0 m, and with low degrees of shear connection. In each case, sudden failure of the stud connectors occurred throughout the 9 m shear span.

There are two problems in using this information. First, there are no comprehensive data on the slip capacity of connectors, and certainly no evidence that for studs it can be assumed to exceed 5 mm, in all of the many situations that occur (e.g., in L-beams, with metal decking, etc.). Secondly, as Dr. Siess wrote in 1960 (24), it is very difficult to make a code more restrictive on the basis of research alone!

DUCTILITY IN SIMPLY-SUPPORTED COMPOSITE BEAMS

Siess's second law could well have been that it is very difficult to make an existing code less restrictive, with or without research. The Australians have this problem now. The 1980 Australian code for composite beams (25) includes a mandatory requirement that the response of beams to sagging bending must be ductile, meaning that strain-hardening of the steel member must precede collapse. The rule is based solely on research (16), and has been found to be restrictive in practice. There is no equivalent requirement in other national codes, nor in the Australian code for reinforced concrete beams. A strong case has been made (26) for relaxing this rule, and its interest to us is that the same research is relevant to current limitations on the use of plastic hinge analysis.

CONCLUSION

It was not thought appropriate on this occasion to make a detailed exposition of new research. One remembers how Randal Wood was always so excited by the challenge of research and how he delighted in its paradoxes.

An attempt has been made to show, by means of a few examples, that in the study of composite structures, there are still plenty of challenges; that research workers and their sponsors should try to make their published work more relevant to designers, or at least to code drafters; and that in our attitudes to codes and research there are paradoxes too. One sometimes feels that industry commissions costly large-scale research, when little is not known (e.g., simply-supported beams and bridge decks), and builds long-span continuous structures, without research, when little is known.

REFERENCES

(1) Wood, R.H: "New techniques in nodal-force theory for slabs". MCR Special Publication, Recent developments in yield-line theory, 31-62. Cement and Concrete Association, May 1965.

(2) Wood, R.H: "Buckling of short elastomeric blocks under shear and bending". Private communication, March 1985.

(3) Wood, R.H: "Composite action in frame buildings". Publications, 15, 247 - 265, Int. Assoc. for Bridge and Struct. Engrg, Zurich, 1955.

(4) Eurocode 4, Composite steel and concrete structures, Report EUR 9886 EN, Commission of the European Communities, 1985.

(5) B.S.5950, Structural use of steelwork in building, Part 3.1, Design of composite beams. In preparation, 1988.

(6) Roeder, C.W, ed: Composite and Mixed Construction. Proceedings of the US/Japan Joint Seminar, Seattle, 1984, pp. 339, Amer. Soc. of Civil Engrs, New York, 1985.

(7) Javor, T, Kozak, J, and Pechar, J, ed: International Symposium, Composite Steel Concrete Structures, 3 vols, pp. 516, Dom techniky CSVTS, Bratislava, 1987.

(8) Buckner, C.D and Viest, I.M, ed: Composite Construction in Steel and Concrete, Proceedings of an Engineering Foundation Conference, pp. 820, Amer. Soc. of Civil Engrs, New York, 1988.

(9) Narayanan, R, ed: Steel-Concrete Composite Structures. pp. 347, Elsevier, London, 1988.

(10) Haaijer, G. et al: "Suggested Autostress procedures for load factor design of steel beam bridges". Bulletin 29, pp. 35, Amer. Iron and Steel Inst., April 1987.

(11) Janovic, K: "Cracking in reinforced and prestressed concrete". Betonwerk und Fertigteil-Technik, 12, 161-169, 1986.

(12) Noakowski, P: "Continuous theory for the determination of crack width under the consideration of bond". Beton und Stahlbetonbau, 80, 7 u.8, 1-18, 1985.

(13) Johnson, R.P. and Allison, R.W: "Cracking in concrete tension flanges of composite T-beams". Struct. Engr., 61B, 9-16, March 1983.

(14) Johnson, R.P. and Smith, D.G.E: "A simple design method for composite columns". Struct. Engr., 58A, 85-93, March 1980.

(15) Vasseghi, A. and Frank, K.H: "Static shear and bending strength of composite plate girders". PMFSEL Rep. 87-4, pp. 122, Univ. of Texas, Austin, June 1987.

(16) Rotter, J.M. and Ansourian, P: "Cross-section behaviour and ductility in composite

beams". Proc. Instn. Civ. Engrs, Part 2, 67, 453-474, June 1979.

(17) Johnson, R.P. and Hope Gill, M.H: "Applicability of simple plastic theory to continuous composite beams". Proc. Inst. Civ. Engrs., Part 2, 61, 127-143, March 1976.

(18) Brett, P.R, Nethercot, D.A and Owens, G.W: "Continuous construction in steel for roofs and composite floors". Struct. Engr., 65A, 355-368, Oct. 1987. Discussion, Struct. Engr., 66, 216-227, 19 July 1988.

(19) Uth, H.-J: "Continuous composite beams in buildings - local instability in negative moment regions" (in German). Dr.-Ing. dissertation, pp. 137, University of Kaiserslautern, 1987.

(20) Manual of Steel Construction, Load and Resistance Factor Design, pp. 1100, Amer. Inst. of Steel Constr., 1986.

(21) Aribert, J.M. and Abdel Aziz, K: "Model for the full-range analysis of continuous composite beams" (in French). Construction Metallique, 23, No. 4, 3-41, Dec. 1986.

(22) Molenstra, N. and Johnson, R.P: "Final report on a study of partial shear connection in composite beams". Engineering Department, University of Warwick, pp. 51, November 1987.

(23) Johnson, R.P: "Limitations to the use of partial shear connection in composite beams". Research Report CE21, pp. 18, University of Warwick, Aug. 1986.

(24) Siess, C.P: "Research, Building Codes, and engineering practice". Journal Amer. Conc. Inst., 56, 1105-1122, May 1960.

(25) AS 2327, Composite construction code, Part 1 - Simply supported beams, Standards Assoc. of Australia, 1980.

(26) Patrick, M. and Bridge, R.Q: "Ductility of simply-supported composite beams". Steel Construction (Australia), 22, No. 2, 3-39, 1988.

7

HILLERBORG'S ADVANCED STRIP METHOD – A REVIEW AND EXTENSIONS

L L Jones
Loughborough University of Technology, UK

SYNOPSIS

After an extensive review and presentation of Hillerborg's 'advanced' strip method some of the limitations imposed by Hillerborg are discussed. In order to remove some limitations and make the method simpler and less restrictive, a general set of equations is developed in terms of the support moments for the parameters γ_1 to γ_5 for Hillerborg's five partial load cases. These values are then used to evaluate the possible reduction in, and the extent of the negative steel over the columns. Equations for additional positive steel which is required in cases which are outside Hillerborg's imposed limits are also presented. Design charts are included for the amount of negative and positive steel and values of the γ parameters.

INTRODUCTION

Hillerborg's 'simple' strip method, which can be used to design slabs supported by beams or undeflecting line supports is well known. In a rectilinear co-ordinate system he assumes the torsional moment $M_{xy} = 0$, so that for a uniform load p the slab equilibrium equation can be subdivided such that

$$\frac{\delta^2 M_x}{\delta x^2} = -\alpha p \quad \text{and} \quad \frac{\delta^2 M_y}{\delta y^2} = -(1 - \alpha)p \quad (1)$$

thus allowing easy calculation of the bending moments and reinforcement. For slabs supported as a whole, or in part, by columns he abandoned the simple strip method and developed what he termed the 'advanced' strip method. The method was presented in English in his book (1) in 1975. In a paper (2) in 1982 he amplified certain points and gave numerous examples of its use. In that paper he states "in most books only the 'simple' strip method is described. The reason for this seems to be that the 'advanced' method has been supposed to be more complicated ..... which may have been caused by the

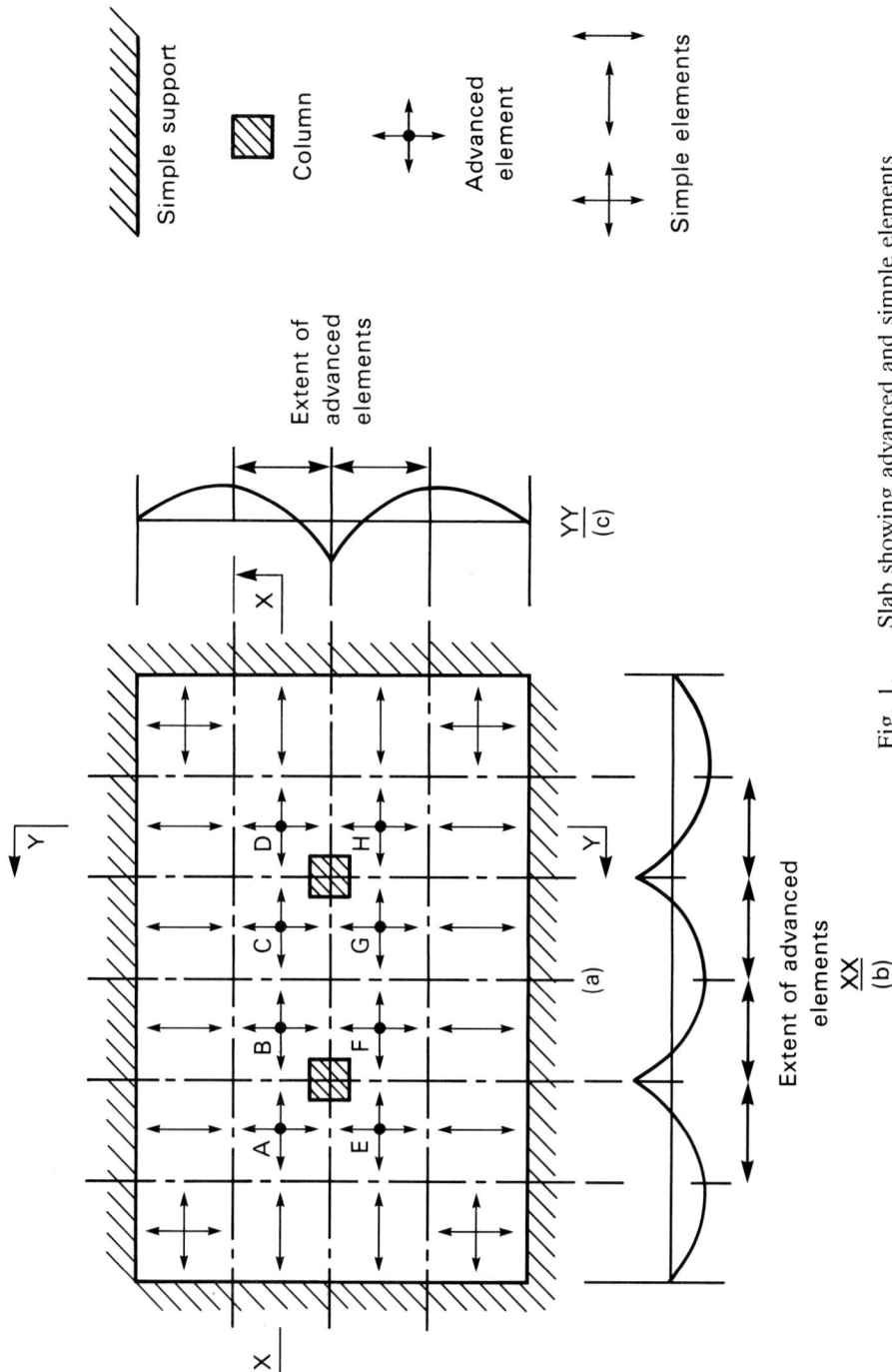

Fig. 1. Slab showing advanced and simple elements

Hillerborg's advanced strip method   147

perhaps too scientific, rigorous and complete treatment given in the book". Whilst this is partly true, the author believes that one drawback which Hillerborg underestimates is that the designer has to work blindly within certain constraints and cannot, without considerable difficulty, calculate the way in which the reinforcement values vary. Generally the designer will not need to know this, but it is felt that this is a detraction from what is otherwise an easy design method.

In 1983 Dr Wood and the author therefore started to work together with the intention of clarifying some of Hillerborg's rules; removing if possible the constraints and extending the method generally.

This paper presents some of that joint work which will be published more fully in the future (3). Because Hillerborg's 'advanced' strip method is not well known his basic theory is outlined first since reference will in any case need to be made to certain features. This introduction should also lead to a greater understanding of the extensions.

HILLERBORG'S BASIC THEORY

The advanced strip method is intended for use for the design of slabs supported as a whole, or in part, by columns. An example of such a slab is shown in Fig. 1. The 'advanced strip' method is used for those slab areas or special elements directly supported by the columns, and any other areas connected to these special elements are designed by his simple strip method with, of course, proper moment continuity between the elements.

In its simplest form the method is extremely easy to use. The designer, merely as a starting point, assumes 'imaginary' supports over the column lines.

At these 'imaginary' supports, moments are chosen and hence normal bending moment diagrams are obtained assuming simple strip action, with the full load p acting on strips, in the x and y directions as shown in Fig. 1b and 1c. The special elements of the slab designed by the 'advanced' strip method extend from the columns to the points of maximum positive bending moment (zero shear), and for this particular slab are marked A to H in Fig. 1. The region outside this area is designed by the simple strip method. The average span and support moments along the edges of the special elements obtained from the hypothetical bending moment diagrams become the average edge moments for the special elements designed by the 'advanced' strip method. The distribution of these average moments may have to be adjusted to satisfy certain constraints but Hillerborg proves in his 'advanced' strip method theory that if the special elements A to H are reinforced, initially across the whole element, for these edge moments then the yield condition will not be exceeded within the elements. It must never be assumed however that the moments within a special element are those of the initial hypothetical bending moment diagram. The sole purpose of this diagram <u>is to obtain suitable average edge moments for the special elements.</u> In reality the special elements are only special in so much that the orthogonal reinforcement is determined from Hillerborg's lower bound moment field and the

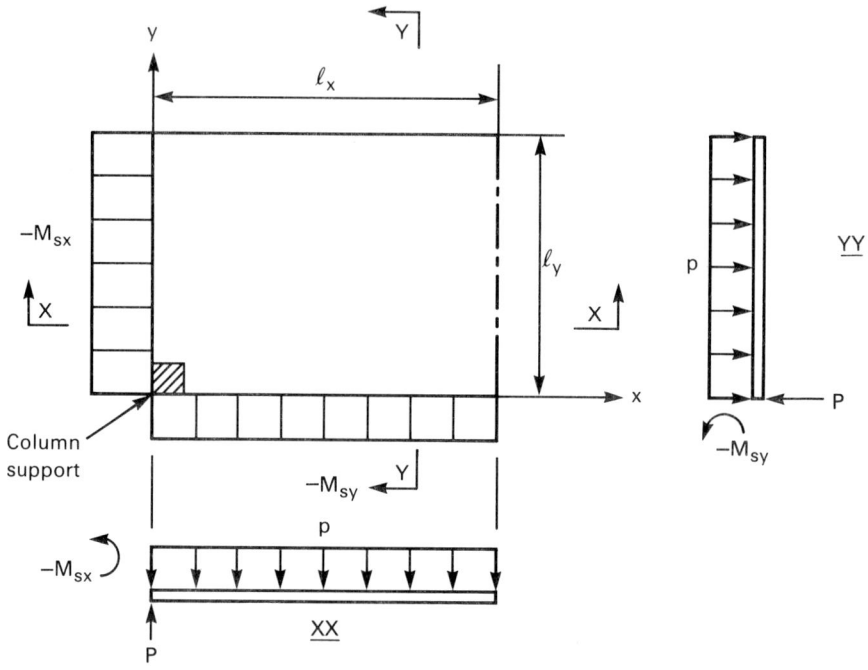

Fig. 2.   Basic distribution element-edge moments and load

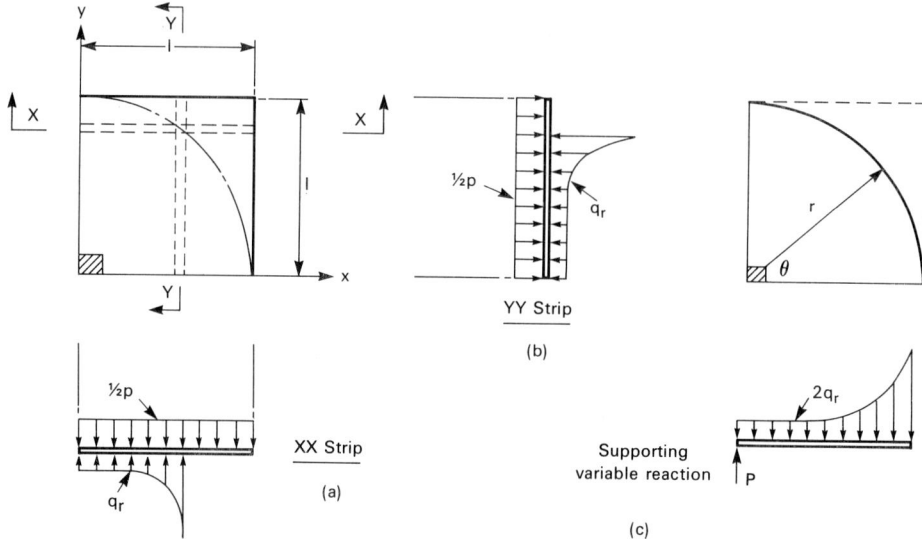

Fig. 3.   Basic distribution element showing radial support reaction

Hillerborg's advanced strip method

Wood-Armer reinforcement rules.

It is now necessary to outline Hillerborg's proof that the yield condition will not be exceeded within his special elements and define his limits for the edge moments. To analyse these special elements Hillerborg (1) introduces the concept of a load-distribution element supported at a corner by a column as shown in Fig. 2 which, as shown in Fig. 1, may be part of a more extensive slab system. Initially it is assumed the element as being under the action of;

i) a downward uniform load p per unit area;

ii) an upward force P (the column) at the corner $x = 0$, $y = 0$;

iii) uniform bending moments $-M_{sx}$ and $-M_{sy}$ only on the $y = 0$ and $x = 0$ edges, which must be in equilibrium with adjacent elements; and

iv) zero shears and moments along the other two edges (but which we will see later have span moments added).

The load distribution element not only carries Hillerborg's lower bound moment field but also any strip actions within the element. The reinforcement is calculated by the contribution from the various moments in accordance with what in this country are termed the Wood-Armer rules but which Dr Wood insisted should really be called the Hillerborg-Wood-Armer (H-W-A rules).

For overall equilibrium of a general element, size l_x by l_y, in Fig. 2:

$$P = pl_xl_y; \quad M_{sx} = pl_x^2/2 \text{ and } M_{sy} = pl_y^2/2.$$

To begin with Hillerborg assumed $l_x = l_y = l$ and as shown in Fig. 3c he postulates as a basis of his lower bound solution that there is a variable polar-symmetrical distribution reaction $2q_r$ where

$$q_r = pl/\pi\sqrt{(l^2 - x^2 - y^2)} \qquad (2)$$

Inside the quarter circle this supports the actual load p and outside there is simple strip action. Half of the actual load p is carried by strips in the x direction and the other half by strips in the y direction. These strips are kept in vertical equilibrium by the variable radial reaction as shown in Fig. 3a and b. For a strip in the x direction vertical equilibrium is maintained since

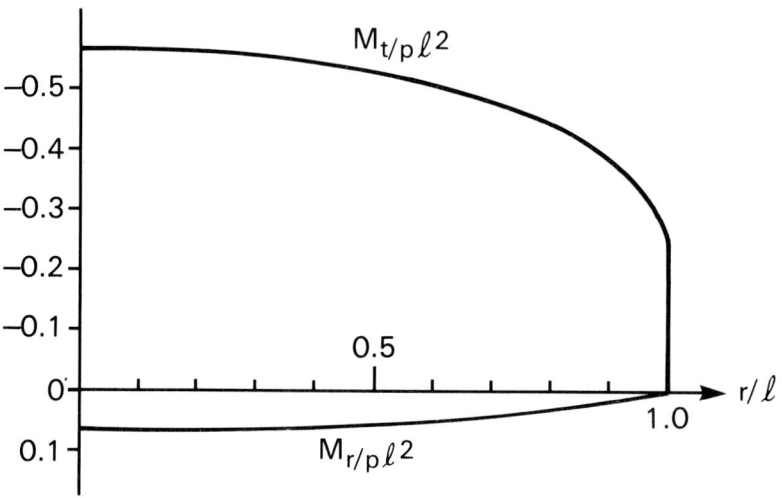

Fig. 4.   Values of M_t and M_r

$$\int_0^{\sqrt{(l^2-y^2)}} q_r dx = \frac{pl}{\pi} \int_0^{\sqrt{(l^2-y^2)}} \frac{dx}{\sqrt{(l^2-x^2-y^2)}} = \frac{pl}{2} \qquad (3)$$

i.e. the same as $p/2$ on a strip of length l.

He next goes on to show that at a point x on a simple x strip the moment M_{x1} due to the load $p/2$ only is

$$M_{x1} = -p(l-x)^2/4 \qquad (4)$$

and that due to the upward radial support reaction

$$M_{x2} = \{x \arcsin[x/\sqrt{(l^2-y^2)}] + \sqrt{(l^2-x^2-y^2)} - \pi x/2\}pl/\pi \qquad (5)$$

Within the quarter circle he next determines the values of the lower bound tangential and radial moments M_t and M_r, which are consistent with $2q_r$. The values of M_t and M_r along any radius are shown in Fig. 4 and it can be observed M_t is always negative and M_r always positive.

Hillerborg's advanced strip method

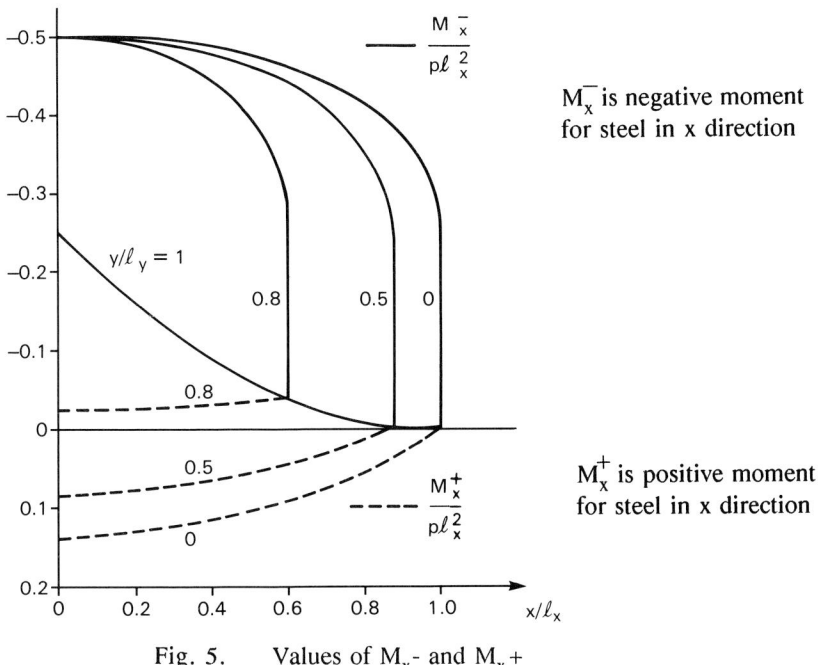

Fig. 5. Values of M_x^- and M_x^+

Earlier in his book Hillerborg shows that in a polar symmetrical field if the value of K in the H-W-A reinforcement rules for positive reinforcement is taken as $|\sin\theta/\cos\theta|$ when $M_r > M_t$ (which it is in this case) then due to M_r and M_t only

$$M_x^+ = M_r \cos^2\theta + M_t \sin^2\theta + |(M_r - M_t) \sin\theta \cos\theta . \sin\theta/\cos\theta| \quad (6)$$

i.e. $M_x^+ = M_r$ \hfill (7)

For the negative reinforcement by a similar process

$$M_x^- = M_t \quad (8)$$

In the load distribution element therefore by adding the various moment effects from equations 4, 5, 7 and 8

$$M_x^- = M_{x1} + M_{x2} + M_t \quad (9)$$

and $M_x^+ = M_{x1} + M_{x2} + M_r$ \hfill (10)

it being noted M_t and M_r are zero when $x^2 + y^2 > l^2$.

Values of M_x and M_x calculated by Hillerborg from equations 9 and 10 are given in Tables 1 and 2. Although the element studied originally was square, by the affinity rules the

x/l_x

y/l_y	0	0.1	0.2	0.3	0.4	0.5	0.6	0.7	0.8	0.9	1.0
0	500	499	497	493	488	479	467	450	424	382	250
0.1	500	499	497	493	487	478	466	449	422	379	0
0.2	500	499	497	493	487	477	464	445	417	369	0
0.3	500	499	497	492	485	475	460	439	407	349	0
0.4	500	499	496	491	483	471	454	429	390	307	0
0.5	500	499	495	489	479	465	444	413	360	3	0
0.6	500	499	494	486	473	455	428	385	260	3	0
0.7	500	498	492	481	464	439	400	317	10	3	0
0.8	500	497	488	473	447	406	290	23	10	3	0
0.9	500	495	478	450	392	63	40	23	10	3	0
1.0	500	203	160	123	90	63	40	23	10	3	0

Table 1. Values of $-1000\, M_x^-/pl_x^2$ for a load-distribution element. The values have a discontinuity of 250 where $(x/l_x)^2 + (y/l_y)^2 = 1$

x/l_x

y/l_y	0	0.1	0.2	0.3	0.4	0.5	0.6	0.7	0.8	0.9	1.0
0	137	135	131	124	114	101	86	68	47	24	0
0.1	135	133	129	122	112	99	84	66	45	23	0
0.2	128	127	122	115	106	93	78	60	41	19	0
0.3	117	116	112	105	95	83	69	51	32	11	0
0.4	101	100	96	90	80	69	55	39	21	1	0
0.5	80	79	75	69	61	50	37	22	6	-3	0
0.6	53	52	48	43	35	26	15	2	-10	-3	0
0.7	17	16	14	9	3	-4	-12	-20	-10	-3	0
0.8	-29	-30	-31	-34	-36	-39	-40	-23	-10	-3	0
0.9	-94	-94	-92	-90	-84	-63	-40	-23	-10	-3	0
1.0	-250	-203	-160	-123	-90	-63	-40	-23	-10	-3	0

Table 2. Values of $1000\, M_x^+/pl_x^2$ for a load-distribution element.

reinforcement is governed by the square of the lengths of the element so more generally the Tables can be labelled with units of pl_x^2. Some of these values are reproduced in Fig. 5 and a similar set exists for M_y, M_y. All along $x = 0$, there is a uniform negative edge value of $-pl_x^2/2$ and if one designed for this edge value then at no point in the element is this negative value exceeded as can be observed from Table 1 or Fig. 5. Comments on the M_x values will be made later.

At this stage two points are worth noting. First his use of K in the H-W-A rules which varies between 0 and infinity, whereas it is well known that locally a K value of 1 gives the minimum value of steel at a point. Using a value of $K = 1$, which is almost universal in computer programs, is however only the most economical over an area if one continuously changes the reinforcement; which is totally impractical. There is therefore scope for more research into the best value of K to use when coupled with practical reinforcement placing as demonstrated by Jones and Wood(3). The second interesting point is that if an isolated square slab footing is considered then yield-line analysis would require bottom reinforcement for the pattern in Fig. 6a of $m = pl^2/2$. For the pattern in Fig. 6b; if we disregard the distributed load on the fan

$$P = 4pl^2 = 2\pi m(1+i) \tag{11}$$

which, with an m value of $pl^2/2$, gives $i = (4-\pi)/\pi = 0.273$ and an im value of $0.137\ pl^2$. Thus yield-line analysis, when taking into account the load direction, would require

$$M_x^- = M_y^- = pl^2/2$$

along the column line; and at the column

$$M_x^+ = M_y^+ = 0.137pl^2$$

which are identical with Hillerborg's lower bound values in Tables 1 and 2. Returning to Hillerborg's theory, he clearly recognised that this particular load distribution element would require top steel over its whole area and it would be preferable to concentrate this into a narrower band near the column. His next step therefore is to postulate that his original column element area can carry its loads to load distribution elements shown hatched in Fig. 7 in five different ways. For simplification in what follows it will be assumed $\alpha = \beta = 1/2$ which gives sensible half element widths for the column strips.

Case 1 in Fig. 7 is his original element. In Case 2 the load outside the shaded area is carried by primary strip action in the x direction onto the load distribution element. In Case 3, initially the load to the right of the shaded area is carried by primary strip action in the x direction as in case 2, then the resulting load on the area 'A', size αl_x by $(1-\beta)l_y$ is carried onto the load distribution element by secondary strip action in the y direction. Cases 4 and 5 and similar to Cases 2 and 3 with the x and y directions transposed. The reinforcement in all cases is then obtained by a summation of simple strip

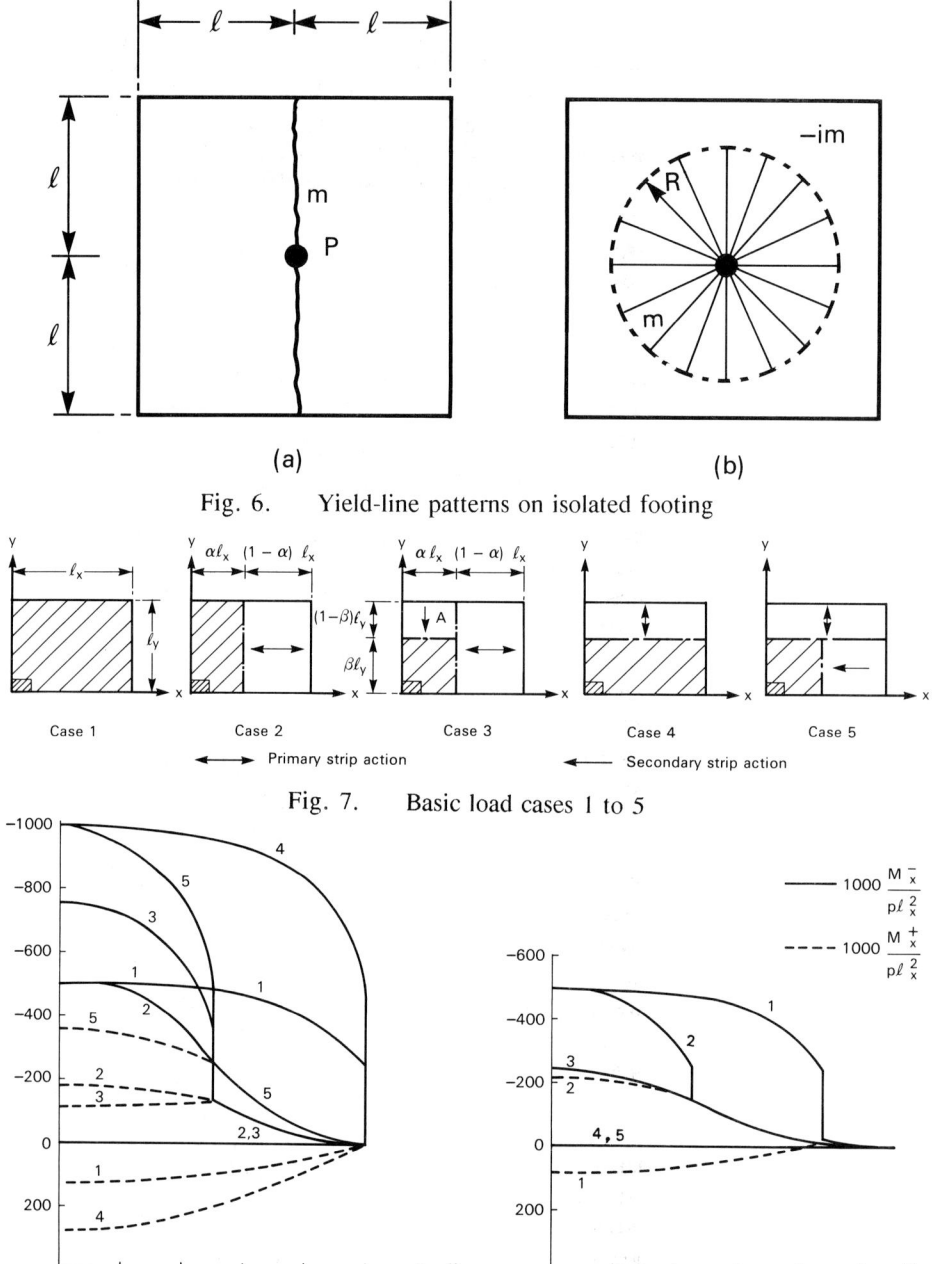

Fig. 6. Yield-line patterns on isolated footing

Fig. 7. Basic load cases 1 to 5

Fig. 8. Variation in M_x^-, M_x^+ for load cases 1 to 5

action and the reinforcement, suitably factored, from Tables 1 and 2 for the load distribution element.

The reinforcement calculations for the five cases are relatively easy and Hillerborg's values for all five cases along $y = 0$ and at $y = l_y/2$ (just outside the distribution element for cases 3, 4 and 5) for both the positive and negative steel are given in Table 3. The author has added an extra value in brackets at $y = 0$, $x = l_x/2$, for the M_x^- values since in cases 2, 3 and 5 there are discontinuities at this point. For pictorial purposes these tabular values are shown in Figs 8a and b.

It will be noted that some of the M_x^+ values are negative which appears nonsense but it must be remembered these values are for elements in which the positive field moment is zero and the field moment M_f has still to be added to the values in Table 3 to find the final reinforcement values.

If the external load p on the column element is divided into five parts, $\gamma_1 p$ to $\gamma_5 p$ with each part being carried as indicated by load cases 1 to 5 then it is possible by addition to achieve any desired negative support conditions. It is necessary of course that $\Sigma \gamma = 1$ and that no γ value is negative. A convenient initial edge moment distribution from the five load cases is of the general form shown in Fig. 9a, to which one can add the field (span) moments in Fig. 9b to achieve the final edge moments in Fig. 9c. Such an element only has negative reinforcement for half the width of the column element and uniform field moments which is suitable from the continuity aspect with other elements. Elements such as those in Fig. 9c will be termed double-intensity elements since the negative reinforcement is twice that of the average negative moment.

It may seem that the designer must know the γ values but this is not the case. As far as the designer is concerned provided he keeps within certain constraints, mentioned later, he need have no knowledge of load distribution elements or how the load is distributed between the five loading cases and the reinforcement is governed only by the edge moments of the element and these values will not be exceeded anywhere within the element. This is clearly evident for the negative moment from Fig. 8 where it can be seen they always decrease below the value at $x = 0$. It is the positive moments which it will be seen later cause difficulties.

As an example consider the symmetrical slab in Fig. 10a. The designer must first choose an average negative support moment over the column. Hillerborg suggests the designer can imagine a line or beam support over the columns and therefore for an x strip XX in Fig. 10b we will choose to make the average support moment the same as the average span moment, i.e. $M_{sx} = M_{fx} = pl_x^2/4$. The extent of the column element in the x direction is

$l_x = l/2 + M_{sx}/pl = 0.586l$ giving $M_{sx} = M_{fx} = p(0.586l)^2/4 = 0.086pl^2$.

If we choose to have a symmetrical double-intensity element we will have $2 \times 0.086pl^2 =$

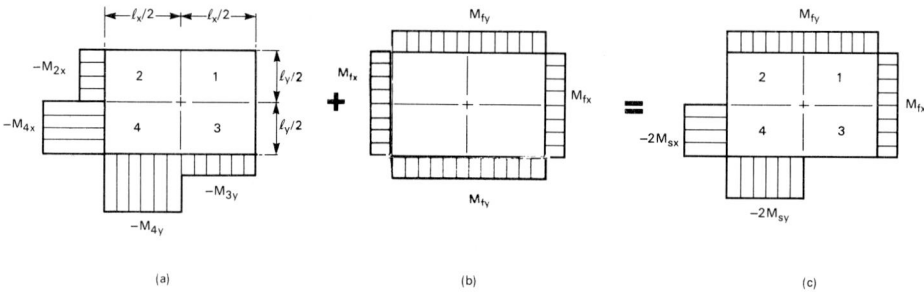

Fig. 9. Edge moments on double-intensity element

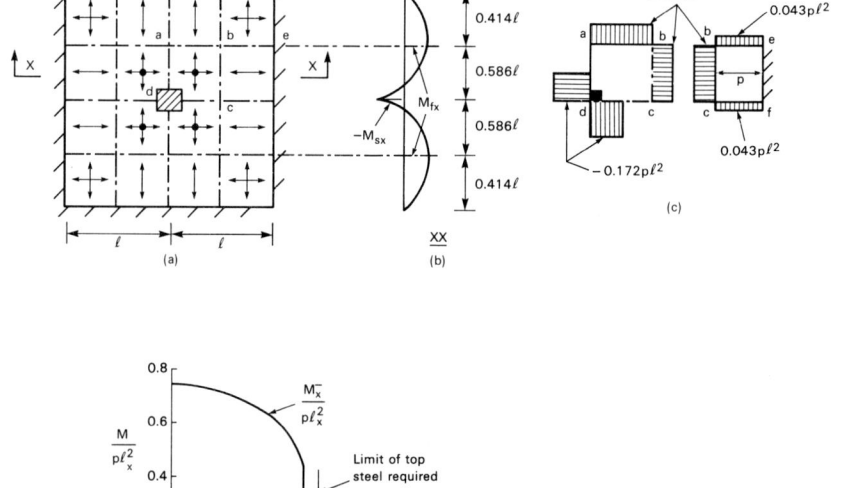

Fig. 10. Design example with double-intensity element

Hillerborg's advanced strip method

0.172pl^2 over half the element width and a uniform span moment $M_{fx} = 0.086pl^2$ as shown in Fig. 10c. The values in the y direction will be similar. At this stage it is assumed these edge reinforcements are carried across the whole column element. Outside the column element the moments are determined by simple strip action and it is easily checked that the field moment along bc is compatible with element bcfe which is chosen to span in one direction and of length 0.414l therefore also giving a span moment of p(0.414l)^2/2 = 0.086pl^2 along bc. If the load in the corners is taken as p/2 in each direction, the reinforcement in the edge strips parallel to the sides will be p(0.414l)^2/4 = 0.043pl^2, and this completes the design.

It is interesting to pursue this example to see how the double intensity solution is achieved and that nowhere inside the element are the edge values exceeded. In general, before any span moments are added, let the edge moments for the four areas 1-4 in Fig. 9a be $-M_{2x}$, $-M_{4x}$, $-M_{4y}$ and $-M_{3y}$ respectively. For an element length l_x in this example we have chosen to make the final average support moment $M_{sx} = M_{fy} = pl_x^2/4$. To ultimately achieve a double-intensity element we initially will require

$$-M_{4x} = -2M_{sx} - M_{fx} \text{ and } -M_{2x} = -M_{fx}$$

so after adding M_{fx}, Fig. 9b we finally have

$$-M_{4x} = -2M_{sx} - M_{fx} + M_{fx} = -2M_{sx};$$

and $\quad -M_{2x} = -M_{fx} + M_{fx} = 0$, Fig. 9c.

Let $\gamma_1 = \gamma_3 = \gamma_5 = 1/3$ and $\gamma_2 = \gamma_4 = 0$, ie $\Sigma\gamma = 1$.

The initial value of $-M_{4x}$ can be obtained by multiplying the five γ values by their relevant coefficients in Table 3a at y = 0 and x = 0, we have

$$-M_{4x} = -(0.5/3 + 0.5 \times 0 + 0.75/3 + 1 \times 0 + 1/3)pl_x^2 = -0.75pl_x^2.$$

The values in Table 3 are for zero field moments so to obtain the final value of $-M_{4x}$ we must add $M_{fx} = pl_x^2/4$ to give

$$-M_{4x} = -pl_x^2/2 \text{ or } p(0.586l)^2/2 = -0.172pl^2 = -2M_{sx}$$

as obtained previously. Also from Table 3b for M_x at y = l/2, x = 0; by the same process

$$-M_{2x} = -(0.5/3 + 0.5 \times 0 + 0.25/3 + 0 \times 0 + 0/3) = -0.25pl_x^2$$

which after adding the field moment $M_{fx} = 0.25\ pl_x^2$ will give zero on the outer edge. It might be noted from Table 3c that the initial positive reinforcement required along edge 4 is $\qquad(0.137 - 0.113 - 0.363)pl_x^2/3 = -0.133pl_x^2$

and from Table 3d that along edge 2 is

$$(0.08 - 0.25 + 0)pl_x^2/3 = -0.057pl_x^2.$$

Both values are negative, indicating that no positive steel other than M_{fx} is required. The variation in the reinforcing moment values across the whole element can be obtained by adding the various rows in Table 3 multiplied by the γ values. To these must be added M_{fx} which can then be used as the base line as shown in Fig. 10d. It will be observed that as Hillerborg intends at no point in the element is the reinforcement greater than the edge values, which is all the designer has to decide. This however is only because the support moment choice was within Hillerborg's limits which are now outlined.

One of the conditions Hillerborg wishes to achieve is that the edge moments are the design moments and nowhere within the element will these be exceeded. From Table 3, or Fig. 8 it is clear this is easily fulfilled for the negative reinforcement. For the positive reinforcement this means in effect that with zero field moment then no positive reinforcement values can be allowed for any combination of the γ values. Examination of Table 3c and d clearly shows that γ_1 and γ_4 produce positive values in the x direction, and γ_1 and γ_2 will therefore produce positive values in the y direction. Hillerborg sets up a

Moment	Basic case	0	0.1	0.2	0.3	0.4	0.5	0.6	0.7	0.8	0.9	
M_x^- $y=0$	1	-500	-499	-497	-493	-488	-479	-467	-450	-424	-382	
	2	-500	-494	-474	-439	-382	-250 (-125)	-80	-45	-20	-5	(a)
	3	-750	-742	-718	-672	-594	-375 (-125)	-80	-45	-20	-5	
	4	-1000	-999	-994	-987	-975	-958	-934	-899	-845	-763	
	5	-1000	-987	-948	-877	-764	-500 (-250)	-160	-90	-40	-10	
M_x^- $y=l_y/2$	1	-500	-499	-495	-489	-479	-465	-444	-413	-360	-3	
	2	-500	-493	-469	-427	-350	-125	-80	-45	-20	-5	
	3	-250	-245	-230	-205	-170	-125	-80	-45	-20	-5	(b)
	4	0	0	0	0	0	0	0	0	0	0	
	5	0	0	0	0	0	0	0	0	0	0	
M_x^+ $y=0$	1	137	135	131	124	114	101	86	68	47	24	
	2	-182	-180	-173	-162	-146	-125	-80	-45	-20	-5	
	3	-113	-114	-116	-119	-123	-125	-80	-45	-20	-5	(c)
	4	273	270	262	247	227	202	171	135	94	49	
	5	-363	-359	-346	-324	-293	-250	-160	-90	-40	-10	
M_x^+ $y=l_y/2$	1	80	79	75	69	61	50	37	22	6	-3	
	2	-210	-207	-200	-186	-167	-125	-80	-45	-20	-5	
	3	-250	-245	-230	-205	-170	-125	-80	-45	-20	-5	(d)
	4	0	0	0	0	0	0	0	0	0	0	
	5	0	0	0	0	0	0	0	0	0	0	

x/l_x

Table 3. Values of $1000\ M_x/pl_x^2$ for load cases 1-5, with zero field moments

Hillerborg's advanced strip method 159

number of equations involving γ_1 to γ_5 to ensure that the summation of any five load cases will not allow M_x^+ to exceed zero at the column (or 5% of the total static moments $pl_x^2/2$ or $pl_y^2/2$ at some distance away from the origin). In the process of doing this, he defines two coefficients κ_x and κ_y which were initially defined as

$$\kappa_x = \beta(M_{4x} - M_{2x})/(pl_x^2/2) \tag{12}$$

and $$\kappa_y = \alpha(M_{4y} - M_{3y})/(pl_y^2/2) \tag{13}$$

By observation of Fig. 9a (where it was chosen to have $\alpha = \beta = 1/2$) these coefficients indicate what proportion of the static moments are carried by the difference in moment between the parts of the edges between 0 to βl_y; and $\beta\lambda_y$ to l_y; and 0 to αl_x; and αl_x to l_x. By considering general values of the edge moments M_{4x}, M_{2x}, M_{4y} and M_{3y} Hillerborg establishes that

$$\kappa_x = \alpha\gamma_3 + \gamma_4 + \gamma_5 \tag{14}$$

and $$\kappa_y = \gamma_2 + \gamma_3 + \beta\gamma_5 \tag{15}$$

From equations 14 and 15 and six other equations limiting the value of M_x^+ and M_y^+, after what must have been extensive calculations, Hillerborg produces a series of charts for upper and lower bound limits for κ_x and κ_y for values for α and β values both varying between 0.1 to 0.7. Provided the choice of κ_x and κ_y are between these limits Hillerborg guarantees in effect that no positive reinforcement (other than the edge values M_{fx} and M_{fy}) will be required and the γ values need not be known. In the particular case we have been considering earlier, namely $\alpha = \beta = 1/2$, by abstracting from his charts, the lower limit of both κ_x and κ_y is 0.3 and the upper limit $\kappa_x + \kappa_y \not> 1.5$. As a simplification he chooses to set the limits to

$$0.3 \leq \kappa \leq 0.75 \tag{16}$$

In the design example earlier, Fig. 10, with a double-intensity element we had $M_{4x} = pl_x^2/2$ and $M_{2x} = 0$ hence from equation 12

$$\kappa_x = (1/2 - 0)pl_x^2/pl_x^2 = 0.5$$

with a similar value for κ_y and hence within Hillerborg's limits of equation 16. It was unnecessary to know the γ values but from the γ values selected, which must satisfy equations 14 and 15 and have $\Sigma\gamma = 1$, it was possible to demonstrate the moments within the element did not exceed the edge moments.

If one considers an edge column, or a corner column where either $M_{sx} = 0$ or $M_{sy} = 0$, or both, then clearly from the definition given by equations 12 and 13 either κ_x or κ_y, or both, are equal to zero and this is outside the limits for κ. In his book(1) for such cases Hillerborg does not then use a constant field moment M_f but adds a value M_{ff}, say, to the inside half edge and deducts it from the outer half, thus maintaining overall equilibrium of the element. However this device begins to cause difficulties of continuity with adjacent elements designed by the simple strip method.

In his paper (2) for the general set of moments given in Fig. 11 where with the author's notation 'i' is used for inner and 'o' for outer, Hillerborg defines his limits as

$$M_{sxo} + M_{fxo} = cpl_x^2/2 \qquad (17)$$

with $\qquad 0.25 \leq c \leq 0.7 \qquad (18)$

Now M_{sx} is the mean of M_{sxo} and M_{sxi} etc. and the static moment equation is

$$(M_{sxo} + M_{fxo})/2 + (M_{sxi} + M_{fxi})/2 = pl_x^2/2 \qquad (19)$$

so that equation 17 may be re-expressed as

$$\kappa_x = [(M_{sxi} + M_{fxi})/2 - (M_{sxo} + M_{syo})/2]/pl_x^2/2 \qquad (20)$$

with the limits

$$0.3 \leq \kappa_x \leq 0.75 \qquad (21)$$

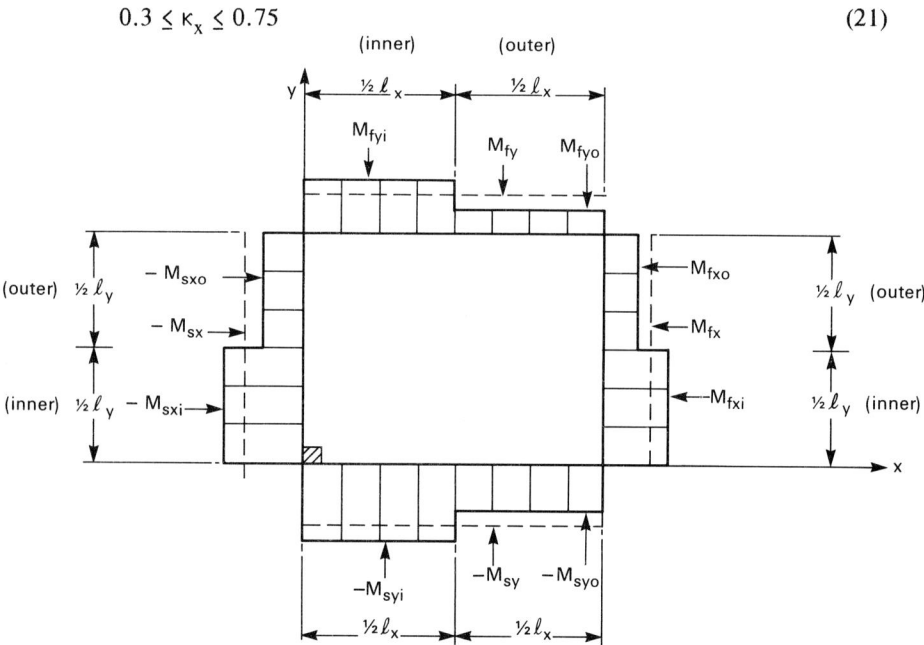

Fig. 11. General set of edge moments

Hillerborg's advanced strip method

i.e. the same limits as equation 16 but now with κ being the difference between the total (arithmetic) edge moments on the inner band and the total on the outer band. The previous definition equation 12 included only the difference between the support moments but in his book Hillerborg clearly uses equation 20 even if it is never expressed in this explicit form. In order to keep within the κ limits when M_{sx} or M_{sy} = 0 Hillerborg must have non-uniform field moments which leads to unnecessary complications in adjoining elements. For example, in the problem (adapted from page 184, Fig. 6.84 of his book (1)) shown in Fig. 12a, in the simple strip element in the area nthm he has to resort to load distributions of 4p/3, -p/3; and 2p/3, p/3 to obtain continuity of moments because of the non-uniform moments along nt (forced in order to keep within the κ limits) and negative steel is required along the whole of th. The loading is shown in area achm and the moments in cdfh. While it is not necessarily a bad solution, if no κ restrictions are imposed a double-intensity element could be used at column c, with uniform field moments, which would give the simple solution shown in Fig. 12b. The difficulty with the latter solution, as will be seen later, is that the positive edge moments are no longer the greatest positive moments in the column element and some additional positive steel near the columns is required since with the moment distribution in Fig. 12b in the case of element abnr $\kappa_x = \kappa_y = 0$ and for element bctr $\kappa_x = 0.55$ and $\kappa_y = 0$. Both elements with these κ values are outside Hillerborg's limits but this example demonstrates how in the solution in Fig. 12a the designer often has to adjust the load distribution on adjoining elements to keep within Hillerborg's limits, and even then in this case a double-intensity element

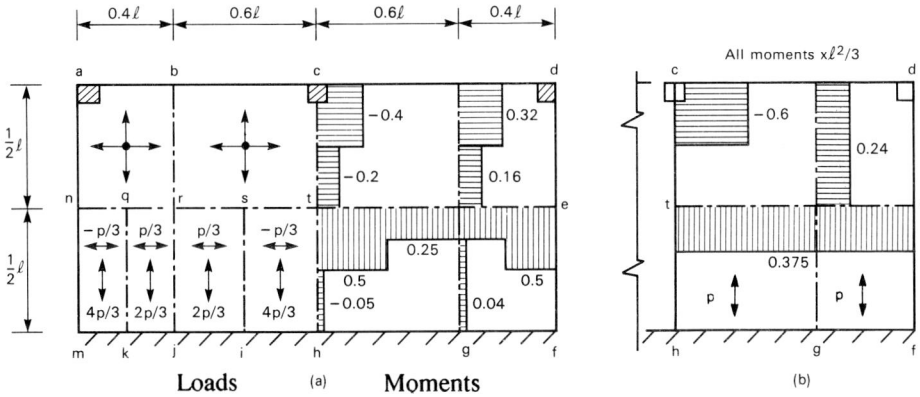

Fig. 12. Design example-alternative designs

cannot be achieved. This seems an unnecessary restriction on the designer and Dr Wood and the author therefore began to examine how to extend Hillerborg's work so that double-intensity elements could be used irrespective of the κ values if the designer so wished, albeit at the cost of extra positive steel. One other problem also needed to be examined, namely the extent of the negative support steel. Hillerborg notes from steel distribution diagrams such as Figs. 8 and 10 that negative steel often is not required beyond about 0.6l and suggests this is a suitable cut-off point. Other diagrams clearly show that some steel is required across the whole element and therefore a more rigorous examination of this problem was needed.

EXTENSIONS TO HILLERBORG'S THEORY

The objectives of the extensions were to find:

(i) by how much the negative support steel could be reduced beyond 1/2 which is the edge of the load distribution element for load cases 2, 3, 4 and 5;

(ii) what was the extent of the reduced steel;

(iii) what additional positive steel was necessary if the designer chose to work outside Hillerborg's κ limits and wished to use double-intensity elements; and

(iv) whether a set of γ values could be defined which could be used for any support conditions thus allowing the actual moment variation across a column element to be found.

a) Relationship between support moments and γ values

Referring back to Fig. 9, double-intensity elements can be achieved if and when the set of edge moments in Fig. 9a have the field moments in Fig. 9b added to them to achieve the moments in Fig. 9c thus giving negative support moments of twice the average value over half the element length. This condition for the initial set of edge moments requires

$$-M_{2x} + M_{fx} = 0;$$

i.e. $\quad M_{2x} = M_{fx}$ \hfill (22)

$$-M_{4x} + M_{fx} = -2M_{sx}$$

i.e. $\quad M_{4x} = M_{fx} + 2M_{sx}$ \hfill (23)

Similarly $\quad M_{3y} = M_{fy}$ \hfill (24)

and $\quad M_{4y} = M_{fy} + 2M_{sy}$ \hfill (25)

Hillerborg's advanced strip method

From Table 3b at $x = 0$ the contributions to the edge moment M_{2x} for γ_1 to γ_5 are $pl_x^2/2$, $pl_x^2/2$, $pl_x^2/4$, 0 and 0 respectively. The values for M_{4x} are found from Table 3a. All four edge moments M_{2x}, M_{4x}, M_{3y} and M_{4y} are shown collectively in Table 4 as a set of γ coefficients expressed non-dimensionally as proportions of $pl_x^2/2$ and $pl_y^2/2$. The values of M_{3y} are obtained from the M_{2x} values by transposing γ_2 for γ_4 and γ_3 for γ_5. Similarly for M_{4y} from M_{4x}.

PARTIAL LOADING CASE

Negative edge moment value	γ_1	γ_2	γ_3	γ_4	γ_5
$M_{2x}/(pl_x^2/2) = [m_{2x}]$	1	1	0.5	0	0
$M_{4x}/(pl_x^2/2) = [m_{4x}]$	1	1	1.5	2	2
$M_{3y}/(pl_y^2/2) = [m_{3y}]$	1	0	0	1	0.5
$M_{4y}/(pl_y^2/2) = [m_{4y}]$	1	2	2	1	1.5

TABLE 4 Contributions to negative edge-moments from partial-loading cases, as multiples of $\gamma_1 \ldots \gamma_5$, when the field moments are zero.

The value of any moment such as $M_{2x}/(pl_x^2/2)$ is found by multiplying the relevant coefficients for that moment in Table 4 by the chosen γ values and adding the products.

Writing non-dimensionally $m_{sx} = M_{sx}/(pl_x^2/2)$ generally, with suffices x and y, then equation 22 and the coefficients for the γ values in Table 4 lead to

$$m_{2x} = m_{fx} = \gamma_1 + \gamma_2 + 0.5\gamma_3 \tag{26}$$

it being noted that the γ_4 and γ_5 coefficients are zero.

Similarly, from equations 23 – 25,

$$m_{4x} = m_{fx} + 2m_{sx} = \gamma_1 + \gamma_2 + 1.5\gamma_3 + 2\gamma_4 + 2\gamma_5 \tag{27}$$

$$m_{3y} = m_{fy} = \gamma_1 + \gamma_4 + 0.5\gamma_5 \tag{28}$$

$$m_{4y} = m_{fy} + 2m_{sy} = \gamma_1 + 2\gamma_2 + 2\gamma_3 + \gamma_4 + 1.5\gamma_5 \tag{29}$$

together with

$$1 = \gamma_1 + \gamma_2 + \gamma_3 + \gamma_4 + \gamma_5 \tag{30}$$

Hillerborg forms similar equations but not quite in this form and states any "desired theoretical distribution of edge moments ... can be obtained by a suitable choice of positive γ values, one of the values being chosen and the others calculated from equations....". This is not true, there are only 3 independent equations, equation 30 and the two total element moment equilibrium equations, namely

$$m_{sx} + m_{fx} = 1 \tag{31a}$$

and $\quad m_{sy} + m_{fy} = 1 \tag{31b}$

It is not therefore possible to solve the five equations directly and one must choose two of the γ values. One way out of this dilemma is to take two arbitrary γ values to the other side of the equations and treat them as 'known' (arbitrary) quantities. The obvious choices are γ_2 and γ_4, since they may often be zero. From equation 26

$$\gamma_3 = 2m_{fx} - 2\gamma_1 - 2\gamma_2$$

which using equation 31a gives γ_3 in terms of support moments, thus:

$$\gamma_3 = 2 - 2m_{sx} - 2\gamma_1 - 2\gamma_2 \tag{26a}$$

Similarly from equations 28 and 31b,

$$\gamma_5 = 2 - 2m_{sy} - 2\gamma_1 - 2\gamma_4 \tag{28a}$$

Substituting these values of γ_3 and γ_5 into equation 30 gives

$$\gamma_1 = [3 - 2m_{sx} - 2m_{sy} - \gamma_2 - \gamma_4]/3 \tag{32}$$

and back substitution of this value of γ_1 into equations 26a and 28a gives

$$\gamma_3 = [2(-m_{sx} + 2m_{sy}) - 4\gamma_2 + 2\gamma_4]/3 \tag{33}$$

$$\gamma_5 = [2(2m_{sx} - m_{sy}) + 2\gamma_2 - 4\gamma_4]/3 \tag{34}$$

It will have been noticed that equations 27 and 29 have not been used. This choice was arbitrary, only two of equations 26-29 can be used and the result is always the same whichever two are chosen.

Equations 32-34 show that if we choose arbitrary positive values for both γ_2 and γ_4 then together with the known values of m_{sx} and m_{sy} at the supports the values of γ_1, γ_3 and γ_5 can be determined, it being noted that only positive values of γ_2 and γ_4 which lead to positive values of γ_1, γ_3 and γ_5 are acceptable. Several important limitations on the values of the support moments can be deduced from these equations. Thus even with $\gamma_2 = \gamma_4 = 0$ from equation 32 if γ_1 is to remain positive then the sum of the support moments

$$m_{sx} + m_{sy} \not> 1.5 \tag{35}$$

Other more general limits can also be deduced but not pursued here. It is now possible to consider some of the previously stated objectives.

b) **Intensity of top reinforcement carried beyond $l_x/2$ and $l_y/2$**

If steel in the x direction is considered first it is necessary to calculate the value of the moments at $l_x/2$ from the column, i.e. along the boundary between areas 4 and 3 in Fig. 9. The final value of the edge moment $2m_{sx}$ is made up from the edge moment m_{4x} value, defined in Table 4, which was partially cancelled (reduced in value) by m_{fx} to leave the double intensity moment $2m_{sx}$. It is the way in which m_{4x} is formed which is important. Now the various m_{4x} values were formed from the basic moments on the distribution elements, and in addition in most cases primary and secondary strip action. Let the moment in the x direction at the interface between area 4 and 3 be M_{3x} and adopting the same notation as before let $m_{3x} = M_{3x}/(pl_x^2/2)$. If any reduction in the required negative reinforcement at y = 0 in Table 1, i.e. from 0.5 at x = 0 to 0.479 at x = $l_x/2$, is ignored when the distribution element extends into area 3, then for partial load cases 1 and 4 Fig. 7 the contribution to m_{3x} (assumed equal to m_{4x}) will, from the coefficients in Table 4, be $1.\gamma_1 + 2.\gamma_4$. For load cases 2, 3 and 5 the distribution elements stop at the boundary between areas 4 and 3 so that their effect is zero beyond $l_x/2$. Consideration of Fig. 7 shows that for load case 2 at $l_x/2$ from the column there is a primary strip moment value

$$|-\gamma_2 \cdot p \cdot l_x/2 \cdot l_x/4| = \gamma_2 p l_x^2/8$$

which after dividing by $pl_x^2/2$ gives a contribution to m_{3x} of $\gamma_2/4$. The primary strip action in load case 3 is the same as in 2 so this also gives a contribution of $\gamma_3/4$. For load case 5 there is primary strip action from areas 1 and 2 onto 3 and 4 so that the secondary strip action from area 3 and 4 is in effect from a load of 2p thus giving due to secondary strip action a contribution to m_{3x} of $\gamma_5/2$. Hence adding the effects of γ_1 to γ_5 the maximum reinforcement C_x carried over (negative) into region 3 in the x-direction can be expressed conservatively in the form of a carry-over coefficient, in units of $pl_x^2/2$, namely

$$C_x = \gamma_1 + 2\gamma_4 + (\gamma_2 + \gamma_3)/4 + \gamma_5/2 - 1 + m_{sx} \tag{36}$$

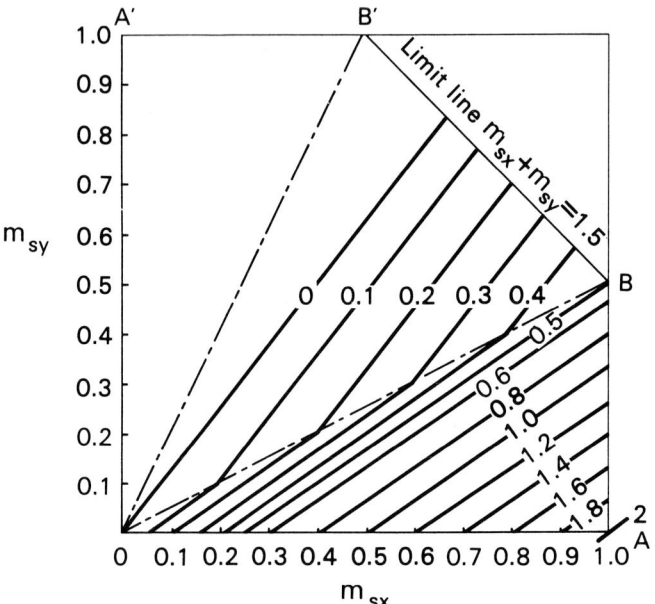

Fig. 13. Contours of $C_x = M/(pl_x^2)$ at $l_x/2$
(Uniform field moments only)

since the total negative moment in region 3 is partly cancelled by $m_{fx} = 1 - m_{sx}$. Substituting for γ_1, γ_3 and γ_5 from equations 32 to 34 leads to

$$C_x = 5m_{sx}/6 - 2m_{sy}/3 - \gamma_2/12 + 7\gamma_4/6 \tag{37}$$

Similarly

$$C_y = -2m_{sx}/3 + 5m_{sy}/6 + 7\gamma_2/6 - \gamma_4/12$$

There is no obvious minimum for C_x, since γ_2 and γ_4 were made arbitrary, in order to determine the γ_1, γ_3 and γ_5 values for any given values for m_{sx} and m_{sy}. Alternative values for C_x are clearly possible, so a search for minimum values of C_x could be laborious. However, a linear contour diagram for C_x might be expected from the nature of the equations, albeit with discontinuities due to different combinations of γ values. Such a linear diagram for C_x as a function of the support moments m_{sx}, m_{sy} shown in Fig. 13 was eventually derived by an intuitive examination of controlling special cases, which search is highly instructive.

Thus at the origin, $m_{sx} = m_{sy} = 0$, C_x is clearly zero if γ_2 and γ_4 are zero in equation 37, as also are γ_3 and γ_5 from equations 33 and 34. This leaves $\gamma_1 = 1$ only, and from Table 4 shows that $m_{4x} = m_{2x} = m_{3y} = m_{4y} = 1$. This refers to the basic distribution element where the support moments are exactly cancelled by field moments, $m_{fx} = 1$, $m_{fy} = 1$, clearly implying a corner column, with free edges on the 'support' lines. There is no negative reinforcement at all, and clearly no carry-over C_x or C_y.

Hillerborg's advanced strip method

At the other extreme where $m_{sx} = 1$, $m_{sy} = 0$, i.e. end A of the base line, Fig. 13, the support moment implies that this is an edge column case with the free edge in the x-direction ($m_{sy} = 0$); and in addition, the maximum possible average support moment about the y-axis ($m_{sx} = 1$). It is not immediately obvious that this implies $\gamma_4 = 1$ only. However, from Table 4 with $\gamma_4 = 1$, $m_{3y} = m_{4y} = 1$, which means that we can add a constant moment $m_{fy} = 1$ and make $m_{3y} = m_{4y} = 0$. Likewise in the x-direction $m_{2x} = 0$, $m_{4x} = 2$, i.e. average $m_{sx} = 1$. Any deviation from $\gamma_4 = 1$, $\gamma_2 = 0$ results in negative values for either γ_1, γ_3 or γ_5. The result is $C_x = 2$, from equation 36, which implies that the doubled-reinforcement $2m_{sx}$ must be carried right across l_x as might be expected. These two special cases, namely $\gamma_1 = 1$ at O and $\gamma_4 = 1$ at A, with all other γ values 0, suggest that all along OA, $\gamma_4 = m_{sx}$, $\gamma_1 = 1 - m_{sx}$, $\gamma_2 = \gamma_3 = \gamma_5 = 0$.

The next important point is B which is on the validity limit line $m_{sx} + m_{sy} = 1.5$ equation 35, shown as BB′. From equation 32 since γ_2 and γ_4 cannot be negative then clearly we must have $\gamma_1 = \gamma_2 = \gamma_4 = 0$ all along this limit line with, since $\Sigma\gamma = 1$, $\gamma_3 + \gamma_5 = 1$. The first term of equation 33 would suggest that line OB, i.e. $m_{sx} = 2m_{sy}$ might well be a discontinuity line so that we may consider area OAB as a whole. At O, A and B we have seen that $\gamma_2 = 0$ so that it is not unrealistic to assume it is zero over this whole region. If γ_4 is assumed to be linear of the form $\gamma_4 = a_4 + b_4 m_{sx} + c_4 m_{sy}$, then since $\gamma_4 = 0$ at $m_{sx} = m_{sy} = 0$ we find $a_4 = 0$; at A $\gamma_4 = 1$ hence $b_4 = 1$ and since $\gamma_4 = 0$ at B then in the area OAB $\gamma_4 = m_{sx} - 2m_{sy}$. Inserting these values of γ_2 and γ_4 in equations 32 to 34 gives, in area OAB

$$\gamma_1 = 1 - m_{sx}; \gamma_3 = 0; \gamma_5 = 2m_{sy}; \text{ with } \gamma_2 = 0; \gamma_4 = m_{sx} - 2m_{sy} \quad (38)$$

The first term of equation 34 again suggests the line OB′, i.e. $2m_{sx} = m_{sy}$, is a discontinuity line. Consequently using similar reasoning to that already employed the following values of γ in area OBB′ were found

$$\gamma_1 = 1 - 2(m_{sx} + m_{sy})/3; \gamma_3 = 2(-m_{sx} + 2m_{sy})/3;$$

$$\gamma_5 = 2(2m_{sx} - m_{sy})/3; \text{ with } \gamma_2 = 0; \gamma_4 = 0 \quad (39)$$

Finally, the same reasoning in area OB′A′ gives

$$\gamma_1 = 1 - m_{sy}; \gamma_3 = 2m_{sx}; \gamma_5 = 0; \text{ with } \gamma_2 = -2m_{sx} + m_{sy}; \gamma_4 = 0 \quad (40)$$

These equations have been used to construct the linear contour diagrams for γ_1 to γ_5 shown in Fig. 14. It can easily be tested that at every value of m_{sx} and m_{sy} that $\Sigma\gamma = 1$, and fascinating that at the five extreme points we have $\gamma_1 = 1$ at O, $\gamma_2 = 1$ at A′, $\gamma_3 = 1$ at B′, $\gamma_4 = 1$ at A and $\gamma_5 = 1$ at B. It should be noted however that these γ values are not unique since the choice of γ_2 and γ_4 is initially arbitrary.

If in area OAB the appropriate γ values are substituted into equation 37 for C_x we find

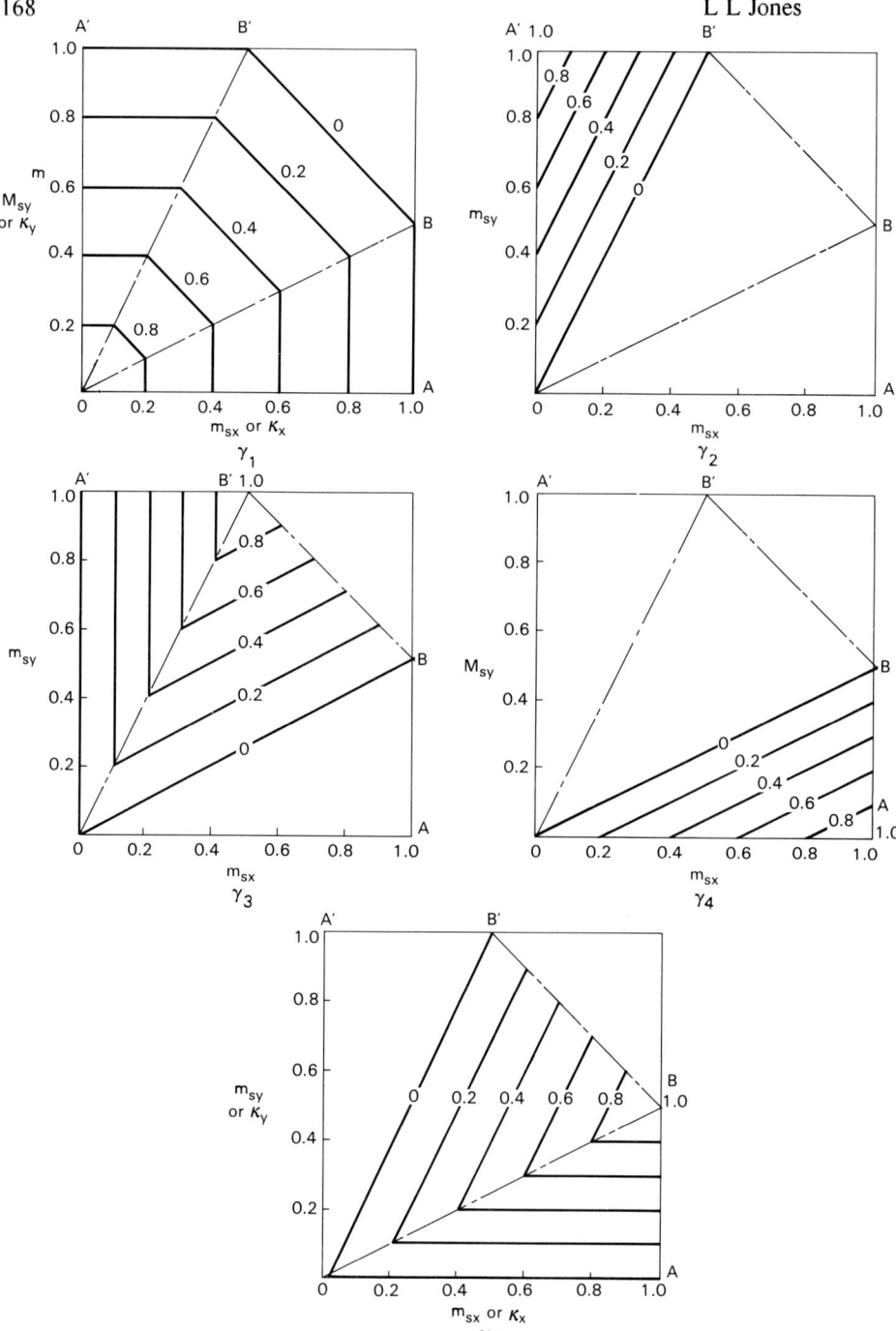

Fig. 14. Contours of $\gamma_1 \ldots \gamma_5$

Hillerborg's advanced strip method   169

$$C_x = 2m_{sx} - 3m_{sy} \text{ for area OAB} \tag{41}$$

and similarly for area OBB' it is found

$$C_x = 5m_{sx}/6 - 2m_{sy}/3 \text{ for area OBB'} \tag{42}$$

Equations 41 and 42 are plotted in Fig. 13. Thus in the process of obtaining a linear diagram for C_x a linear diagram for the five γ values has also been produced. What is perhaps even more interesting is that exhaustive tests of all other possible γ_2 and γ_4 values which give non-negative values of γ_1, γ_3 and γ_5 have shown that a minimum C_x is produced by these linear γ values in area OAB. In area OBB' this is not quite true but the differences are trivial. Values of C_x can be read by designers from Fig. 13, being the intensity of negative reinforcement resistance moment (in $pl_x^2/2$ units) to be carried over from area 4 into area 3, i.e. at $l_x/2$. To find C_y the axes m_{sx} and m_{sy} are transposed and the units are in terms of $pl_y^2/2$.

Examination of the general equations for κ_x and κ_y showed that a similar set of equations for the γ values could have been obtained in terms of κ_x and κ_y rather than m_{sx} and m_{sy}. Therefore more generally the charts in Fig. 14 might have been labelled κ_x and κ_y rather than m_{sx}, m_{sy}. Fig. 13 can also be labelled κ_x and κ_y provided the field moment is uniform, but if it is not the final term in equation 36 for C_x needs adjusting.

As an example of the use of Fig. 13, in the problem shown in Fig. 10 $\kappa_x = \kappa_y = m_{sx} = m_{sy} = 0.5$ and therefore from Fig. 13 we find $C_x \approx 0.08 \times pl_x^2/2 = 0.014 \, pl^2$. This small amount of steel required beyond $l_x/2$ is easily observed in Fig. 10d. In this case it would hardly be practical to make the reduction in view of the fact that all the steel can be stopped off at about $0.58 \, l_x$.

One of the initial assumptions in calculating C_x was that the $\overline{M_x}$ moment due to γ_1 and γ_4 remained constant up to $l_x/2$ at the edge value. However, the reduction can be included by writing 0.479 instead of 0.5 in the γ_1 and γ_4 coefficients in equation 36 to give

$$C_x = 0.958\gamma_1 + 1.916\gamma_4 + (\gamma_2 + \gamma_3)/4 + \gamma_5/2 - 1 + m_{sx} \tag{36a}$$

and substitution for the γ values in the region OBB' would give

$$C_x = -0.042 + 5.168 \, m_{sx}/6 - 3.832 \, m_{sy}/6 \tag{42a}$$

which is slightly different to equation 42. A similar allowance can be made to equation 41. The effect of such revised equations is to reduce the values of C_x slightly.

Before concluding this section it must be recalled Fig. 13 for C_x applies only to elements having a uniform field moment which, for continuity, is desirable. Should a designer wish to use different inner and outer field moment m_{fxi} and m_{fxo} then the last terms of equation

36 needs changing from $-m_{fx}$ to $-m_{fxi}$ and κ_x and κ_y included rather than m_{sx} and m_{sy}. Such equations have been developed since the γ values are still valid but they are not shown here.

c) Extent of C_x

The length of top-reinforcement carried beyond $l_x/2$ can now be found by considering equation 36 for C_x. If it is assumed that part due to the distribution element moments, $\gamma_1 + 2\gamma_4$ is assumed conservatively to stay constant. That part due to strip action from γ_2, γ_3 and γ_5 will rapidly decrease, parabolically. The actual intensity of top-reinforcement (moment), at a distance $\lambda_x l_x$ from the corner-column will be M_λ and writing

$$m_\lambda = M_\lambda/(pl_x^2/2)$$

then $\quad m_\lambda = \gamma_1 + 2\gamma_4 + 4(1 - \lambda_x)^2(\gamma_2/4 + \gamma_3/4 + \gamma_5/2) - m_{fx}$

The negative steel can be completely stopped when $m_\lambda = 0$. For region OBB' of the contours for C_x, if all the appropriate γ values are inserted it is eventually found that

$$m_{sy}/m_{sx} = 3(1 - \lambda_x)^2 + 1/2 \tag{43}$$

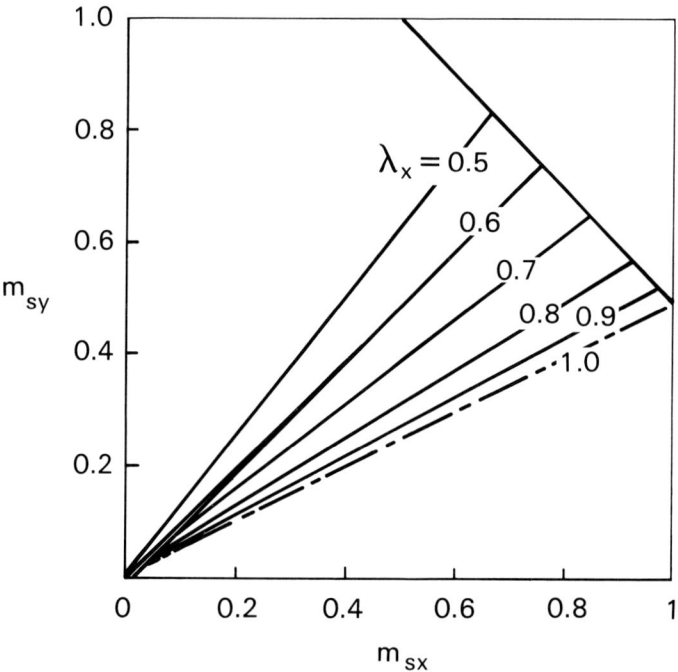

Fig. 15. Extent $\lambda_x l_x$ of C_x

Hillerborg's advanced strip method 171

These contours of λ_x, the length of the negative carry-over reinforcement, are shown in Fig. 15. These λ_x values are very conservative. They are based on the worst possible extent of negative support moment, namely along y = 0 in Table 1, and in addition not allowing for the reduction of reinforcement on the distribution elements. It serves however to demonstrate that contours can be obtained once a set of $\gamma_1 - \gamma_5$ values have been established. Values allowing for the continuous reduction in γ_1 and γ_4 are being prepared.

As an example of the use of Fig. 15, in the problem in Fig. 10 with $m_{sx} = m_{sy} = 0.5$ from Fig. 15 $\lambda_x = 0.6$, i.e. the steel C_x would extend to 0.6 x 0.586l = 0.35l from the column. It can be observed from Fig. 10d that the value of λ_x allowing for a reduction in γ_1 gives $\lambda_x \simeq 0.58$.

d) Intensity of bottom reinforcement

Hillerborg makes a special feature of carrying all the m_{fx} and m_{fy} reinforcement right across the whole element. The reason is primarily on account of the heavy twists near the column, in which positive reinforcement contributes, just as negative reinforcement does, to the dissipation of energy in any fan mechanisms at failure. It is essential hereabouts that the positive reinforcement is present at full strength, because it has reduced the required negative reinforcement by the same amount according to the H-W-A. rules. Any reduction in positive reinforcemnt would, near the column, directly affect the magnitude of the collapse load by small fan mechanisms. If the designer chooses to work outside Hillerborg's limits of $0.3 \leq \kappa \leq 0.75$ then extra positive steel will be required near the column, indeed the lower limit was calculated by Hillerborg specifically to avoid this. The author takes the view that while it is convenient not to add extra positive steel, if circumstances demand it then a method of calculating it is required.

Corner and edge columns are the most serious cases for investigating the value of any extra positive reinforcement. The significance is best demonstrated by yield-line theory locally, when with symmetrical conditions a tiny fan of yield lines could form, of small enough radius to neglect the distributed load, Fig. 6b. The radial yield lines are governed by a square mesh, $-2m_s$ in a double intensity element. The boundary yield line is $+(m_f + \varepsilon)$ where ε is the extra local value of resistance moment due to extra isotropic reinforcement in pl^2/2 units. From equation 11

$$pl^2 = \pi(m + im)/2$$

but $\quad m = 2m_s.pl^2/2$ and im is $(m_f + e) pl^2/2$ so that

$$pl^2 = \pi pl^2(2m_s + m_f + \varepsilon)/4 \tag{44}$$

Since $m_f = 1 - m_s$ this leads to $\varepsilon = 0.273 - m_s$

Hence at the column a symmetrical double-intensity element requires no extra positive reinforcement if $m_s > 0.273$. As this applies to much of the practical range the conclusion is that the increased negative reinforcement has eliminated or reduced the need for extra positive reinforcement. Only a corner column ($m_s = 0$) needs the extra reinforcement in full, both ways.

Reverting back to Hillerborg's moment field it will be noted from Table 1 that the maximum positive moment at $x = 0$, $y = 0$ for the distribution element part only is $0.137\,pl_x^2$ or $0.273.pl_x^2/2$ where p and l_x are the quasi-load intensity on the distribution element from strip action and l_x is the length of the element in the x direction. The contribution from the various load cases is therefore

$$\varepsilon_d = 0.273(\gamma_1 + \gamma_2/2 + \gamma_3 + 2\gamma_4 + \gamma_5)$$

$$\varepsilon_d = 0.273(1 - \gamma_2/2 + \gamma_4); \text{ since } \Sigma\gamma = 1 \qquad (45)$$

From this we need to deduct the strip moments along $y = 0$ for load cases 2, 3 and 5 so that the total extra positive steel at $x = 0$, $y = 0$ in units of $pl_x^2/2$ is

$$\varepsilon_x = 0.273(1 - \gamma_2/2 + \gamma_4) - (\gamma_2/2 + \gamma_3/2 + \gamma_5) \qquad (46)$$

Similarly $\quad \varepsilon_y = 0.273(1 + \gamma_2 - \gamma_4/2) - (\gamma_3 + \gamma_4/2 + \gamma_5/2) \qquad (47)$

There is a check on equation 46 since if in turn each γ value is taken to be unity, the others being zero, the ε_x value should correspond to M_x^+ at $x = 0$ in Table 3c, which it does (after multiplying by $pl_x^2/2$). By inserting the various γ values in equation 46 it is found

$$\varepsilon_x = 0.273 + 0.273m_{sx} - 2.546m_{sy} \text{ for area OAB} \qquad (48)$$

$$\varepsilon_x = 0.273 - m_{sx} \text{ for area OBB}' \qquad (49)$$

and $\quad \varepsilon_x = 0.273 + 0.273m_{sx} - 0.637m_{sy} \text{ for area OB}'\text{A}' \qquad (50)$

These values of ε_x are plotted in Fig. 16 from which the extra positive reinforcement is easily found. Over the practical range the extra reinforcement will be small, except with corner columns (the origin), and edge columns (the base line). The designer obtains ε_y by changing axes, i.e. transposing m_{sx} for m_{sy}, and using $pl_y^2/2$ as a multiplier.

Since ε_x contains only γ values, which are valid for κ, the axes can be labelled κ_x and κ_y and therefore Hillerborg's limits of $0.3 \leq \kappa \leq 0.75$ have been drawn in as dotted lines. It can be seen the author's zero ε_x lines corresponds closely to the κ lower limit which Hillerborg sets and which includes other limits but which he admits is slightly questionable in view of the computation involved.

Hillerborg's advanced strip method 173

As an example of the use of Fig. 16, in the problem in Fig. 12 for element abrn for the simplified reinforcement in Fig. 12b, $\kappa_x = \kappa_y = 0$, hence from Fig. 16 $\varepsilon_x = \varepsilon_y = 0.273$. The extra ε_x steel at the column would be $0.273\,(0.4l)^2/2 = 0.066$ in $l^2/3$ units and $\varepsilon_y = 0.273(0.5l)^2/2 = 0.1$ in $l^2/3$ units.

Although not shown here graphs of the length and width of the extra positive steel can also be constructed. These indicate that the maximum width and extent of positive steel generally need not exceed $l_y/2$, $l_x/2$.

GENERAL COMMENTS AND CONCLUSIONS

The extensions which have been outlined allow greater flexibility in using Hillerborg's 'advanced' strip method. The designer begins as is normal by postulating hypothetical bending moment diagrams to determine the average support and field moments for the advanced elements. Where possible the support steel would then be adjusted to give a concentration of top steel closer to the column usually of twice the average moment over half the element width. Should this adjustment cause the κ values to be outside Hillerborg's limits the designer now has the option of further adjustment or accepting the values. In the latter case additional positive steel as indicated in Fig. 16 in excess of the field moment value needs to be added.

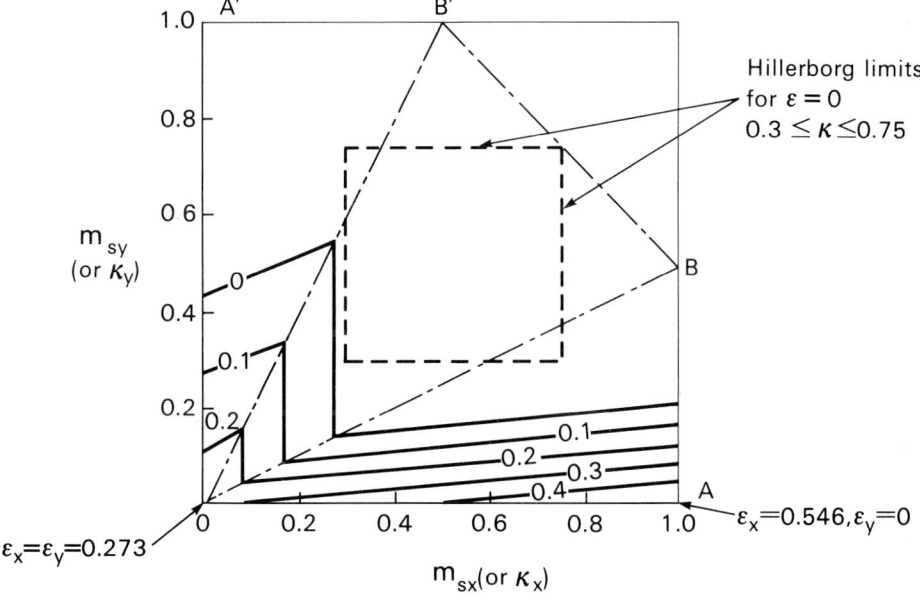

Fig. 16. Contours of ε_x (extra positive x-steel) in units of $pl_x^2/2$

Allowing the designer to work outside Hillerborg's limits will also often enable the designer to retain constant field moments and avoid the problem of adjusting the load distribution on adjacent elements to ensure moment continuity across the element boundaries. In such cases with uniform field moments Figs. 13 and 15 may be used to determine the reduction in steel at the half element point and the cut-off point of this steel.

The set of γ values which have been formulated in terms of m_{sx} and m_{sy} now enable the designer, should he wish to, to evaluate the variation of steel across an element thus introducing a greater degree of confidence into the use of the 'advanced' strip method.

ACKNOWLEDGEMENT

The majority of the work presented in this paper was originally carried out jointly with Dr R H Wood and the author for their book (3).

REFERENCES

(1) Hillerborg A: "Strip method of design". Viewpoint Publication, 1975, Cement and Concrete Association.

(2) Hillerborg A: "The advanced strip method – a simple design tool",. Mag. Concrete Research. Vol. 34, No. 121, Dec. 1982, pp. 175-181.

(3) Jones L L and Wood R H: "Yield-line and strip method of analysis of slabs". Chapman and Hall (to be published 1989).

8

PLASTIC FLOW RULES FOR USE IN THE ANALYSIS OF COMPRESSIVE MEMBRANE ACTION IN CONCRETE SLABS

K O Kemp, J R Eyre and H M Al-Hassani
University College, London, UK

SYNOPSIS

A controversy exists over the appropriate flow rule to use in the plastic analysis of membrane action in concrete slabs. The two flow rules, one based on incremental strains and the other on total strains are considered in relation to the material properties of reinforced concrete, particularly the discontinuities produced by tension cracking.

The conclusion reached is that a total strain flow rule should be used when the membrane compression is increasing and an incremental strain flow rule when the membrane compression is decreasing. The former stage will usually include the important maximum collapse load.

The application of the combined flow rule is illustrated by considering a rigid-plastic analysis of a reinforced concrete slab axially restrained by rigid supports with end gaps.

NOTATION

	d	Depth to the tension reinforcement
	f_b	Yield strength of concrete in bending compression
	h	Depth of the section
	L	Span of the slab
	M	Yield bending moment on the section
	M_o	Value of M in pure flexure
	m	Non-dimensional yield bending moment M/M_o
	N	Yield axial compressive force on the section
	n	Non-dimensional yield axial compression N/T_o
	P	Collapse load
	P_o	Collapse load in pure flexure

T_o Yield tensile strength of the reinforcement in the section
w Deflection at the centre of span
w_i Value of w at beginning of membrane action
α Coefficient of n in yield criterion
β Coefficient of n^2 in yield criterion
Δ Magnitude of the end gap at the rigid restraint
ε Plastic axial strain at centre of section
η Distance of the strain neutral axis from the centre of the section.
η_i Value of η for incremental strains
η_t Value of η for total strains
θ Rotation in the collapse mechanism
κ Plastic curvature strain
λ Scalar coefficient
μ Distance of the stress neutral axis from the centre of the section
μ_o Value of μ in pure flexure

INTRODUCTION

The application of rigid perfectly plastic theory to the flexural behaviour of reinforced concrete slabs in now standard, following the pioneering work of Johansen (1) on yield line theory and Hillerborg (2) on the strip method and their many followers. The beneficial effects of compressive membrane action in axially restrained concrete slabs however remains largely unexploited in design.

R.H. Wood (3) in his classic text devoted a chapter to membrane action and in his plastic analysis of circular slabs adopted a total strain plastic flow rule but offered no justification for his choice. Neither did Kemp (4), Park (5) or Roberts (6) for their use of the same flow rule in the plastic analysis of membrane action.

In contrast, Morley (7) and Janas (8) in their studies of membrane action adopted an incremental strain flow rule in accordance with the normality rule of plasticity, but again without real justification. These different choices of flow rule were examined by Al-Hassani (9) in a Ph.D. thesis in 1978. He concluded that normally a combination of these two flow rules was required. In 1980, Braestrup and Morley (10,11) discussed the question of plastic flow rules including Al-Hassani's ideas but came down in favour of the incremental flow rule. In 1985, Eyre (12) in a Ph.D. thesis rejected Braestrup and Morley's conclusions and generally, though not precisely, supported Al-Hassani's approach.

There remains therefore a theoretical controversy over the nature of the plastic flow rule to be used in the plastic analysis of membrane action in reinforced concrete slabs or beams. It is the purpose of this paper to explore this question. There is little doubt that Randal Wood, in whose memory this conference has been organised would have been

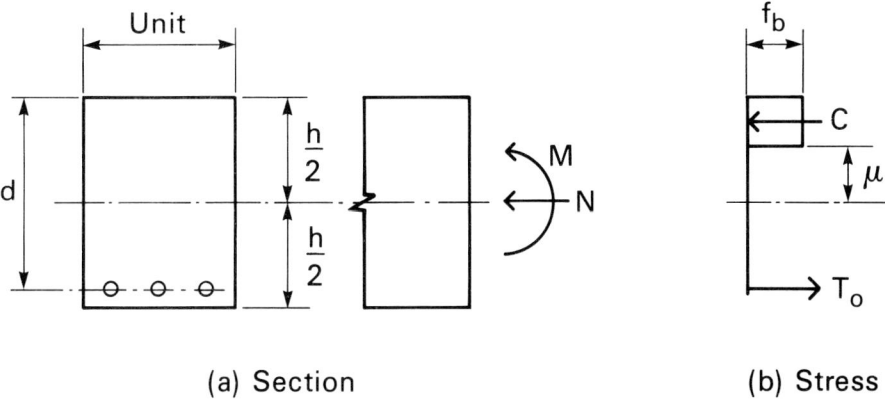

Fig. 1. Reinforced concrete section under axial compression and moment

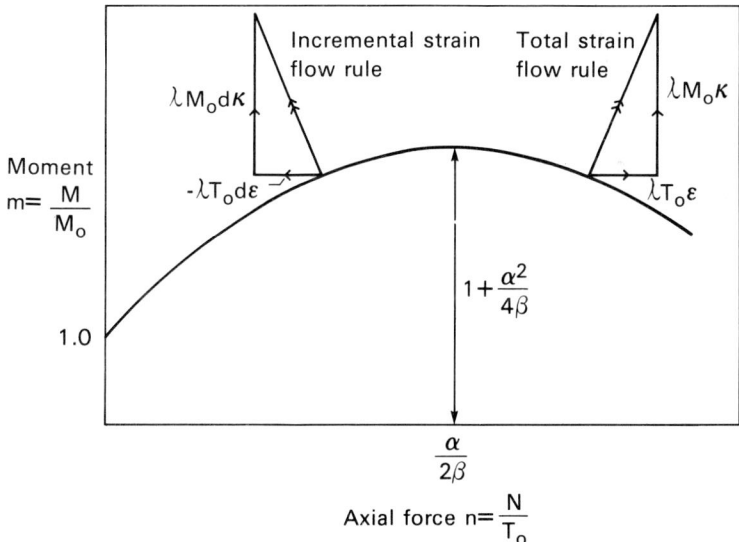

Fig. 2. Yield criterion

intrigued by this question since he was always interested in theoretical rigour as well as in practical applications.

THE YIELD SURFACE AND ASSOCIATED PLASTIC STRAINS

Consider the reinforced concrete section of unit width shown in Figure 1 subjected to a moment M and a compressive axial force N applied at the centre of the section. Both materials are assumed to be rigid perfectly plastic with the concrete having a compressive yield stress in bending f_b and zero tensile strength. Wood (3) showed that the yield criterion could be expressed in non-dimensional form as:

$$m = 1 + \alpha n - \beta n^2 \qquad (1)$$

$$\text{with } n = \frac{\alpha}{2\beta} - \frac{\mu}{2\beta} \frac{T_o}{M_o}$$

where $m = M/M_o$ and M_o is the pure flexural plastic moment

$n = N/T_o$ and T_o is the tensile yield force of the reinforcement

μ is the distance of the stress neutral axis from the centre of the section measured positive toward the compressive face

α and β are coefficients defined by the section properties where,

$$\alpha = \mu_o T_o/M_o, \qquad \beta = \tfrac{1}{2}(h/2-\mu_o)'T_o/M_o$$

and μ_o is the pure flexural value of μ

The yield criterion Equation (1) is shown diagramatically in Figure 2 in which sagging moments and compressive axial forces are considered positive. Equation (1) is only valid for a certain range of positive values of n but it will suffice for the present needs.

The plastic strains associated with the generalised forces M and N will be a curvature strain κ considered positive when sagging and an axial stain ε at the centre of the section considered positive when compressive. Since plastic hinges will be assumed to form over infinitely small lengths of the member, the curvature strains will be non-dimensional rotations and the axial strains will have the dimensions of length.

Plastic strain diagrams are shown in Figure 3, in which for each increment of plastic deformation, the incremental strains vary linearly with distance from the incremental strain neutral axis (ie plane sections remain plane). The neutral axis for incremental strains occurs at a distance η_i from the centre of the section and an incremental plastic flow rule

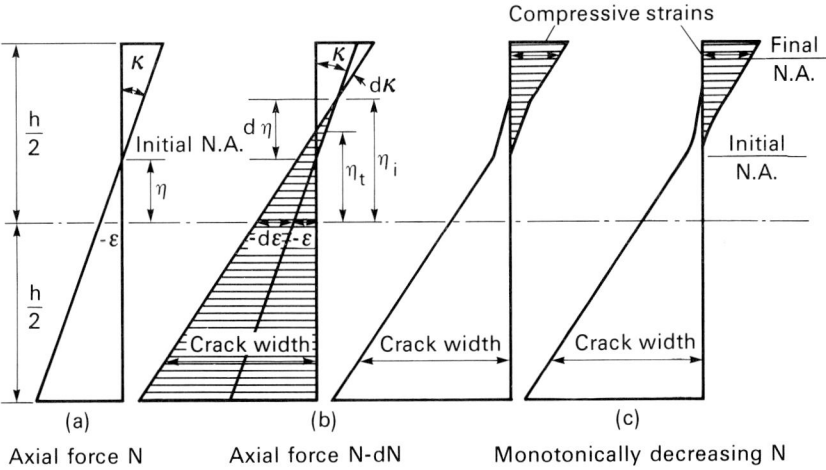

Fig. 3. Strain: Decreasing axial compression

(Figure 3b) defines that the plastic incremental strains are related by;

$$\eta = -\frac{d\varepsilon}{d\kappa} = \eta_i$$

The neutral axis for total strains occurs at a distance η_t from the centre of the section and the total strain plastic flow rule (Figure 3a) defines the relationship between the total or accumulated plastic strains to be;

$$\eta = -\frac{\varepsilon}{\kappa} = \eta_t \qquad (3)$$

The distances η_i and η_t will only be equal if the stress state remains constant on the yield surface when $\eta_i = \eta_t = \eta$. This is illustrated in Figure 3b and is also shown by differentiating Equation 3 to give;

$$-\frac{d\varepsilon}{d\kappa} = \eta_t + \kappa \cdot \left(\frac{d\eta_t}{d\kappa}\right) = \eta_i$$

Since $\mu = \eta_i$ or $\mu = \eta_t$ defines the stress state, the two flow rules will in general predict different stress states. The total strain flow rule is dependent on the stress history whereas the incremental flow rule is not.

For the non-dimensional yield criterion, Equation (1), shown in Figure 2, the associated generalised plastic strains are ($M_o d\kappa$, $T_o d\varepsilon$) so that $m.M_o d\kappa$, $n.T_o d\varepsilon$, define plastic work. The normal to the yield surface has the slope $-\frac{1}{\eta} M_o/T_o$ so defines the relative magnitudes of either the generalised incremental strains or the generalised total strains as indicated on Figure 2.

THE CHOICE OF A PLASTIC FLOW RULE

In choosing an appropriate plastic flow rule for reinforced concrete subjected to axial compression and moment, it is essential to be clear about the assumed behaviour of the materials involved. The steel reinforcement presents no problem since it is a ductile material which can be assumed to act in a perfectly plastic manner. Concrete in compression is strictly a strain softening material but experience has shown that it can be considered to be perfectly plastic provided a suitable yield stress in bending f_b is chosen, less than the cube or cylinder strength. The concrete will, however, be assumed to remain a continuum in compression.

The tensile strength of concrete is normally neglected in strength analysis since the cracking stress is relatively low. In plastic analysis, however, two different assumptions could be implied by a zero tensile stress. The first could be that the concrete is a ductile material in tension with a zero yield stress in which true plastic tensile strains occur and the material remains a continuum. The second could be that the concrete is brittle in tension and cracking immediately follows the imposition of any tensile stress. In this latter case, the material becomes discontinuous and the opening of the tension crack is considered to be the equivalent of plastic tensile strains. They are though quite different in nature from true plastic strains in that discontinuities are introduced and such pseudo-plastic tensile strains do not reduce any previously existing compressive plastic strains in the same zone.

If the first of the assumptions is adopted there is no reason to doubt that an incremental flow rule would be appropriate as in other continuum materials. The true nature of concrete is however much closer to the second assumption and this will be adopted here.

Suppose that the reinforced concrete section shown in Figure 1 is subjected first to a stress state (M,N) which induces total plastic strains (κ,ε) as indicated in Figure 3a. The neutral axis for stress (μ) and the zero total strain position (η_t) will be coincident at this stage, plane sections will remain plane and below the neutral axis the concrete will be cracked, the width of the crack increasing linearly with depth below the neutral axis.

Plastic flow rules

Case (A) Decreasing Axial Compression

If now the section is subjected to a new stress state in which N is reduced by an increment dN, as in Figure 3b, the new neutral axis position for stress will have to move incrementally upwards by dη. Since it is moving into the compression zone where the material is continuous there are no restrictions on its position. The next incremental strains (dκ, dε) will therefore occur as indicated in Figure 3b, obeying an incremental plastic flow rule where;

$$\eta + d\eta = -d\varepsilon/d\kappa = \eta_i = \mu$$

The concrete below the new incremental strain neutral axis will be cracked but the crack width will be bi-linear with depth as shown in Figure 3b. Compressive plastic strains will exist in the concrete above the initial neutral axis position and will also vary in a bi-linear manner with height above the initial neutral axis as shown in Figure 3b. It will be noted that the new position of the incremental strain neutral axis is no longer coincident with the current position of zero total strain.

If the section were subjected to stress states (M,N) with a monotonically decreasing axial compression N, the neutral axis would continue to move upwards into the continuous compression zone and each successive increment of strain (dκ, dε) would be correctly defined by an incremental strain flow rule. The resulting plastic strains are indicated in Figure 3c. The tension crack extends up to the final neutral axis position and its width varies non-linearly with depth between the initial and final neutral axis positions and linearly below the initial neutral axis. Compressive plastic strains exist above the initial neutral axis and vary non-linearly with height between the initial and final neutral axis positions and linearly above the latter. The total strain diagram, however, remains linear above and below the current position of zero total strain.

The conclusion is that where the stress state involves a monotonically decreasing axial compression, an incremental plastic flow rule should be adopted.

Case (B) Increasing Axial Compression

Suppose now that the same section after first being subjected to the stress state (M,N) producing total plastic strains (κ,ε) Figure 4a, is then acted upon by a new stress state in which N is increased by dN as indicated in Figure 4b.

It is important to recall that below the initial neutral axis the section is cracked - there is a material discontinuity. Under the new stress state, the neutral axis will have to move downwards into the cracked zone.

Compressive stresses will be required to exist above the new neutral axis and it is apparent that this cannot happen unless the tensile crack closes above the new neutral axis. The total strain therefore must become zero at this new neutral axis position together with an incremental rotation dk as shown in Figure 4. The incremental axial strain is then composed of two parts;

$$d\varepsilon = -\eta d\kappa + \kappa d\eta$$

and $-(\varepsilon + d\varepsilon)/(\kappa + d\kappa) = \eta_t$

which is consistent with a total strain flow rule.

It should also be noted that the tension crack begins at the new position of zero total strain and its width varies linearly with depth below. Similarly the compressive strains begin at the same position and increase linearly with height above it. The position of zero total strain therefore defines the new stress neutral axis $\eta_t = \mu$.

If the section were to be subjected to stress states with a monotonically increasing axial compression, the neutral axis would continue to move downwards into the cracked tensile zone. At each stage, compressive stresses must exist down to the current neutral axis so the tensile crack must close above that position. The total strain is always zero at the current neutral axis and the strain diagram for both compressive and tensile strains remains linear.

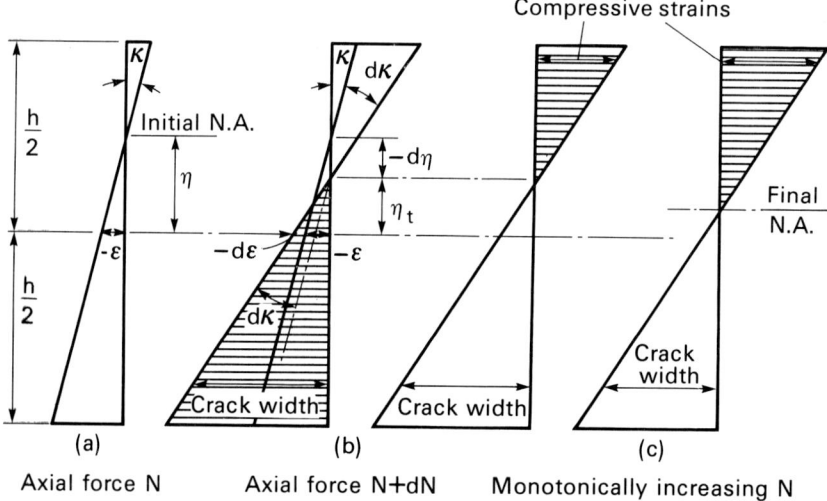

Fig. 4. Strain: Increasing axial compression

Plastic flow rules

The conclusion is that for a monotonically increasing axial compressive force, a total strain flow rule should be adopted.

Case (C) Monotonically Increasing Axial Compression Followed by Monotonically Decreasing Axial Compression

This sequence of loading will occur in compressive membrane action of axially restrained concrete slabs and beams. From the physical arguments presented a total strain flow rule should be used for the increasing axial compression and if this is followed by decreasing axial compression an incremental strain flow rule should then be applied.

Space does not permit a discussion of the reverse loading situation, decreasing axial compression followed by increasing axial compression, except to point out that the analysis may then become more difficult. The reason being that the tension crack width will vary non-linearly with depth below the highest neutral axis position reached under the decreasing axial compression.

As an illustration of the application of the proposed plastic flow to a problem with the loading sequence of Case (C), the example of a rigid perfectly plastic slab between rigid axial restraints with end gaps will now be considered.

Example: Rigid Plastic Analysis of a Simply Supported R.C. Slab Strip Axially Restrained by Rigid Supports With End Gaps

The simply supported reinforced concrete slab of unit width is shown in Figure 5a. It is singly reinforced in the bottom face providing a yield tensile force To but there is no top reinforcement at the supports. The axial restraints at the ends are assumed to be rigid but there are gaps Δ at each end.

Initially the collapse mechanism, Figure 5b, will be purely flexural with a positive plastic hinge at the centre and a collapse load $P_o = 4M_o/L$. After some deformation when the end sections come into contact with the end restraints, negative plastic hinges will form at each end producing the collapse mechanism shown in Figure 5d involving compressive membrane action.

Yield Criteria

Adopting the same notation and sign conventions as were used previously the yield criteria for the centre and end sections will be:

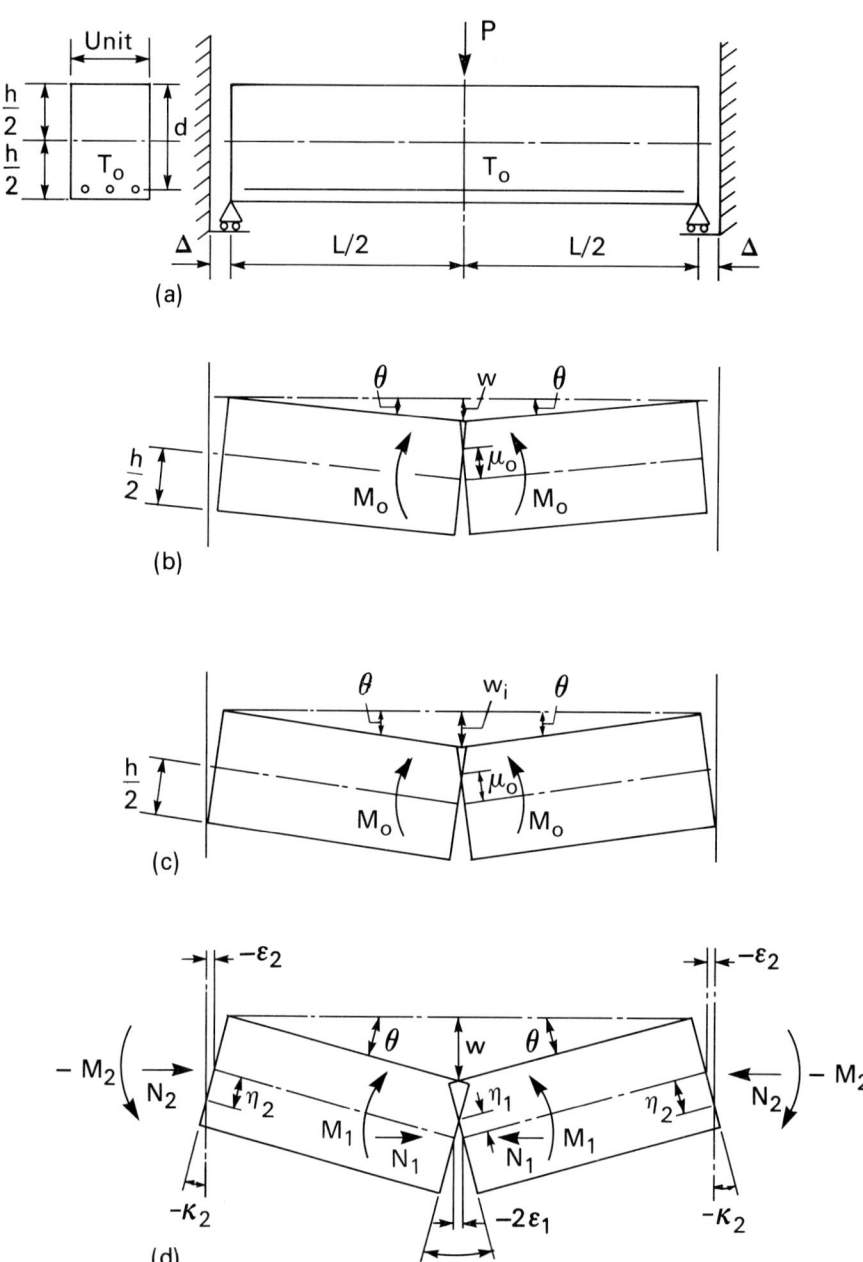

Fig. 5. Reinforced concrete slab strip: Simply supported, axially restrained by rigid supports with gaps Δ

Plastic flow rules

Centre section $\quad m_1 = 1 + \alpha_1 n_1 - \beta_1 n_1^2$ (4a)

$$n_1 = \frac{\alpha_1}{2\beta_1} - \frac{\mu_1}{2\beta_1}\frac{T_o}{M_o}$$

where $m_1 = \dfrac{M_1}{M_o}$, $n_1 = \dfrac{N_1}{T_o}$, $\alpha_1 = \mu_o\dfrac{T_o}{M_o}$, $\beta_1 = \frac{1}{2}(h/2 - \mu_o)\dfrac{T_o}{M_o}$

End sections $\quad -m_2 = \alpha_2 n_2 - \beta_2 n_2^2$ (4b)

$$n_2 = \frac{\alpha_2}{2\beta_2} - \frac{\mu_2}{2\beta_2}\frac{T_o}{M_o}$$

where m_2 and n_2 are non dimensionalised by the centre section values M_o and T_o respectively so that;

$$m_2 = \frac{M_2}{M_o}, \quad n_2 = \frac{N_2}{T_o}, \quad \alpha_2 = \frac{h}{2}\frac{T_o}{M_o}, \quad \beta_2 = \frac{1}{2}(h/2 - \mu_o)\frac{T_o}{M_o} = \beta_1$$

Compatibility Equations And Flow Rules

For the collapse mechanism shown in Fig. 5d, the compatibility equation for axial deformations after a mechanism rotation θ is:

$$L + 2\Delta = \frac{2L}{2}\cos\theta - 2\varepsilon_1 - 2\varepsilon_2$$

and using the first order approximation $\cos\theta \approx 1 - \frac{1}{2}\theta^2$, the compatibility equation becomes;

$$\varepsilon_1 + \varepsilon_2 = \frac{-L.\theta^2}{4} - \Delta \qquad (5)$$

If a strain increment flow rule is adopted then by differentiating Equation (5) with respect to θ;

$$\frac{d\varepsilon_1}{d\theta} + \frac{d\varepsilon_2}{d\theta} = -\frac{L.\theta}{2}$$

Noting that $\kappa_1 = \theta$, $\kappa_2 = -\theta$ and $w = \dfrac{L\theta}{2}$

$$\dfrac{d\varepsilon_1}{d\kappa_1} - \dfrac{d\varepsilon_2}{d\kappa_2} = -w$$

and since $\dfrac{d\varepsilon_1}{d\kappa_1} = -\mu_1, \quad \dfrac{d\varepsilon_2}{d\kappa_2} = +\mu_2$

$$\mu_1 + \mu_2 = w \qquad \text{(Incremental strain)} \qquad (6)$$

It is important to note that the end gap Δ is not represented in this compatibility equation based on an incremental strain flow rule. If now a total strain flow rule is adopted, the compatibility equation (5) becomes:

$$\dfrac{\varepsilon_1}{\theta} + \dfrac{\varepsilon_2}{\theta} = -\dfrac{L\theta}{4} - \dfrac{\Delta}{\theta}$$

Noting again that $\kappa_1 = \theta$, $\kappa_2 = -\theta$, $w = \dfrac{L\theta}{2}$

$$\dfrac{\varepsilon_1}{\kappa_1} - \dfrac{\varepsilon_2}{\kappa_2} = -\dfrac{w}{2} - \dfrac{\Delta L}{2w}$$

and since $\dfrac{\varepsilon_1}{\kappa_1} = -\mu_1, \quad \dfrac{\varepsilon_2}{\kappa_2} = +\mu_2$

$$\mu_1 + \mu_2 = \tfrac{1}{2} w + \dfrac{\Delta L}{2w} \qquad \text{(Total strain)} \qquad (7)$$

where in this compatibility equation based on a total strain flow rule, the end gap Δ is represented.

Initial Deflection

The slab will continue to collapse in pure flexure with a single central plastic hinge and at a constant load $P_o = 4M_o/L$ until the bottom fibre of the end sections just comes into

Plastic flow rules

contact with the end restraints. This deformation stage is depicted in Figure 5c in which

$$\mu_1 = \mu_0 \, , \, \mu_2 = h/2$$

Introducing these values of μ into the total strain compatibility equation (7) leads to;

$$w_i^2 - 2\,(h/2 + \mu_0)\,w_i + \Delta L = 0$$

where w_i is the initial deflection marking the end of the pure flexural mechanism and the commencement of membrane action.

Hence $\qquad w_i = (h/2 + \mu_0) - \sqrt{(h/2 + \mu_0)^2 - \Delta L}$

It is significant that the compatibility equation for incremental strain cannot be used to determine the initial deflection w_i since Δ is not represented. In effect $w_i = 0$ is predicted for all Δ.

Equilibrium of Axial Forces

The two collapse mechanisms before and after membrane action commences are depicted in Figures 5b and 5d. Up to the initial deflection w_i when membrane action begins;

$$N_1 = N_2 = 0 \text{ with } P_0 = 4M_0/L$$

After membrane action begins compressive axial forces N will be induced assumed to be acting at the centre of the section as previously.

Equilibrium of axial forces requires that;

$N_1 = N_2 = N$ and substituting for N from the yield criteria. Equations 4a and 4b gives;

$$\frac{\alpha_1}{2\beta_1} - \frac{\mu_1}{2\beta_1}\frac{T_0}{M_0} = \frac{\alpha_2}{2\beta_2} - \frac{\mu_2}{2\beta_2}\frac{T_0}{M_0}$$

and noting that $\beta_1 = \beta_2 = \beta$

$$\mu_1 - \mu_2 = \frac{M_0}{T_0}(\alpha_1 - \alpha_2) \qquad\qquad (8)$$

Using the compatibility equation (7) based on a total strain flow rule in association with the equilibrium equation (8) defines the values of μ as;

$$\mu_1 = \frac{w}{4} + \frac{\Delta L}{4w} + \frac{1}{2} \frac{M_o}{T_o} (\alpha_1 - \alpha_2)$$

$$\mu_2 = \frac{w}{4} + \frac{\Delta L}{4w} - \frac{1}{2} \frac{M_o}{T_o} (\alpha_1 - \alpha_2)$$

Substituting for μ_1, μ_2 into either of the yield equations for axial force gives;

$$n = \frac{(\alpha_1 + \alpha_2)}{4\beta} - \frac{w}{8\beta} \cdot \frac{T_o}{M_o} - \frac{\Delta L}{8\beta} \cdot \frac{T_o}{M_o} \cdot \frac{1}{w}$$

and writing $\alpha_1 + \alpha_2 = \bar{\alpha}$, $\beta_1 = \beta_2 = \beta$

$$n = \frac{\bar{\alpha}}{4\beta} - \frac{1}{8\beta} \cdot \frac{T_o d}{M_o} \cdot \frac{w}{d} - \frac{\frac{\Delta L}{d\,d}}{8\beta} \cdot \frac{T_o d}{M_o} \cdot \frac{1}{w/d} \quad \text{(Total strain)} \quad (9)$$

When $w = w_i = (h/2 + \mu_o) - \sqrt{(h/2 + \mu_o)^2 - \Delta L} = \frac{M_o}{T_o} \bar{\alpha} - \sqrt{\left(\frac{M_o}{T_o} \cdot \bar{\alpha}\right)^2 - \Delta L}$

the value of n is zero, but for greater values of w, the axial force n will be increasing as shown in Figure 6a and the neutral axis will be moving into the cracked tensile zones at both the centre and end sections. The end gaps are not of course strictly tensile cracks but they will have exactly the same effect.

Now it was argued earlier that a total strain flow rule was valid when n was increasing monotonically so Equation (9) will be valid up to $n = n_{max}$. By differentiating Equation (9) it can be shown that n_{max} occurs at;

$$w = \sqrt{(\Delta L)}$$

and the maximum value of n is;

$$n_{max} = \frac{\bar{\alpha}}{4\beta} - \frac{\sqrt{(\Delta L)}}{4\beta} \cdot \frac{T_o}{M_o}$$

For values of $w > \sqrt{(\Delta L)}$, the axial force will be decreasing and then it has been argued that an incremental flow rule should be adopted. If an incremental flow rule is employed, then by solving Equations (6) and (8);

Plastic flow rules

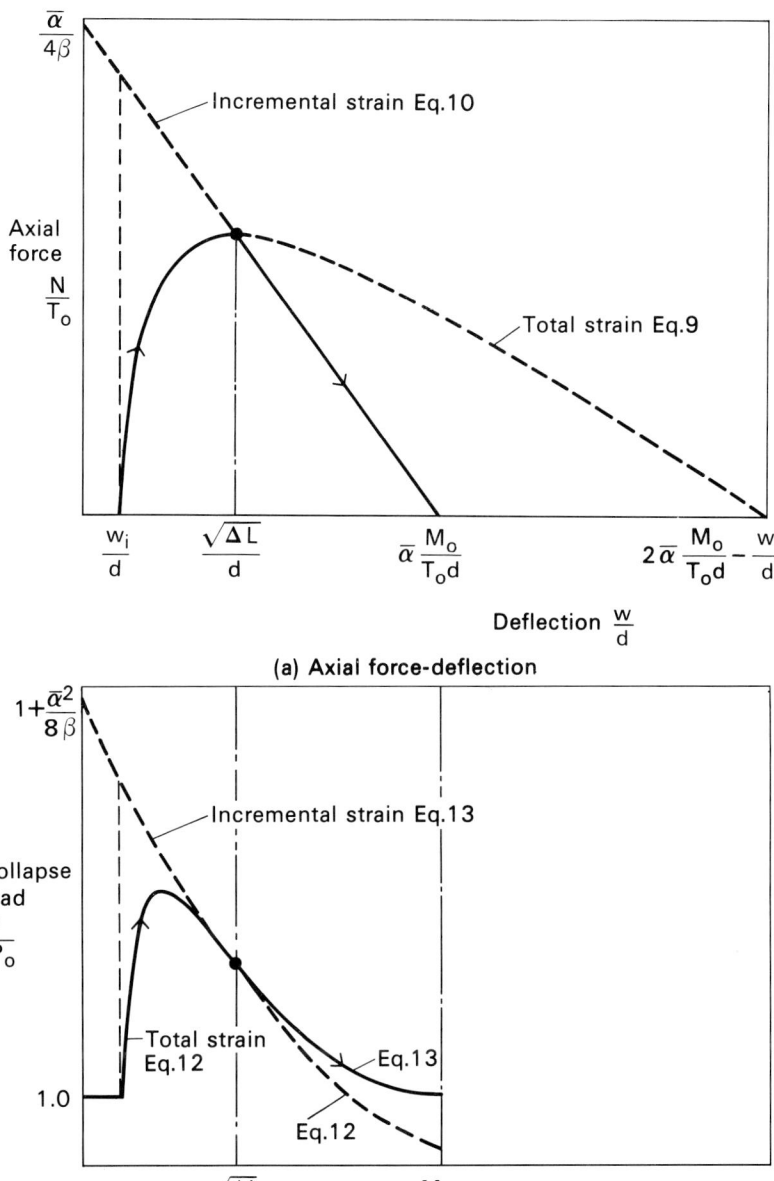

(a) Axial force-deflection

(b) Collapse load-deflection

Fig. 6. Collapse load behaviour

$$\mu_1 = \frac{w}{2} + \tfrac{1}{2}\frac{M_o}{T_o}\cdot(\alpha_1 - \alpha_2)$$

$$\mu_2 = \frac{w}{2} - \tfrac{1}{2}\frac{M_o}{T_o}\cdot(\alpha_1 - \alpha_2)$$

and substituting into either of the yield equations for axial force gives:

$$n = \frac{\bar{\alpha}}{4\beta} - \frac{1}{4\beta}\cdot\frac{T_o d}{M_o}\cdot\frac{w}{d} \quad \text{(Incremental strain)} \tag{10}$$

This equation is also plotted in Figure 6a and shows the axial force to be a maximum at zero deflection, (where in reality there can be no membrane force), and then decreases linearly with w. The line intersects the total strain axial force equation at $w = \sqrt{(\Delta L)}$ where the total strain axial force is a maximum.

It could be argued physically that Equation (10) should only apply for values of $w > w_i$ but then the axial force would have to increase instantaneously at $w = w_i$ to a maximum and then decrease linearly.

Moment Equilibrium with Membrane Action

For the mechanism shown in Figure 5d, moment equilibrium requires:

$$PL/4 = M_1 - M_2 - N.w$$

and non-dimensionalising with respect to P_o.

$$P/P_o = m_1 - m_2 - n\cdot\frac{T_o}{M_o}\cdot w \tag{11}$$

Using the yield equations for m_1 and m_2 and the total strain value for n from Equation (9) leads to:

$$P/P_o = 1 + \frac{\bar{\alpha}^2}{8\beta} + \frac{\frac{\Delta L}{d\,d}}{16\beta}\cdot\left(\frac{T_o d}{M_o}\right)^2 - \frac{\bar{\alpha}}{4\beta}\cdot\frac{T_o d}{M_o}\cdot\frac{w}{d}$$

$$+ \frac{3}{32\beta}\left(\frac{T_o d}{M_o}\right)^2\left(\frac{w}{d}\right)^2 - \frac{\left(\frac{\Delta L}{d\,d}\right)^2}{32\beta}\cdot\left(\frac{T_o d}{M_o}\right)^2\cdot\frac{1}{(w/d)^2} \quad \text{(Total strain)} \tag{12}$$

Plastic flow rules

which satisfies the requirement that $P/P_0 = 1$ at $w = w_i$

The maximum value of P/P_0 occurs when;

$$w^4 - \frac{4}{3}\bar{\alpha}\frac{M_0}{T_0}w^3 + \frac{(\Delta L)^2}{3} = 0$$

for which there is no explicit solution.

A plot of Equation (12) is however shown in Figure (6b) where it will be observed that $(P/P_0)_{max}$ occurs at a value of w less than where n is a maximum $w = \sqrt{(\Delta L)}$.

For values of $w > \sqrt{(\Delta L)}$ with n decreasing and the neutral axis at both the centre and end sections moving into the continuous compressive zones, an incremental flow rule should be applied. Substituting again into the equilibrium Equation (11) for m_1 and m_2 from the yield criteria and using the incremental strain value for n from Equation (10) gives;

$$P/P_0 = 1 + \frac{\bar{\alpha}^2}{8\beta} - \frac{\bar{\alpha}}{4\beta}\left(\frac{T_0 d}{M_0}\right)\frac{w}{d} + \frac{1}{8\beta}\left(\frac{T_0 d}{M_0}\right)^2\left(\frac{w}{d}\right)^2 \quad \text{(Incremental strain)} \quad (13)$$

If this equation is compared with the total strain equation (12), it will be seen that the first two terms and the coefficient of the w term are identical, the remaining terms are different and the end gap Δ is missing.

At $w = \sqrt{(\Delta L)}$, the axial force n is identically predicted by the total strain and the incremental strain equations and from the moment equilibrium equation (11) it is clear that P/P_0 will also be identically predicted as shown on Figure 6b.

At $w = \sqrt{(\Delta L)}$, $P/P_0 = 1 + \frac{\bar{\alpha}^2}{8\beta} - \frac{\bar{\alpha}}{4\beta}\frac{T_0 d}{M_0}\sqrt{\frac{\Delta L}{d d}} + \frac{\frac{\Delta L}{d d}}{8\beta}\left(\frac{T_0 d}{M_0}\right)^2$

from either Equation (12) or (13).

For values of $w > \sqrt{(\Delta L)}$, the value of n and P/P_0 will be correctly predicted by the strain increment equations as indicated in Figure 6b.

CONCLUSIONS

The total strain and incremental strain plastic flow rules have been examined and their relevance to a reinforced concrete section under axial compression and bending discussed. Previous researchers working on the plastic analysis of membrane action in concrete slabs have used one or the other of these two flow rules except for Al-Hassani (9) and Eyre (12).

The material has been assumed to be rigid perfectly plastic in compression but to be brittle in tension with cracking following the imposition of any tensile stress producing discontinuities in the material. The alternative assumption of a material which is rigid perfectly plastic in tension with a zero tensile yield stress and which remains a continuum has been rejected.

The conclusion reached here on physical arguments is that a total strain flow rule is correctly applied when the compressive axial force increases monotonically and the neutral axis is moving into the discontinuity produced by the tensile crack. When the compressive axial force decreases monotonically and the neutral axis moves into the continuous compression zone, then an incremental strain flow rule is required.

As an illustration of the application of these rules, a rigid plastic analysis has been presented of a simply supported reinforced concrete slab strip, having rigid axial restraints but with end gaps Δ. In this example the membrane axial force initially increases to a maximum value and then declines. A total strain flow rule has therefore been applied up to the deflection corresponding to the maximum axial force and an incremental strain flow rule thereafter. The equation for the collapse load is nevertheless continuous both in value and slope through the change point. The equation for the membrane force is however only continuous in value at the transition point.

The total strain flow rule produces a more realistic result for the initial part of the load-deformation behaviour with a gradually increasing membrane force up to a maximum value which is dependent on the magnitude of the end gap. In contrast, the incremental strain flow rule leads to load-deformation and membrane force-deformation relationships which are independent of Δ which is unreal. They would only be valid for the special case of $\Delta = 0$ when the membrane compressive force declines from a maximum value at $w = 0$.

The significance of the conclusions for future work is that it is correct in plastic analysis to use a total strain flow rule to predict the maximum collapse load due to compressive membrane action in a concrete slab or beam since the maximum load will occur in advance of the maximum membrane axial force. This offers particular advantages in elastic-plastic analysis of membrane action in slabs and beams since the analysis is simplified.

REFERENCES

(1) Johansen K.W: "Brudlinieteorier". Copenhagen 1943; English edition, "Yield Line Theory" Cement and Concrete Association, London (1962).

(2) Hillerborg A: "Strip Method of Design", Viewpoint Publication, Cement and Concrete Association, London (1975).

(3) Wood R.H: "Plastic and Elastic Design of Slabs and Plates". Thames and Hudson, London, (1961).

(4) Kemp K.O: "Yield of a square reinforced concrete slab on simple supports allowing for membrane forces" The Structural Engineer, V45, July 1967, pp 235-240.

(5) Park R: "Ultimate strength of rectangular concrete slabs under short term uniform loading with edges restrained against lateral movement" Proc Inst Civ Engrs, V28, June 1964, pp125-150.

(6) Roberts E.H: "Load carrying capacity of slab strips restrained against longitudinal expansion". Concrete, V3, Sept 1969, pp369-378.

(7) Morley C.T: "Yield line theory for reinforced concrete slabs at moderately large deflexions", Mag Conc Res, V19, December 1967, pp211-222.

(8) Janas M: "Large plastic deformations of reinforced concrete slabs", Int J Solids and Structs, V4, Jan 1968 pp 61-74.

(9) Al-Hassani H.M: "Behaviour of axially restrained concrete slabs" PhD thesis, Univ of London, (1978).

(10) Braestrup M.W: "Dome effect in R.C. slabs: Rigid Plastic Analysis" Proc ASCE, J Struct Div, V106, ST6, June 1980, pp1237-1253.

(11) Braestrup M.W, Morley C.T: "Dome effect in R.C. slabs: Elastic-Plastic Analysis", Proc ASCE, J Struct Div, V106, ST6, June 1980, pp1255-1262.

(12) Eyre J.R: "Strength enhancement in reinforced concrete slabs due to compressive membrane action": PhD thesis, Univ of London, (1985).

9

NODAL FORCES IN SLABS AND THE 'EQUILIBRIUM METHOD'

C T Morley
University Engineering Department, Cambridge, UK

SYNOPSIS

The paper surveys 'equilibrium methods' for finding least upper bounds on the collapse loads of plastic structures with assumed mechanisms, concentrating particularly on nodal force theory for slabs and highlighting the contributions of R H Wood. Equilibrium methods are presented for various types of structure including plane frameworks and walls loaded in plane, and the methods turn out to be broadly similar. However, as yet there seems to be no comparable method for plane strain problems.

INTRODUCTION

Nodal forces in slabs, and particularly the question of when they can legitimately be used to find collapse loads, were one of R H Wood's great interests. In his wide-ranging 1961 book (1) on plastic and elastic analysis of concrete slabs, an entire 45 page chapter was devoted to 'equilibrium methods' for finding the most critical member of a family of yield line collapse mechanisms, raising queries about the rationale of the method and pointing out various circumstances in which the standard nodal-force formulae seemed not to work. Yield line theory for slabs was clearly, in the terms of plasticity theory, a kinematic 'mechanism' approach giving an upper bound on the collapse load from a work equation, with no equilibrium considerations at all. Yet paradoxically Johansen (2) had been extremely successful with an equilibrium approach in finding the critical collapse pattern and the corresponding least upper bound on the collapse load. He wrote equilibrium equations for slab portions between yield lines, including in the equations certain concentrated transverse forces ('nodal forces') at the junctions between yield lines. But in some special cases application of Johansen's standard formulae for nodal forces gave incorrect or anomalous results. This rather unsatisfactory situation was elucidated by Wood in his usual forthright style.

In 1965 several workers (3) realised that the various paradoxes could be resolved, and the 'equilibrium method' for slabs put on a sound footing, if restrictions were imposed so that the standard nodal-force formulae were only used in certain defined circumstances. The aim of the equilibrium method was to find the most critical member of a specified family of yield-line patterns with variable parameters, and so the nodal forces would be related to just how a yield line changed position between one member of the family and the next. In particular, if a yield line remained in the same place whatever the pattern parameters, one could not expect to know the nodal forces at its ends. Conversely, if a yield line were completely free to move, one would expect the usual formulae to apply for the nodal forces at its ends (except in special cases where the end lay on a supported boundary or too many yield lines met at the same point, when an extended formula for nodal forces would be needed).

Wood contributed a paper to reference 3 in which restrictions and extended formulae for nodal forces were derived by direct partial differentiation of the work equation rather than by considering equilibrium fields of moment within the slab (the method favoured by the other authors). And in 1967 in collaboration with L L Jones, Wood published another book on slabs (4) in which three chapters were devoted to a comprehensive exposition of the equilibrium method and nodal forces, complete with alternative derivations of the restrictive rules and the extended formulae.

However, the aim of this paper is not to set out the nodal force method for slabs in full detail. Rather it is intended to explore, in a general way with examples, the possibility of developing useful 'equilibrium methods' for finding least upper bounds on plastic collapse loads of other types of structure. The nodal force method for slabs is by far the best developed equilibrium method - but is it a special case or does it fit within a wider group of similar methods?

BASIC CONCEPTS

Rigid, perfectly-plastic materials are considered, forming structures for which small-deflection theory is appropriate. The typical problem is that of a given structure with given boundary conditions and a given pattern of applied loads W. The loads increase proportionally with factor λ, and we wish to find the value λ_c at which plastic collapse occurs, under constant loads.

In the kinematic method, a mechanism or mode of collapse is considered, with displacements Δ and compatible deformations ε in a small increment of motion in the assumed mode. From the material yield criterion, the plastic energy dissipation per unit volume is some known function $D(\varepsilon)$ of the deformations. If the work done by external loads on a small increment of motion is equated to the internal energy dissipation, then according to plasticity theory this gives an upper bound $\lambda_u \geq \lambda_c$, i.e.

$$\lambda_u \Sigma\ W.\Delta = \int_V D(\varepsilon).dV \qquad (1)$$

Notice that so far there has been no consideration whatsoever of equilibrium, which in strict plasticity theory plays a part only in lower-bound calculations.

However, if we write the dissipation as a scalar product of two tensors $D(\varepsilon) = \sigma_p.\varepsilon$, where σ_p are the stresses at the point on the convex yield surface at which the normal is parallel to ε, equation (1) becomes

$$\Sigma(\lambda_u W).\Delta = \int_V \sigma_p.\varepsilon.dV \qquad (2)$$

Since Δ and ε are geometrically compatible, this begins to look like (though strictly is not) a virtual work equation, expressing some equilibrium relation between $\lambda_u W$ and σ_p.

Now suppose that the chosen collapse mechanism is a member of a family of mechanisms, with one or more variable parameters. The parameters are to be varied to find the least upper bound λ_{lu} for that family of mechanisms. Since λ_{lu} is clearly rather special, and since (2) looks like a virtual work equation, one wonders whether some special equilibrium condition might be satisfied at λ_{lu}. This seems to be the basis of the 'equilibrium method' - which aims to find λ_{lu} by writing equilibrium equations, avoiding lengthy differentiations of λ_u from equation (1) with respect to the pattern parameters.

Notice that we are not interested in whether λ_{lu} for this assumed family of mechanisms is indeed the true collapse load factor λ_c; and the equilibrium statements we have in mind are not related to the statical approach which gives a lower bound on λ_c.

SOME SIMPLE EXAMPLES

(a) Consider first the partially-loaded uniform propped cantilever shown in Fig. 1a. The (obviously incorrect) family of mechanisms shown in Fig. 1b is postulated, with one plastic hinge fixed at B and one at a variable position D. Anticipating some results (clearly related to the restrictions on nodal forces in slabs mentioned above) we write equilibrium equations with zero shear force at the movable hinge D, but allowing for an unknown shear force at the fixed hinge B. Taking moments for the left and right parts shown in Fig. 1c about their outer ends.

$$2M_p = W_{lu}.\frac{x_c}{2L}.\frac{x_c}{2}$$

$$M_p = W_{lu}\frac{(2L - x_c)}{2L}\frac{(2L - x_c)}{2} \qquad (3)$$

which combine to give $x_c = 1.172L$, $W_{lu} = 5.828 M_p/L$, exactly the results obtained by differentiation from the work equation to minimise W.

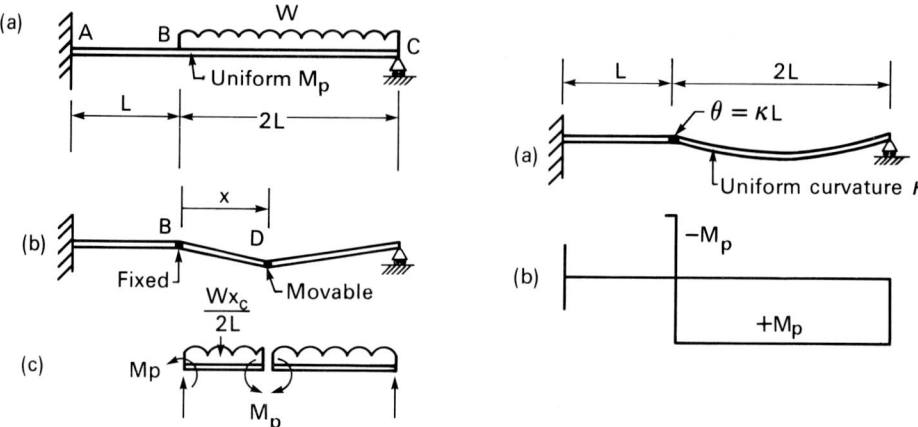

Fig. 1. Propped cantilever   Fig. 2. Alternative mechanism

It must be pointed out here that it is easy to think of collapse mechanisms, perfectly legitimate from the geometrical point of view, for which it is impossible to satisfy equilibrium. Thus for the example of Fig. 1a an alternative fully-compatible mechanism is shown in Fig. 2, with uniform sagging curvature κ over length 2L, hinge rotation κL, and central deflection $\kappa L^2/2$. The work equation gives

$$W_u \cdot \frac{2}{3} \cdot \frac{\kappa L^2}{2} = M_p \cdot \kappa L + M_p \cdot \kappa \cdot 2L$$

i.e. $W_u = 9M_p/L$ \hfill (4)

This is indeed an upper bound on the collapse load (compare above), but this mechanism and the yield criterion produce uniform moments $+ M_p$ all along the loaded part of the beam, with a sudden jump to $- M_p$ at B. Obviously this moment diagram (Fig. 2b) cannot satisfy the various equilibrium equations.

Thus it seems that if an 'equilibrium method' is to have any success it can only be applied to a restricted class of mechanisms which allow at least some equilibrium equations to be written down for large rigid pieces of material between yielding zones. These circumstances arise in simple yield-line theory for slabs (which perhaps explains the success of nodal force methods) and also in plastic-hinge theory for frameworks.

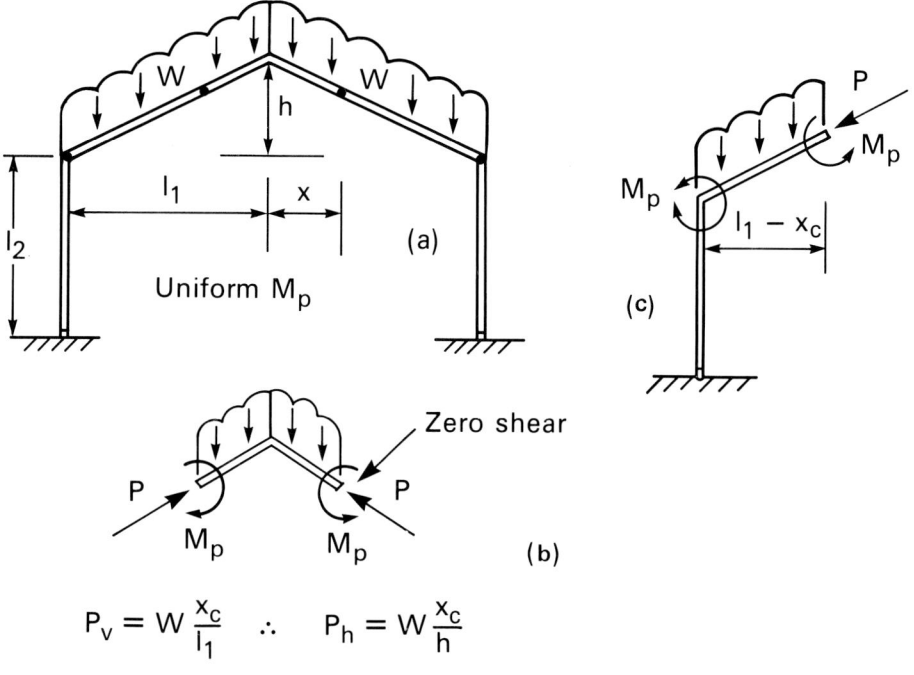

$$P_v = W \frac{x_c}{l_1} \quad \therefore \quad P_h = W \frac{x_c}{h}$$

Fig. 3.   Portal frame

(b)   So we consider next the symmetrical pitched-roof portal frame, with pinned feet and uniform M_p, shown in Fig. 3a. The postulated mechanism has fixed hinges at the corners, and movable hinges in the rafters, with variable parameter B. If axial force has no effect on M_p we write equilibrium equations with zero shear force (transverse to the rafter) at the movable hinges. Equilibrium of the top cap (Fig. 3b) gives the vertical and horizontal components of the rafter force P at the movable hinge. Moments for the rafter about its bottom end (Fig. 3c) then give

$$2M_p = W_{lu} \cdot \frac{(l_1 - x_c)}{l_1} \cdot \frac{(l_1 - x_c)}{2} \tag{5}$$

and moments about the bottom pin for the entire left part give

$$P_v(l_1 - x_c) + W_{lu} \frac{(l_1 - x_c)^2}{2l_1} = M_p + P_h(l_2 + h \frac{(l_1 - x_c)}{l_1}) \tag{6}$$

On eliminating W_{lu} these combine to give

$$4l_1 l_2 x_c = h(l_1 - x_c)^2 \tag{7}$$

which is exactly the expression for the critical hinge position x_c obtained by differentiation from the work equation to minimise W.

(c)   Another type of example is shown in Fig. 4. A beam made of special material with yield stress σ_Y in tension, $k\sigma_Y$ in compression, spans 2L between fixed abutments which prevent both rotation and displacement. If $k > 1$ this might be a rudimentary model of compressive membrane action in a one-way concrete slab. Compatibility requires that the neutral axes in the postulated end and midspan hinges be at the same level y; but what is the value of y at the least upper bound?

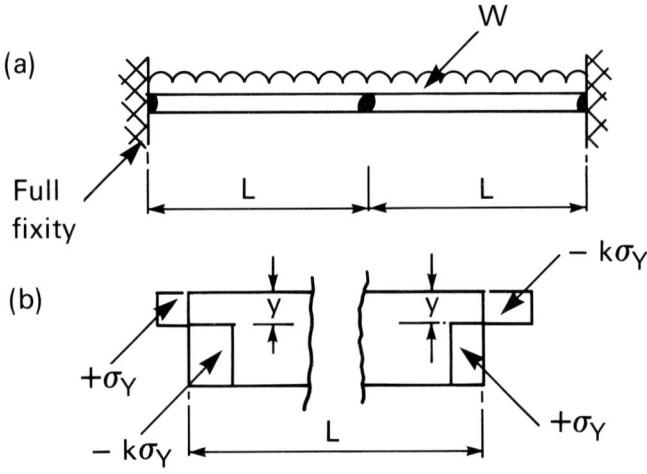

Fig. 4.   Beam with lateral restraint

The work equation is

$$W_u \cdot \frac{\Delta}{2} = (M_e + M_c) \cdot \frac{2\Delta}{L} \tag{8}$$

where the end and central moments about the neutral axis are given by

$$M_e = \frac{\sigma_Y b}{2} [y^2 + k(d-y)^2]$$

$$M_c = \frac{\sigma_Y b}{2} [(d-y)^2 + ky^2] \qquad (9)$$

Thus

$$W_u = \frac{2\sigma_Y b}{L} (1+k) [y^2 + (d-y)^2] \qquad (10)$$

from which by differentiation the least upper bound occurs at $y = y_c = d/2$. But for this y (and not otherwise) half of the beam is in equilibrium under the horizontal (membrane) forces developed at the end and midspan hinges. Writing this horizontal equilibrium equation would give the critical value of y, without differentiation.

It may seem obvious here that one should select y to satisfy equilibrium - but we see now that in doing so one is minimising the load obtainable by varying y. There is however a contrast between this and the previous examples. There at least some of the hinges moved (and occurred in different material) as the pattern parameters changed - and the equilibrium equations were written using forces (e.g. shear forces) which did not follow immediately from the yield criterion. Here the same material yields (though in a different way) as the pattern parameter changes; and the forces in the (useful) equilibrium equation follow directly from the yield criterion.

Case A: Same material yielding

This leads us to distinguish two cases of the equilibrium method for least upper bounds, according to whether or not the same actual material yields as the parameters of the collapse mode change. Dealing first with case A, in which the same material does yield, to make an equilibrium method possible we consider a situation where yielding is limited to the material close to certain defined surfaces of area A_p within the structure (rather than spread throughout the volume). If q now represents the deformations across these surfaces, compatible with displacements Δ in the assumed collapse mode, the work equation becomes

$$\lambda_u \Sigma W.\Delta = \int_{A_p} D(q).dA \tag{11}$$

where $D(q)$ is the energy dissipation per unit zone area. If the parameters of the collapse mode change slightly, so that the displacements become $\Delta + \delta\Delta$ etc., the work equation for the new mode is

$$(\lambda_u + \delta\lambda) \Sigma W.(\Delta + \delta\Delta) = \int_{A_p} D(q + \delta q).dA \tag{12}$$

Since A_p is fixed, we can subtract (11) from (12) to obtain

$$\delta\lambda \Sigma W.(\Delta + \delta\Delta) + \Sigma(\lambda_u W).\delta\Delta = \int_{A_p} \delta D.dA \tag{13}$$

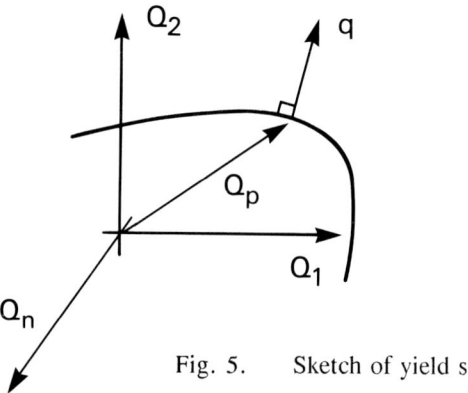

Fig. 5.   Sketch of yield surface

If the dissipation $D(q)$ is now expressed as $Q_p.q$ (Fig. 5), where Q_p are the appropriate stresses at yield with deformations q and the given yield criterion, the new dissipation is

$$D(q + \delta q) = (Q_p + \delta Q_p)(q + \delta q) \tag{14}$$

so that the increase δD in dissipation is

$$\delta D = \delta Q_p.q + Q_p.\delta q + \delta Q_p.\delta q \tag{15}$$

To the first order of small quantities the first term on the right of (15) is zero, by the normality rule of plasticity theory, and if (as is usual) flat zones on the yield surface are not significant, equation (15) reduces to

$$\delta D = Q_p.\delta q \tag{16}$$

Equation (13) then becomes

$$\delta\lambda \Sigma W.(\Delta + \delta\Delta) = \int_{A_p} Q_p.\delta q.dA - \Sigma(\lambda_u W).\delta\Delta \tag{17}$$

Here δq and $\delta\Delta$ certainly form a geometrically compatible system, with deformation only on A_p. So if we can show that $\lambda_u W$ and Q_p on A_p are in equilibrium (at least to the extent tested by δq, $\delta\Delta$) the principle of virtual work will show that the right hand side of (17) vanishes. Then we would have $\delta\lambda = 0$, and so would have obtained a value of λ_u stationary as pattern parameters change, presumably the least upper bound for this family of mechanisms.

Thus in case A where the same material yields, satisfying equilibrium (in an overall way for rigid portions of the structure rather than in detail for each element) leads to stationary λ_u, presumably λ_{lu}. This seems to explain the success of the horizontal equilibrium equation in the membrane-force example (c) above.

Case A: Further examples

(a) In (3) Wood considered at length so-called 'ring slabs', simply supported on inner and outer boundaries, as examples of cases where more than three yield lines of differing sign meet at one point (e.g. A in Fig. 6). Special extended formulae were needed for nodal forces at points such as A, where a single Mohr's circle for moment cannot be drawn and there must be a discontinuity.

Strictly speaking there is a prior problem for yield-line theory here. This is that we normally assume in the theory that the moments developed on the yield lines correspond to zero membrane force. To ensure this, the neutral axes within each yield line must strictly be at an appropriate level within the slab (so that compressive and tensile forces balance). The slab portions must move appropriately in their own planes (to get the neutral axes right) as well as deflecting out of plane. Often there are not enough in-plane degrees of freedom to allow all the neutral axes to be correctly placed, so that non-zero membrane forces must be set up.

The square slab on a central support shown in Fig. 6 is a case in point. There are three distinct types of yield-line (CA, AA' and AB) and because of symmetry only two lateral displacement quantities u_1 and u_2 to be varied. So membrane forces must be set up, although there is no lateral restraint at any of the simply-supported boundaries. For yield lines at a fixed position on the slab surface, each set of lateral displacements u_1, u_2 (as multiples of the transverse displacement Δ_A) gives an upper bound on the collapse load. And according to the Case A Principle (above) the least upper bound (for variation of u) occurs when the slab portions are in horizontal equilibrium under the membrane forces. Two equilibrium equations are available, to settle u_1 and u_2.

The numerical consequences of these membrane forces are of course quite small (5) and neglect of them will be entirely justified in practice. But strictly speaking they should be considered for each fixed position of A, since they affect the magnitudes of the moments developed on the yield lines, before one goes on to allow the position of A to vary, using the extended nodal-force formulae.

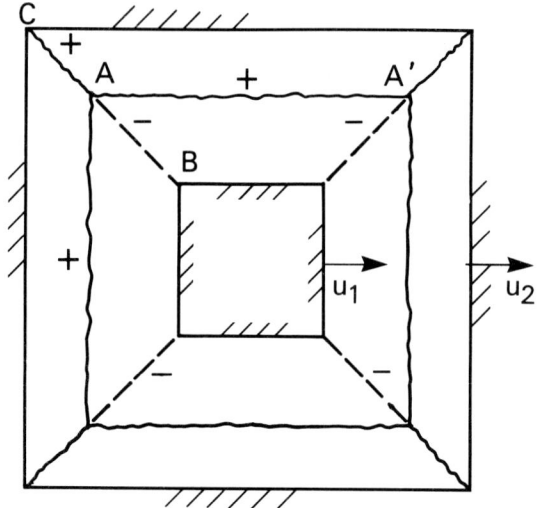

Fig. 6.　Square slab on central pier

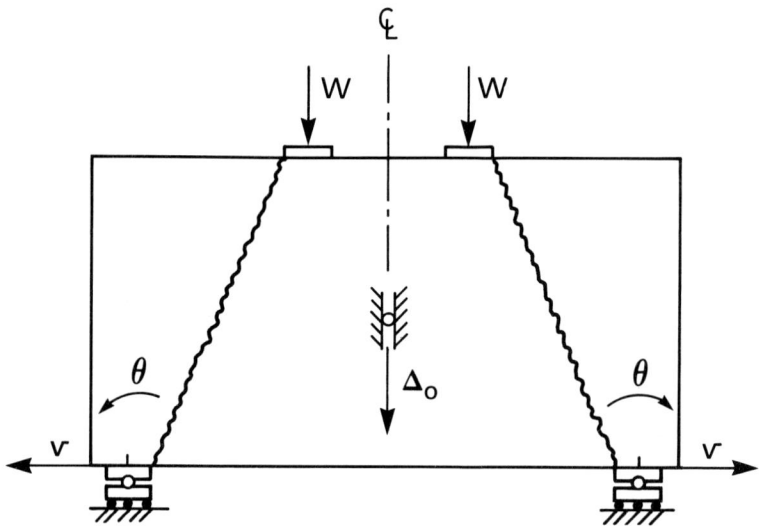

Fig. 7.　Deep beam with straight yield zones

(b) Again, it perhaps seems obvious in the above example that equilibrium ought to be satisfied. A rather different example is shown in Fig. 7 where a simply-supported deep beam carries symmetrically-positioned point loads. Straight (but narrow) yield zones are postulated, connecting the inner edges of the support and load plates as shown.

For a given vertical translation Δ_0 of the central portion, a range of collapse mechanisms can be generated by varying the outward displacement v (at the support) and rotation θ of the outer rigid portions. Since the same material yields - in the postulated straight yield zones - the Case A Principle applies, and one expects to find the least upper bound by writing overall equilibrium equations for the deep-beam portions in their own plane. There are three quantities to find, v and θ (as multiples of Δ_0) and the collapse load W_{lu}; and three independent equilibrium equations (vertical forces on the central zone, horizontal forces and moments about the simple support for an outer zone).

Least upper bounds obtained in this way do not seem to appear much in the literature, perhaps because of the complexities involved in finding the strains in the yield zone from Δ_0, v, θ and integrating the corresponding stresses to find forces to go into the equilibrium equations. In such cases the 'equilibrium method' may give no advantage over writing the work equation and differentiating.

However, Mohamed (6) developed a numerical technique for plastic analysis of concrete walls loaded in plane (e.g. deep beams with or without holes). A pattern of yield zones is specified by their end-points, dividing the structure into a series of rigid portions each with variable displacement in the collapse mechanism. The yield zones are rectangular hyperbolae, on axes with origin at the instantaneous centre of relative rotation of adjacent blocks, the optimum shape (7) of yield zone for concrete with a square yield criterion in principal stress space. For a given set of pattern parameters (essentially the displacements of the blocks subject to any boundary conditions) the various terms in the work equation are evaluated numerically by computer. An automatic minimisation technique is then implemented, to vary the pattern parameters and find the least upper bound on the collapse load.

For a few of his examples, Mohamed checked whether at the computed least upper bound the rigid slab portions were in fact in equilibrium under the integrated forces predicted on the yield zones from the yield criteria. An example is shown in Fig. 8 (this is essentially the left outer block of Fig. 7). At the least upper bound, the predicted forces C_e, T_e and V_e on the left block are effectively coincident above the centre of the reaction plate, and their magnitudes satisfy equilibrium. So in this example the rigid blocks are indeed in overall equilibrium at the least upper bound - though of course the 'equilibrium method' was not used to find the critical values of the parameters. However, Gurley (8) has recently developed an equilibrium method for finding collapse loads for beams and deep beams in shear.

Strictly speaking the example in Fig. 8 does not fall precisely within category A, since

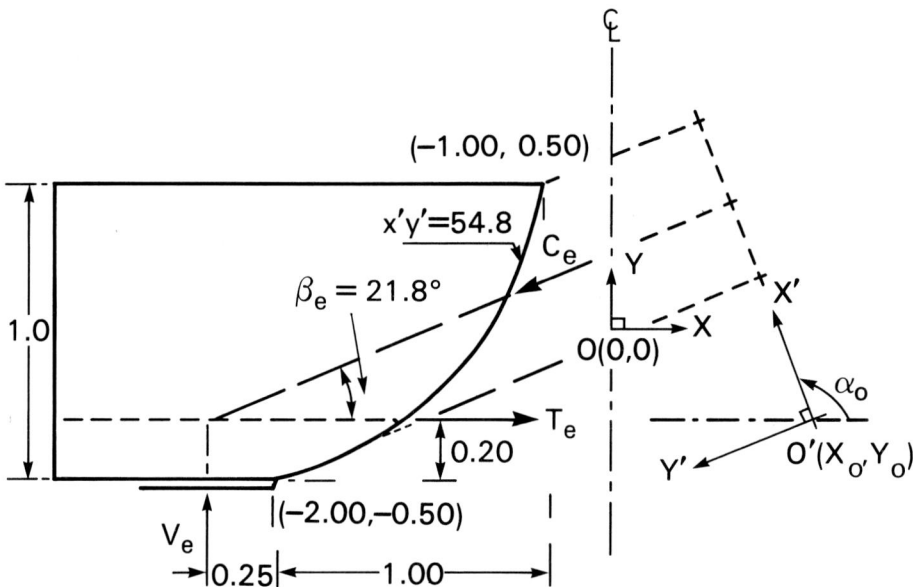

Fig. 8.   End block yield zone for deep beam (Fig. 7)

the yield zones were fixed only by their end points, and different hyperbolic curves between the same end points (with yielding in different material) were considered, depending on the pattern parameters. However, suppose that a particular pattern of hyperbolae is fixed upon: one could then vary the block displacements and rotations and seek the least upper bound for those fixed yield zones: the Case A Principle would apply, and equilibrium of the blocks would give this least upper bound. So we would expect equilibrium to be satisfied at Mohamed's least upper bounds - but the question remains open whether some of the forces in the equilibrium equations should have special values (as nodal forces in slabs do).

Case B: Different material yielding

Let us now turn to what might be called Case B, where different material yields if the parameters of the collapse pattern are changed. Care is now necessary to ensure that integrals of internal energy dissipation or virtual work are taken over properly comparable regions. We must restrict attention to cases where equilibrium is possible (contrast Fig. 2) - one example being transversely-loaded slabs with straight yield lines separating extensive rigid portions. For such a slab, with a collapse mechanism 1 with displacements Δ and yield lines L_1 with rotations θ, the work equation is

$$\lambda_u \, \Sigma \, W.\Delta = \int_{L_1} m_p.\theta.ds' \qquad (18)$$

where s′ is the distance along a yield line. If direction n is normal to the yield line in the slab plane, only the normal bending moment per unit length $M_n = m_p$ is determined by the yield criterion (since only M_n affects the energy dissipation for rotation θ).

Now suppose that the parameters of the mechanisms are changed slightly, to give yield-line pattern 2. It is assumed that, at least in principle, an equilibrium system for pattern 2 can be found, satisfying the yield criterion on the yield lines of pattern 2 but not necessarily anywhere else, and giving normal bending moments $M_{n'}$ on the lines of pattern 1. The virtual work equation can be written, for the displacements of pattern 1 and the equilibrium system of pattern 2, to give

$$(\delta\lambda + \lambda_u) \Sigma \, W.\Delta = \int_{L_1} M_{n'} .\theta.ds' \qquad (19)$$

Since the integrals in (18) and (19) are over the same yield-lines, the equations can be subtracted to give

$$\delta\lambda \, \Sigma \, W.\Delta = \int_{L_1} (M_{n'} - m_p)\theta.ds' \qquad (20)$$

Thus if we can find pattern parameters such that $M_{n'} = m_p$ along every yield line, from (20) we shall have $\delta\lambda = 0$, i.e. a stationary load factor, presumably the least upper bound on the collapse load for this family of mechanisms. A principle of this sort seems to underlie successful 'equilibrium methods' where different material yields as pattern parameters change.

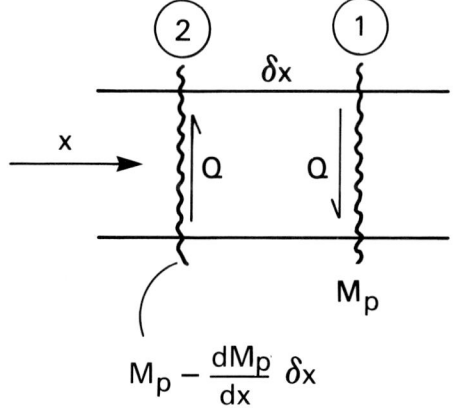

Fig. 9.   Beam element

Application to frameworks

Frameworks can be regarded as a special case of slabs spanning one way, and arranging that $M_{n'} = m_p$ is a matter of getting the correct shear force Q transverse to the framework member containing the movable hinge under consideration. Thus for a straight member in which axial force is assumed to have no effect on yield, the situation at a movable plastic hinge is shown in Fig. 9.

The fully-plastic moment at section 1 is M_p, but if the beam is not uniform the fully-plastic moment at section 2 will differ slightly. The moment $M_{n'}$ at section 1 in the equilibrium system for pattern 2 is then

$$M_{n'} = M_p - \frac{dM_p}{dx}.\delta x + Q.\delta x \qquad (21)$$

so that the requirement for stationary λ is

$$Q = \frac{dM_p}{dx} \qquad (22)$$

If the beam has uniform M_p, stationary load factor is given by taking zero shear force at movable hinges - explaining the success of 'equilibrium methods' for the examples of Figs. 1 and 3.

Condition (22) can conveniently be adapted to cases where the assumed yield criterion predicts that axial (or membrane) forces have some effect on yield. Then for a hinge fixed at say section 1 in Fig. 9 the neutral axis position at that section must be found, to give stationary λ_u, presumably by relying on Principle A and writing equilibrium equations involving the axial force at section 1 (see Fig. 4). Similar remarks apply if the hinge is fixed at section 2 in Fig. 9. So when we come to compare the load factor for the hinge at 1 and the hinge at 2, searching for stationary λ_u, the axial forces at the two hinge positions will both be in equilibrium with the same external loading $\lambda_{lu}W$. If as often happens the axial force is the same at sections 1 and 2, the derivative in condition (22) can be interpreted as the partial derivative of yield moment with respect to position, for a constant axial force in equilibrium with the applied load. For a uniform beam this derivative will be zero, again giving $Q = 0$.

A slightly more complex example is shown in Fig. 10a, of a propped cantilever subject to uniformly distributed load w per unit length, both longitudinally and transversely. If the axial force affects yield, the critical position of the sagging hinge in the mechanism of

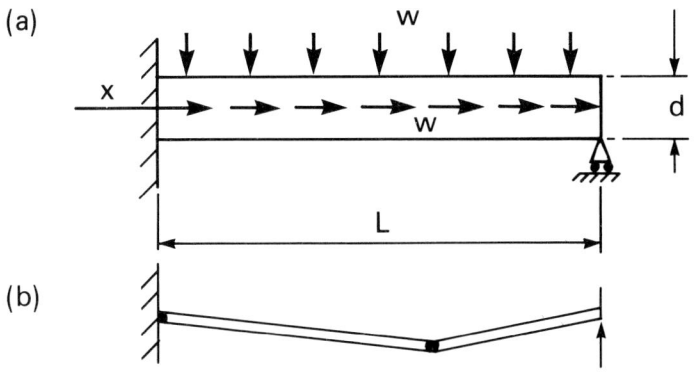

Fig. 10. Beam with longitudinal load

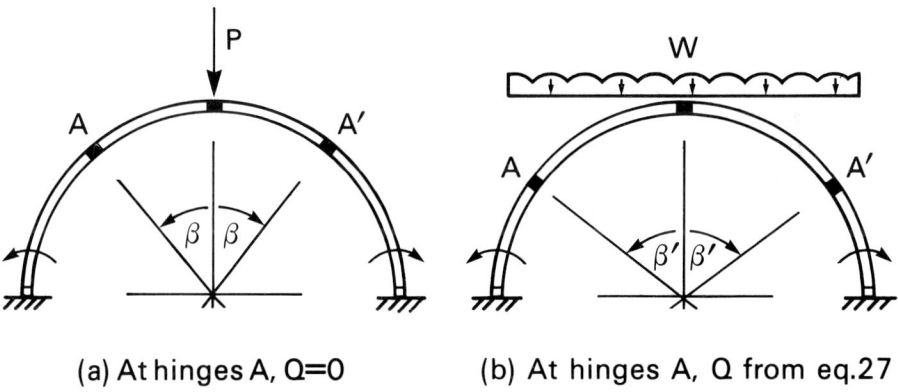

(a) At hinges A, Q=0　　　(b) At hinges A, Q from eq.27

Fig. 11. Typical arch problems

Fig. 10b is changed somewhat. For tensile axial force N and rectangular cross-section b by d in material of yield strength $\pm\sigma_Y$, the reduced plastic moment M_p is given by

$$M_p = M_o - \frac{N^2}{4\sigma_Y b}$$

$$M_o = \frac{\sigma_Y b d^2}{4} \; ; \quad -\sigma_Y bd < N < +\sigma_Y bd \tag{23}$$

So for a uniform beam

$$\frac{dM_p}{dx} = -\frac{N}{2\sigma_Y b} \cdot \frac{dN}{dx} \tag{24}$$

But for axial equilibrium at the least upper bound

$$\frac{dN}{dx} = -\lambda_{lu} w \tag{25}$$

and condition (22) for stationary λ_u becomes

$$Q = \frac{N(\lambda_{lu} w)}{2\sigma_Y b} \tag{26}$$

where w is the applied axial load per unit length. The solution now proceeds along the lines given earlier for Fig. 1, with two equilibrium equations containing two unknowns, the critical hinge position x_c and λ_{lu}. In this example solution by the equilibrium method is rather easier than by the work equation, which has extra terms allowing for axial stretching at the hinges and work done by the longitudinal load.

For arches of radius R, the equivalent of equation (25) for tangential equilibrium contains the shear force Q, and eventually equation (26) (for the same uniform rectangular section) must be replaced by

$$Q\left[1 + \frac{N}{2\sigma_Y bR}\right] = \frac{N}{2\sigma_Y b} (\lambda_{lu} w) \tag{27}$$

So the shear force is again zero at stationary load factor, if the local tangential load w vanishes. Condition (27) enables solutions to be found by the equilibrium method for typical arch examples with axial force affecting yield, such as those in Fig. 11 which both have symmetrical mechanisms with one hinge on each side at an unknown angle β.

Application to slabs

We turn now to the application of Principle B (equation 20) to slabs with zero membrane forces (for which it was first developed). Consider a typical yield line (Fig. 12) of pattern 1, with an adjacent yield line of pattern 2 obtained by making small changes in the pattern parameters. The slab is assumed to be homogeneous over reasonably large areas, with only a small number of discontinuities in strength. However, within each area the slab may be anisotropically reinforced, with the yield moment m_p dependent on the orientation ϕ of the yield line relative to some fixed direction.

For a yield line AB through a single strength area, suppose that at a typical point D on line 1 the normal yield moment is m_p in direction n'. At an adjacent point C on line 2 the yield moment is

$m_p - (dm_p/d\phi).\delta\phi$ in direction n.

In the equilibrium system for pattern 2, the moment M_n at D is then

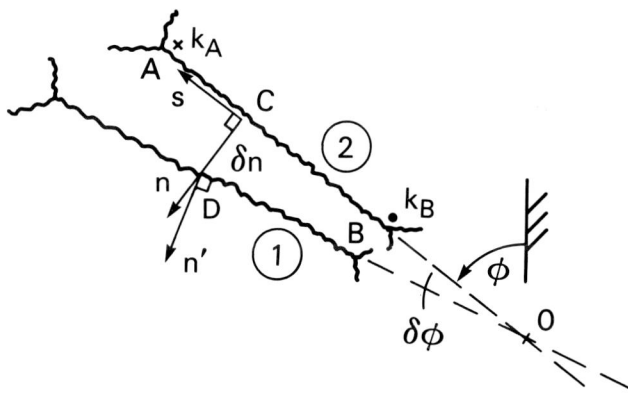

Fig. 12. Movable yield line

$$(M_n)_D = m_p - \frac{dm_p}{d\phi}.\delta\phi + (\frac{\partial M_n}{\partial n})_C.\delta n \tag{28}$$

This must be transformed round to direction n' using Mohr's circle, giving (to the first order)

$$(M_{n'})_D = m_p - \frac{dm_p}{d\phi}.\delta\phi + (\frac{\partial M_n}{\partial n})_C.\delta n + 2(M_{ns})_C.\delta\phi \tag{29}$$

Thus to have $M_{n'} = m_p$ at D requires at C

$$\frac{\partial M_n}{\partial n}.\delta n + (2M_{ns} - \frac{dm_p}{d\phi})\delta\phi = 0 \tag{30}$$

If the yield line does not move as pattern parameters change, both δn and $\delta\phi$ are zero, so that (30) is satisfied irrespective of M_{ns} etc. – and we obtain no information about twisting moments, nodal forces etc.

If the yield line does move, Kemp (3) pointed out that it will effectively rotate about some point O (Fig.12) which may be taken as the origin for s, so that $\delta n = s\delta\phi$. The twists and reactions on line 2 may now be represented by 'part nodal forces' k at the line ends A and B, by taking moments about O to give

$$k_A s_A - k_B s_B = (M_{ns})_A s_A - (M_{ns})_B s_B + \int_B^A V_n s.ds \tag{31}$$

where

$$V_n = \frac{\partial M_n}{\partial n} - 2\frac{\partial M_{ns}}{\partial s} \tag{32}$$

Substituting (30) and (32) into (31) gives eventually

$$k_A s_A - k_B s_B = \left[\frac{dm_p}{d\phi} - (M_{ns})_A\right] s_A - \left[\frac{dm_p}{d\phi} - (M_{ns})_B\right] s_B \tag{33}$$

If point O in Fig.12 is itself free to move, and takes up different positions as the pattern parameters change in different ways, then s_A and s_B are independent and (33) may be replaced by

$$k_A = \frac{dm_p}{d\phi} - (M_{ns})_A$$

Nodal forces in slabs ...

$$k_B = \frac{dm_p}{d\phi} - (M_{ns})_B \qquad (34)$$

The nodal forces which result from equations (34) may be used to write equilibrium equations for slab portions, provided that if point O is in fact fixed the nodal forces are only used to take moments about an axis through O.

In equations (34) $dm_p/d\phi$ is assumed to be known from the reinforcement layout and the assumed yield criterion (and for isotropic slabs is zero). The twisting moment M_{ns} at the end of the yield line can be found from Mohr's circle for the moments on the various intersecting yield lines. The total nodal force on a slab portion at a node is then the sum of two appropriate part nodal forces. In this way all the main formulae of Johansen's nodal force theory can be recovered - for example that nodal forces are zero at a junction of yield lines all of the same sign; or the formulae for 'edge forces' where a yield line meets a free edge (compounded of equation (34) and the usual free-edge concentrated force formulae). 'Strength-step' forces can also be derived for points where a yield line crosses a discontinuity in strength.

But the nodal forces thus derived can only be applied in restricted circumstances, depending on how the yield lines move as pattern parameters change. And there is a further restriction to do with Mohr's circle of moment at a node. Using equation (20) we have obtained stationary λ_u by making the various terms ($M_{n'} - m_p$) uniformly zero along every yield line. However, if too many lines of different sign meet at a node, a single Mohr's circle for that node cannot be drawn; local discontinuities and high shears must then occur and would have to be taken into account in equations such as (29). So the standard nodal force formulae will not apply in such a case - hence the restrictions and extended formulae mentioned above.

The usual nodal force formulae are of 'sufficient' nature only: <u>if</u> we can find an equilibrium system with the usual nodal forces, and hence $M_{n'} - m_p = 0$ on all yield lines, <u>then</u> equation (20) shows that the load factor λ_u will be stationary. But there may well be cases where equilibrium conditions cannot be properly satisfied with the usual nodal forces, which are not 'necessary' for stationary λ_u.

Special cases; extended formulae and fans

In (3) Wood pointed out firmly that in the process of minimising the load factor there are only a limited number of unknowns to be found - the critical values of the pattern parameters (essentially the positions of the movable nodes) and λ_{lu} itself. So if the equilibrium method is to be properly equivalent it must lead to the same number of equations as unknowns, and no more. Thus one should expect to know only a limited number of nodal forces when writing the equilibrium equations.

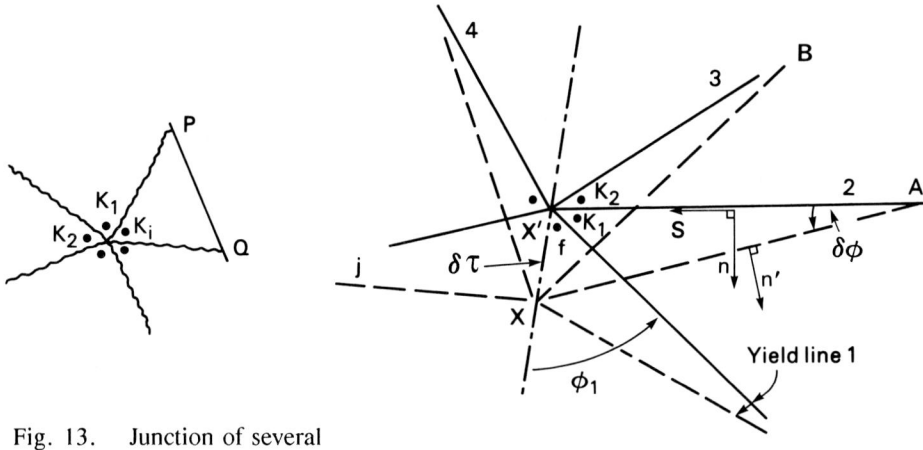

Fig. 13. Junction of several yield lines

Fig. 14. Displacement of Node at X to X'

A critical case occurs when many (say j) yield lines meet at a node (Fig. 13). The maximum number of unknowns associated with the node is three, two giving its position in plan and one related to λ_{lu}. But if every nodal force at the node was known, j equilibrium equations could be written, one for every slab segment meeting at the node, taking moments about remote axes such as PQ. To reconcile numbers, j - 3 nodal forces must remain arbitrary at the least upper bound, with three equations giving the other nodal forces.

So the standard nodal force formulae are only valid for j = 3, even at a movable node (and Johansen (2) ruled out nodes with j > 3). Extended formulae are needed for j > 3 and Wood (3) derived some formulae for the case j = 4 by direct differentiation of the work equation.

It is possible (9) to derive formulae for arbitrary j by making the integral in equation (20) vanish - but since the individual $M_{n'}$ - m_p terms are no longer zero, the formulae involve the rotations θ in the yield lines. The node is assumed to displace a small distance δτ in a given direction from X to X' in Fig. 14, where K_i are the upward nodal forces representing shear and torsion on the yield lines and any local high shears at the node, and f similarly represents XX' and line 1. Taking moments for the small triangle XAX' about XA effectively gives ∫ $M_{n'}$ ds' along XA. Similar equations are written for the other triangles such as XBX' and substituted appropriately into (20). The terms in f and in m_o (the moment on XX') eventually vanish, and for an isotropic slab the condition for stationary load factor becomes

Nodal forces in slabs ...

$$\sum_{i=1}^{j-1} K_i \left(\sum_{r=1}^{i} \theta_r \sin\phi_r \right) = \sum_{i=1}^{j} \theta_i m_i \cos\phi_i \tag{35}$$

One (and only one) further independent equation, similar in form to (35), can be obtained by considering movement of the node in a different direction; and the sum of the nodal forces must be zero. For $j = 3$ these equations reduce to the usual Johansen formulae, and for $j = 4$ they agree with Wood's extended formulae.

Not much use is made of these extended formulae in solving examples. It is more usual to avoid their use by restricting the mechanisms considered, to give nodes with $j = 3$ (or some form of symmetry). However, they do explain difficulties encountered with standard nodal force formulae at the centres of fan mechanisms, where large numbers of yield lines meet at one point. If this point does not move one cannot expect to know the nodal forces on the elements. If the central point does move the extended formulae (35) should apply, with j very large. It is much simpler to limit the attack using nodal forces to cases where the centres of fans are fixed.

Successful examples for slabs

It should be emphasised that Johansen and others have had great success in solving examples using nodal forces - indeed one suspects that Johansen was intuitively well aware of the various restrictions established later. One of his great successes in 1943 was to establish, using his nodal force formulae for $j = 3$, the optimum boundary curve for a fan mechanism with fixed origin and movable boundary (Fig. 15).

He obtained the shape

$$r = \frac{c}{\cosh[\frac{c}{h}(\theta + \gamma)]} \quad ; \quad h^2 = \frac{6(m + m')}{p} \tag{36}$$

where c and γ are constants - a conclusion confirmed by Mansfield (10) in 1957 using calculus of variations.

Also many examples of successful application of nodal forces are given in the book by Jones and Wood (4), which was written in full knowledge of the various restrictions mentioned above. A typical rectangular slab example is shown in Fig. 16, with two unknowns, the variable parameter z and the collapse load p. The standard formulae predict zero nodal force at the movable junction B, with a restriction to take moments only about axes through A, since yield line AB is constrained to pass through A. This is precisely what we do, taking moments for the slab segments about AD and AC to obtain the required two equations.

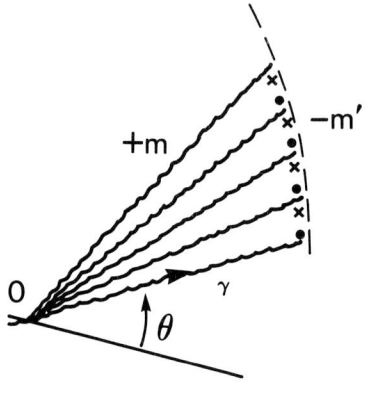

Fig. 15. Typical fan mechanism

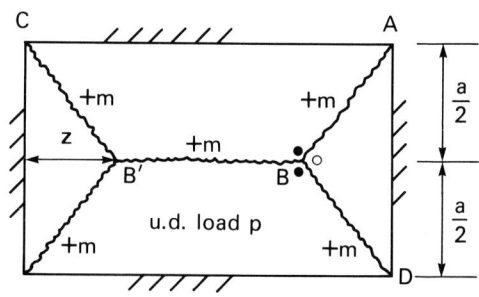

Fig. 16. Simply supported rectangular slab

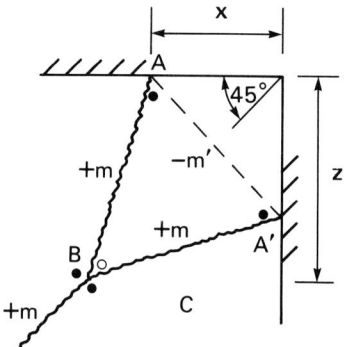

Fig. 17. Corner lever

Nodal forces in slabs ...   217

A further example (one corner of a symmetrical pattern) is the 'corner lever' of Fig. 17, with three unknowns x, z and λ_{lu}. The only nodal force knowable at A is that indicated (the others involve supported edges with unknown reactions) and the standard formula can be used, with zero moment on the simply supported edge. The nodal forces at B are zero, and the three required equilibrium equations emerge: vertical forces for ABA', moments about AA' for ABA', and moments about the edge for the slab segment C.

Speculation: Plane strain plasticity

Thus the equilibrium method is very successful for slabs in bending. However there are some well-known analogies (11,12) between plastic bending problems for plates and the metal-forming type of problem in plane strain plasticity. One wonders therefore whether there is an analogous equilibrium method for finding least upper bounds in plane strain problems.

As explained above, attention must presumably be restricted to cases where it would be possible to satisfy equilibrium - so mechanisms with plastic strain distributed over large areas will not be considered. We are left with mechanisms with rigid blocks of material separated by straight narrow zones of material yielding in shear. For example, in the extrusion problem of Fig. 18a we might assume the mechanism shown, with three types of rigid block and one variable parameter (the horizontal coordinate x of Q). In the soil wedge problem of Fig. 19 there are only two rigid blocks separated by a single yield zone, and one variable parameter ϕ.

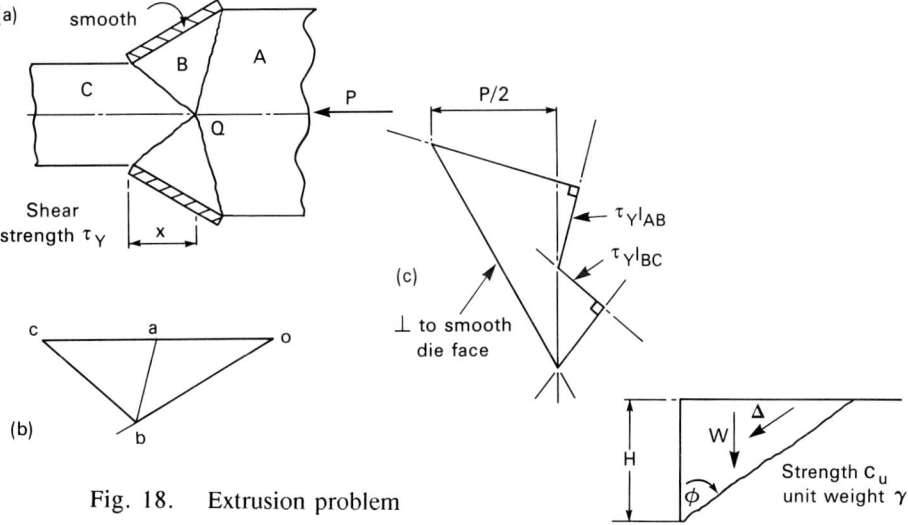

Fig. 18.   Extrusion problem

Fig. 19.   Soil wedge

In yield line theory for uniform isotropic slabs, for each assumed mechanism a 'hodograph' (11) or diagram of angular velocities can be drawn, showing rotations of slab segments and rotations in yield lines (the relative rotations of adjacent segments). In plane strain plasticity an analogous diagram (eg Fig. 18b) can be drawn, giving the shear in the yield zones as the relative movement of adjacent blocks. In each case a work equation is then written, with uniform m_p or τ_Y in the yield zones, as the case may be.

So perhaps one could, for examples like Figs. 18 and 19, (i) find the analogous slab problem, (ii) use nodal force theory to find its critical parameters, and (iii) apply these to the plane strain problem. If sufficient examples were solved in this way, there might perhaps emerge some rules (hopefully relatable to Principle B above) about equilibrium of blocks, which could be applied directly in the plane strain situation.

There is a technique (13) of drawing force-plane diagrams (eg Fig. 18c) for plane strain problems, and recently Avitzur et al (14) have considered rigid block mechanisms with velocity discontinuities of the sort postulated here. They show that if equilibrium equations for the blocks are written, or alternatively the force-plane diagram is drawn, the same result emerges (for the load factor at collapse) as from the work equation. This is for any position of the yield zones - there is no question yet of searching for the most critical member of a family of mechanisms - and it is not strictly an application of Principle A (as was applied above to deep beams in shear). This last is for two reasons; firstly with the yield zones fixed in position there are no variable parameters (because of the imposed requirement of no volume change) and so no least upper bound to be found; and secondly (and connected with the first) the yield criterion specifies only the shear stress in the yield zones but not the direct stress across.

Suppose now that the yield zones do move as pattern parameters change, in analogy with movable yield lines in slabs, and we search for the least upper bound. Are there some special force magnitudes, or special forms of the block equilibrium equations or the force-plane diagram, at stationary load factor? What are the force-type quantities (if any) analogous to the nodal forces in slab theory? These questions seem not yet to have been answered in the literature, though one suspects that there may perhaps be answers to be found, giving a corresponding 'equilibrium method' for plane strain plasticity problems.

CONCLUSIONS

The paper has surveyed 'equilibrium methods' for finding least upper bounds on collapse loads of various plastic structures, including plane frameworks, slabs loaded in plane, and pre-eminently slabs loaded transversely. To this latter topic R H Wood made some very significant contributions. The equilibrium methods seem to form a more or less coherent whole, with two main divisions depending on whether or not different material yields as the parameters of the assumed collapse mechanism are changed. However, there appears as yet to be a gap in the theory, and possibly a missing equilibrium method, for the case of plane strain plasticity.

REFERENCES

(1) Wood, R.H: "Plastic and elastic design of slabs and plates". Thames and Hudson, London, 1961, 344 pp.

(2) Johansen, K.W: "Brudlinieteorier", Doctoral theseis 1943. (Teknisk Forlag, Copenhagen, 1952).

(3) "Recent developments in yield-line theory". MCR Special Publication, May 1965, 74 pp.

(4) Jones, L.L and Wood, R.H: "Yield-line analysis of slabs". Thames and Hudson, London, 1967, 405 pp.

(5) Morley, C.T: "Yield-line theory for reinforced concrete slabs at moderately large deflections". Magazine of Concrete Research, 19 (61), Dec. 1967, p.211-222.

(6) Mohamed, Z.B: "Shear strength of reinforced-concrete wall-beam structures". Ph.D. Thesis, Cambridge, 1987, 228 pp.

(7) Jensen, J.F: "Plastic solutions for reinforced concrete disks and beams " (in Danish), Technical University of Denmark, Dept. of Structural Engineering, Copenhagen, Report no. R141 1981, 153 pp (See also Magazine of Concrete Research 34 (119), June 1982, pp.100-103).

(8) Gurley, C.R: "Shear in reinforced beams". Transactions of the Institution of Engineers, Australia. CE29(2), 1987, 10 pp.

(9) Morley, C.T: "The ultimate bending strength of reinforced concrete slabs". Ph.D. thesis, Cambridge, 1966, 212 pp.

(10) Mansfield, E.H: "Studies in collapse analysis". Proc. Roy. Soc. (A), 241, 1957, p.311.

(11) Johnson, W: "Upper bounds to the load for the transverse bending of flat rigid-perfectly plastic plates." Int.J.Mech.Sci., 11, 1969, pp.913-938.

(12) Collins, I.F: "On an analogy between plane strain and plate bending solutions in rigid/perfect plasticity theory ." Int.J.Solids and Structures, 7, 1971, pp.1057-1073.

(13) Johnson, W and Mamalis, A.G: "Some force-plane diagrams for plane-strain slipline fields". Int.J.Mech.Sci., 20, 1978, pp.47-56.

(14) Avitzur, B, Choi, J.C and Kim, J.M: "The unity of upper-bound approaches to plane-strain deformation problems." J.Metal Working Technology, 15, 1987, pp.297-307.

10

SELECTED PUBLISHED WORKS 1948 - 1986

1948   "Some notes on vibrations in structures". R.I.B.A. Journal 1948, V 55, (12), pp. 553-5.

1951   "A special type of group displacement for use in the relaxation technique". Quart. Journal Mech. and Applied Math., 1951, V 4, (4), pp. 432-438.

"An economical design of rigid steel frames for multi-storey buildings". National Building Studies, Research Paper No. 10, HMSO, London 1951.

1952   "Degree of fixity methods for certain sway problems". With E Goodwin. The Structural Engineer Guard Sheet, 1952, V 30, (7), July, pp. 153-162.

"Studies in composite construction: Part I. The composite action of brick panel walls supported on reinforced concrete beams". National Building Studies, Research Paper No. 13, HMSO, London 1952.

1953   "A derivation of maximum stanchion moments in multi-storey frames by means of nomograms". The Structural Engineer, 1953, V 31, (11), pp. 316-28.

1954   "Some recent foundation research and its application to design". With G G Meyerhof. The Structural Engineer, May 1954, V 32, (5), pp. 156-7.

1955   "A note on the problem of rapid design of multi-storey frames". The Structural Engineer, 1955, V 33, No. 7, pp. 223-224.

"A preliminary study of composite action in framed buildings". International Association for Bridge and Structural Engineering, V 15, Zurich 1955, pp. 247-265.

"Studies in composite construction: Part II. The interaction of floors and beams in multi-storey buildings". National Building Studies, Research Paper No. 22, HMSO, London 1955.

"Stress measurements in the steel frame of the new government offices, Whitehall Gardens". With R J Mainstone. The Institution of Civil Engineers, 1955, pp. 74-106.

"The absolute and the intuitive in dimensional analysis". Euler Society Note No 8, C & CA 1955.

1958 "Composite construction". The Structural Engineer, Jubilee Issue, July 1958, pp. 135-139.

"The stability of tall buildings". Proceedings Institution of Civil Engineers, V 11, pp. 69-102, September 1958.

1960 "Stiffness of a crane jib". With J F Eden. The Engineer, 29 July 1960, pp. 181-187.

1961 "Plastic and elastic design of slabs and plates with particular references to reinforced concrete floor slabs". Thames and Hudson, London 1961.

1964 "Plastic design of slabs using equilibrium methods". Proceedings of the International Symposium on Flexural Mechanics of Reinforced Concrete, Miami, 10-12 November 1964, pp. 319-36.

1965 "New techniques in nodal-force theory for slabs". Magazine of Concrete Research Special Publication: May 1965, 'Recent developments in yield-line theory', pp. 31-62.

1966 "Nuevas tecnicas en la teoria de fuerzas nodales para losas". Sobretiro de la Revista IMCYC, V 4, No. 21, July/August 1966, Mexico, D.F.

"Nuevas tecnicas en la teoria de fuerzas nodales para losas". Sobretiro de la Revista IMCYC, V 4, No. 22, September/October 1966, Mexico, D.F.

1967 "Yield-line analysis of slabs". With L L Jones. Thames and Hudson, Chatto and Windus, London 1967.

1968 "The reinforcement of slabs in accordance with a pre-determined field of moments". Concrete, February 1968, V 2, (3), pp. 69-76.

"The theory of the strip method for design of slabs". With G S T Armer. Proceedings of the Institution of Civil Engineers, October 1968, V 41, pp. 285-311.

"Test of a multi-storey rigid steel frame". With F H Needham and R F Smith. The Structural Engineer, April 1968, V 46, (4), pp. 107-119.

Selected published works 1948 - 1981

"Some controversial and curious developments in the plastic theory of structures". Published in Engineering Plasticity, Cambridge University Press, March 1968, pp. 665-691.

1969 "A partial failure of limit analysis for slabs, and the consequences for future research". Magazine of Concrete Research, 1969, V 21, (67), pp. 79-90.

1971 "Slab design: Past, present and future". Cracking Deflection and Ultimate Load of Concrete Slab Systems, ACI Publication SP-30, pp. 203-221.

"The importance of shear in the yield criterion for bending of slabs". Journal of Strain Analysis, 1971, V 6, (1), pp. 13-19.

1974 "Effective lengths of columns in multi-storey buildings Part 1: Effective lengths of single columns and allowances for continuity". The Structural Engineer, V 52, July 1974, pp. 235-244.

"Effective lengths of columns in multi-storey buildings Part 2: Effective lengths of multiple columns in tall buildings with sidesway". The Structural Engineer, V 52, August 1974, pp. 295-302.

"Effective lengths of columns in multi-storey buildings Part 3: Features which increase the stiffness of tall frames against sway collapse, and recommendations for designers". The Structural Engineer, V 52, September 1974, pp. 341-346.

"A new approach to column design with special reference to restrained steel stanchions". HMSO, London 1974.

1975 "Effective lengths of columns in multi-storey buildings". Discussion. The Structural Engineer, June 1975, V 53, No. 6, pp. 235-241.

"A graphical method of predicting side-sway in the design of multi-storey buildings". With E H Roberts. Proceedings of the Institution of Civil Engineers, Part 2, 1975, V 59, pp. 353-372.

"Non linearity and limit-state design of complete composite structures". Symposium on structural analysis. TRRL Crowthorne, 1975, Supplementary Report 164 UC.

1977 "Torsional buckling at the limit state of collapse in braced multi-storey I-section steel column design". With B Chakrabarti. Journal of Strain Analysis, 1977, V 12, (3), pp. 233-250.

1978	"Plasticity, composite action and collapse design of unreinforced shear wall panels in frames". Proceedings Institution of Civil Engineers, Part 2, 1978, V 65 (June), pp. 381-411. Discussion in Proceedings Institution of Civil Engineers, Part 2, V 67, (March), pp. 237-45. ICE Paper 8110.
1979	"Developments in the variable-stiffness approach to reinforced concrete column design". With M R Shaw. Magazine of Concrete Research, 1979, V 31, (108), pp. 127-41.
1981	"A simplified method for evaluating the natural frequencies and corresponding modal shapes of multi-storey frames". With E H Roberts. Structural Engineer, 1981, V 59B, (1), pp. 1-9, Correspondence; 1981, V 59B, (4), pp. 64-5.
1982	Discussion of paper: "The advanced strip method - a simple design tool" by Arne Hillerborg. Magazine of Concrete Research, V 34, No. 121, December 1982.
1986	"Yield-line analysis of edge supported uniformly loaded slabs by computer". With L L Jones. International Conference on Computer Applications in Concrete, Singapore, March 1986.
	Discussion of paper: "Equilibrium design solutions for torsionless grillages or Hillerborg slabs under concentrated loads" by C T Morley. Proceedings of Institution of Civil Engineers, Part 2 September 1986.

11

Paper No. 6280

THE STABILITY OF TALL BUILDINGS*

by

Randal Herbert Wood, D.Sc., Ph.D., A.M.I.C.E.
Principal Scientific Officer, Building Research Station, D.S.I.R.

For discussion at an Ordinary Meeting on Tuesday, 21 October, 1958, at 5.30 p.m., and for subsequent written discussion

SYNOPSIS

The benefits arising from plasticity in structures are well established, but it is not so well known that the beneficial effects may be curtailed in multi-storey frames because of simultaneous deterioration of elastic stability. This study of frame instability is perhaps the most perplexing and intriguing research subject of the moment in frame design, and an attempt is here made to clarify what is involved. Part of the present difficulty arises from a neglect in the past to study the stiffening effects of the cladding of tall buildings.

Within this setting an account of the work of the Building Research Station is given, and in relating this work to that of other research schools it is convenient to present the subject in the form of a brief historical account of the development of research. The emphasis throughout is on the necessity of producing rapid design methods. The great distinction between "no-sway" designs and designs involving side-sway is brought out, together with the effects which frame instability and composite action have on this issue. In spite of the tremendous numerical work obviously involved a mathematical treatise is purposely avoided.

It appears that, whereas it is imperative to make provisions for probable loss of carrying capacity of frames due to instability, the general behaviour of practical frames favours "collapse" design, particularly if some simple, and in the meantime modest, contribution from composite action can be devised.

NOTATION

A denotes cross-sectional area
C (with suffix 1 and 2) denotes out-of-balance fixed-end moments
c is a stability function
DC denotes double-curvature
E ,,   Young's modulus
K (with appropriate suffix) denotes stiffness
L denotes length
L/r ,,   slenderness ratio
M (with appropriate suffix) denotes bending moment

* Crown copyright reserved.

m, n, and o are stability functions
P (with appropriate suffix) denotes load
s is a stability function
SC denotes single-curvature
X and Y are alternative stability functions to s, c, m, n, o
λ (with appropriate suffix) denotes load factor

General Survey of Design Problems

The designer is nowadays well acquainted with the fact that first yield of the most highly stressed member is not usually followed immediately by collapse.[1] The behaviour of various structures may be divided into three principal types, according to their load/deflexion characteristics. Thus in structures normally envisaged in simple "plastic" design, a variation in moment redistribution takes place between first yield and attainment of a complete collapse mechanism at constant load. If, however, there is a favourable change of geometry (e.g. if the shape of the deflected surface of a slab leads to membrane action), then the ideal plastic-collapse load is exceeded. Conversely if the deflexions result in unfavourable change of geometry, such as in the buckling of struts, then an instability-type of failure may occur, where a peak load is reached short of the ideal plastic-collapse load. Multi-storey frames unfortunately may come into this third class.

2. The instability problem encountered in multi-storey frames may be conveniently subdivided as follows:—

   (a) Instability of individual stanchions by bending about the minor axis. Strictly speaking this is a special case of item (e) below.
   (b) Torsional instability of beams or stanchions when failure is combined with twisting.
   (c) Local crinkling of flanges.
   (d) Instability due to the type of stress/strain curve of the materials themselves, e.g. plastic hinges in reinforced concrete have a limited rotation before the moment of resistance falls off.
   (e) "Frame" instability, consequent upon unfavourable deflexions developing in the frame as a whole, the mutual interaction of all members being the important feature here.

3. Various aspects of the instability problem have been the object of intensive research of recent years, particularly when the aim was to provide rapid design methods.[2, 3, 4] Without such research certain slender stanchions and light sections could not have been used with safety. It does not necessarily follow that such sections are being used under uneconomical conditions for it is the ratio of the ideal plastic-collapse load factor λ_p to the corresponding elastic critical (overall buckling) load factor λ_{crit} which really decides whether members are being used economically. Moreover, the whole concept of the "equivalent length" (of stanchions) is here involved. Perhaps the best procedure is to give a brief historical account of the work of the various research centres since the

---

[1] The references are given on p. 101.

The stability of tall buildings

time of the Steel Structures Research Committee (1929–36). In doing so it is as well to remember that any new design methods must face up to the same high standards of rapidity demanded of that Committee by designers—a task of some magnitude.

Early post-war rapid design methods for rigid-jointed multi-storey frames

4. The earliest post-war rapid design methods[5, 6] relied considerably on the design of stanchions proposed by the Steel Structures Research Committee (S.S.R.C.). These methods, and the "Draft Rules" themselves, depended on the side-sway of the frame being eliminated by separate means (as is made clear in the introduction to the Draft Rules).[7]

5. Although seemingly restrictive, this was a wise decision, for it is impossible for designs for multi-storey building frames, where side-sway is allowed, to equal, in point of economy, modern no-sway designs. Moreover, many types of city buildings are exempt from an analysis of wind stresses.

6. The no-sway design methods appearing immediately after the 1939–45 war were:—

(a) A combination of the S.S.R.C. method of designing stanchions, together with a new method of designing beams using nomograms as devised by Horne.[5]

(b) A modification of the basic S.S.R.C. method of designing stanchions, together with a new graphical method of designing beams as devised by Wood.[6, 8]

In both cases the difficulty was to predict rapidly either maximum support moments or alternatively central-span moments in beams, since either might control. Horne found that if the far ends of the adjacent beams were assumed pinned, and the far ends of the adjacent stanchions fixed, then this simple substitute frame gave accurate predictions of maximum beam moments, which could be read from nomograms. More extensive substitute frames were similarly employed by Wood, and graphs were used to predict the influence of whole groups of beam loads (Fig. 1*). There is little to choose between these methods, and together they made elastic (no-sway) design simple, for no mathematical analysis whatever was needed in the design process itself.

7. In spite of the considerable proportion of steel that was saved in the beam design, the stanchion design lagged behind in terms of economy, although to effect any improvements has subsequently meant long and arduous research carried out at various centres.

---

* Fig. 1 is taken from B.R.S. Notes A:47 and A:58.
L, l_1, and l_2, denote fixed-end moments for live loads in central, L.H., and R.H. bays.
D, d_1, and d_2, denote fixed-end moments for dead loads in central, L.H., and R.H. bays.

$$\text{Stiffness} = \frac{\text{Moment of inertia}}{\text{length}}$$

Worst possible loading arrangement when beam support moment is required at point 1. (For support moment at point 2, turn the whole diagram (and the corresponding loading arrangement) to opposite hand).
Add together the three contributions.

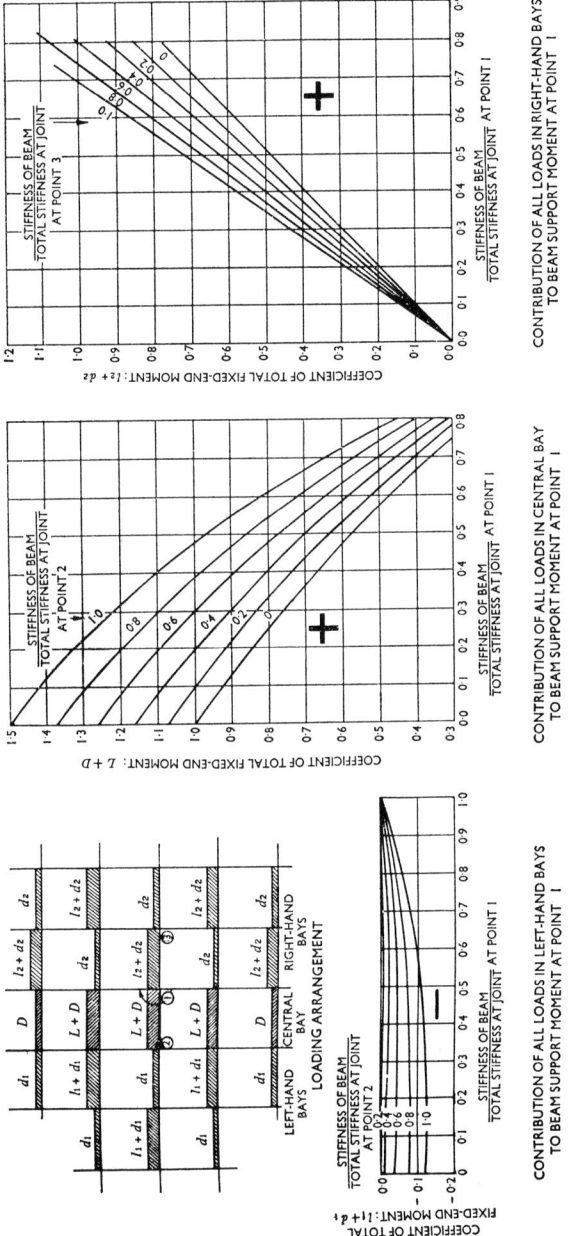

Fig. 1.—Maximum support moments in beams

Modern research work on frame behaviour

Adaptation of the moment distribution procedure of Hardy Cross to include the effects of direct loads in stanchions

8. Baker and Holder, in the final report of the S.S.R.C., pointed out that the original stability functions due to Berry[9] could be re-arranged (as X and Y functions) to fit very neatly into the well-known slope/deflection equations. In America similar functions were discovered independently by James,[10] who in 1935 adapted the moment-distribution methods of Hardy Cross to include the effects of direct loads. It is interesting to note that, whereas in this country the S.S.R.C. decided that the changing end moments in stanchions (with increase of direct load) could be neglected in the much more pressing search for a rapid, conservative design method, elsewhere the work of James was extended by Lundquist[11] in search of ways of determining the elastic-collapse loads of frameworks (the "critical" load). It is not everywhere appreciated that there exists a collapse load factor, leading to indefinitely large displacements, even if the structure remained always elastic, as depicted in Fig. 2 (facing p. 88). Since, however, this elastic critical load usually occurs only at a very high load factor the concept of elastic critical loads has led to a certain amount of confusion. The degree of importance (or otherwise) of the elastic critical load, the way in which it enters into multi-storey frame design, and how it is affected by composite action will be examined. In what is probably the best existing treatise on critical loads of elastic structures Chandler[12] makes the observation that Lundquist's work was "the foundation stone in the concept of stability, but in regard to the numerical evaluation of critical loads left much to be desired".

9. Certainly the concept of stability must be approached by a study of the elastic behaviour in relation to the critical load, and afterwards by the deterioration of stability as plastic zones develop. Merchant, Livesley and Chandler[13] have recently produced the most extensive tabulation of stability functions s, c, m, n, and o, for use in elastic moment-distribution analysis. Broadly speaking, s and c provide the effective stiffness and carry-over coefficients when there is no sway; n and o are used instead when there is unrestricted side-sway (no-shear); and m is used when side loads are present. In what follows, these functions have been used to trace the elastic behaviour of frames, and to predict the approximate behaviour of the frame between the formation of successive plastic hinges.

Developments in degree-of-restraint methods

10. First developed in this country by Shepley[14] and extended in various forms by Wood,[6, 15] Beaufoy and Diwan,[16] and Allen,[17] degree-of-restraint methods are essentially a means of rapidly calculating the effective stiffness of any member to account for all other attached members. Successive stiffness modifications appear rather like carry-over coefficients. Their first use for design purposes was for rapid comparison of the effects of many alternative loading arrangements, but the methods, have now had a new lease of life for determining elastic critical loads, for probably the best definition of the critical "load" is "that multiple of working loads at which the sum of all the effective stiffnesses of members meeting at any joint becomes 0." (At high loads, stanchions can acquire a negative stiffness, i.e. for a clockwise end rotation an anti-clockwise restraint moment may be required). This method is particularly

valuable for tracing the deterioration of the critical load when plastic hinges develop.

The concept of "frame" instability

11. The surprise to which modern references to frame instability have given rise would never have occurred had elastic design of structures been properly based on the load-factor concept right from the start. This is well brought out by the recent studies of frameworks by Merchant.[18] For the effects of direct load, of side-sway, and of the stability functions s, c, m, n, and o mean that even a completely elastic multi-storey frame is sufficiently non-linear in behaviour that ultimately a peak load factor is reached where the deflexions increase out of all proportions (i.e. it "collapses"), as depicted by the demonstration model in Fig. 2a. The fact that the stanchions are here made deliberately slender ($L/r = 660$ approx.) does not detract from a faithful representation of the behaviour of a more orthodox frame near to its critical load if it remains elastic, for the critical load factor depends only on the various ratios of P/P_{Euler} in the stanchions, and the various stiffness ratios of members throughout the frame. Consequently for any stanchion length:

$$\frac{P}{P_E} = \frac{PL^2}{\pi^2 EAr^2} = \frac{1}{\pi^2 E} \cdot \frac{P}{A}\left(\frac{L^2}{r}\right) \qquad \ldots \ldots (1)$$

and the slenderness ratio L/r may be increased and the direct stress P/A decreased for demonstration purposes, keeping the effect the same. Except where the direct loads in stanchions might be altered by corresponding shear in the beams the overall critical load is not affected by different fixed-end moments due to different beam-loading arrangements: fixed-end moments only alter the resultant shape.

12. It is tempting to dismiss the elastic critical load as fictitious, since infinite displacements cannot occur at finite stresses. Its importance, however, lies in the enhancement of stress long before the critical load is reached. Furthermore, instability in the more general elasto-plastic range is best demonstrated by deterioration of stiffness rather than enhancement of stress, for at collapse a flat load/deflexion curve implies zero stiffness even at finite deflexions.

13. Thus the unsymmetrical frame of Fig. 3a, where the stanchion AB of stiffness K is restrained by surrounding members of total effective stiffness ΣK_A at end A, and ΣK_B at end B may be considered (these can include modifications for continuity, etc.). Now if direct load effects were ignored, the effective rotational stiffness[15] of the stanchion AB, at end A, with no side-sway, also taking into account everything attached at end B would be:

$$K'_A = K\left(1 - \tfrac{1}{4} \cdot \frac{1}{1 + \frac{\Sigma K_B}{K}}\right) \qquad \ldots \ldots (2)$$

Thus for a pin end at B, $\Sigma K_B = 0$, $K'_A = \tfrac{3}{4}K$; and for a fixed end at B, $\Sigma K_B = \infty$, $K'_A = K$, both being results well known to the designer. When, however, direct load is included there is a neat transformation:

$$K'_A = K \cdot \tfrac{1}{4}s\left(1 - c^2 \frac{1}{1 + \frac{\Sigma K_B}{K \cdot \tfrac{1}{4}s}}\right) \qquad \ldots \ldots (3)$$

This expression for the effective stiffness at A becomes negative with a pin end at B whenever $P > P_E$; or with a fixed end at B whenever $P > 2 \cdot 05 P_E$.

The stability of tall buildings

14. Evidently the critical (elastic collapse) load factor λ_{crit} will be reached when:

$$K'_A + \Sigma K_A = 0 \quad \ldots \ldots \quad (4)$$

The question arises as to whether or not this same critical load will make the total effective stiffness at the opposite end B simultaneously 0, however unsymmetrical the frame may be. On this point hangs the whole meaning of "frame" instability.

15. Re-arranging equation (4) using equation (3) leads immediately to:

$$\left(1 + \frac{\Sigma K_A}{K.s/4}\right)\left(1 + \frac{\Sigma K_B}{K.s/4}\right) - c^2 = 0 \quad \ldots \quad (5)$$

and from the symmetry of this expression it is obvious that

$$K'_B + \Sigma K_B = 0 \quad \ldots \ldots \quad (6)$$

simultaneously, a result originally obtained in a different way by Lundquist, as discussed by Chandler.[12] (It will be recognized that expression (5) implies that the determinant of the stiffness matrix becomes 0 at the critical load, and that successive modification of stiffnesses is probably the fastest way of evaluating this determinant, at least for simple frames, as has been discussed by Owen and Bolton.[19])

16. A structure will recover from any accidental disturbance so long as any overall elastic stiffness still remains. A small increase in the load factor itself is one such "disturbance", and is the popular idea of the test for collapse—but it is too restrictive. When the critical load is approached with a symmetrical frame and balanced loads, a small horizontal force will bring about a sway-type collapse (Fig. 6b). This raises, of course, the whole question of worst loading arrangements, and the sudden change of configuration known as "unwrapping".

17. The outcome is that there is really no such thing as individual stanchion collapse. It has always been "frame" collapse. Composite action of frame and cladding, is therefore included in a more general statement of the stability of tall buildings. Similarly a pin-ended stanchion buckles at the Euler load because the remainder of the frame consists of two pins. Attempts to find out, by laboratory tests of isolated stanchions, the exact state of a stanchion at collapse can miss the decisive test for instability. A stanchion does not collapse, either elastically or plastically, just because certain end moments or end rotations (or ratios of these) have been reached at a certain direct load. It collapses only if the rate of change of end moments, for a disturbance that may imply end rotation and side-sway, is no longer counterbalanced by the opposing rate of change of adjoining beam moments—once more a question of zero collective stiffness even in the elasto-plastic range. It is possible to tabulate all elasto-plastic states of stanchions in terms of moment-rotation characteristics. A specimen for rectangular stanchions at the Euler load,* is shown in Fig. 4. Any or all of these states—completely elastic, partially plastic, or with full plastic hinges—could take part in the collapse of a frame, or may represent states before, at, or after collapse. No part of the diagram represents specifically collapse states. Collapse is decided by the overall interaction of all such members. In the example given later (Fig. 21) there is a whole variety of stanchion states in a frame collapse. The exact states occurring at collapse depend

---

* Assuming reversibility of stress and strain; while Horne has shown that it is possible to account for non-reversible states.

primarily on the ratio between the simple plastic-collapse load, and the elastic critical load, and on local loading arrangements.

18. In Fig. 3 are shown the critical loads of two simple, but unsymmetrical, basic systems of beams and stanchions, one system with no side-sway, the other with unrestricted side-sway. The range of P_{crit}/P_E for no-sway conditions is 4·0 down to 1·0; the range for unrestricted side-sway is 1·0 down to 0. Over a wide range of beam/column stiffnesses there is a drop of 4:1 or more in the critical load due to side-sway. This distinction grows in importance when further

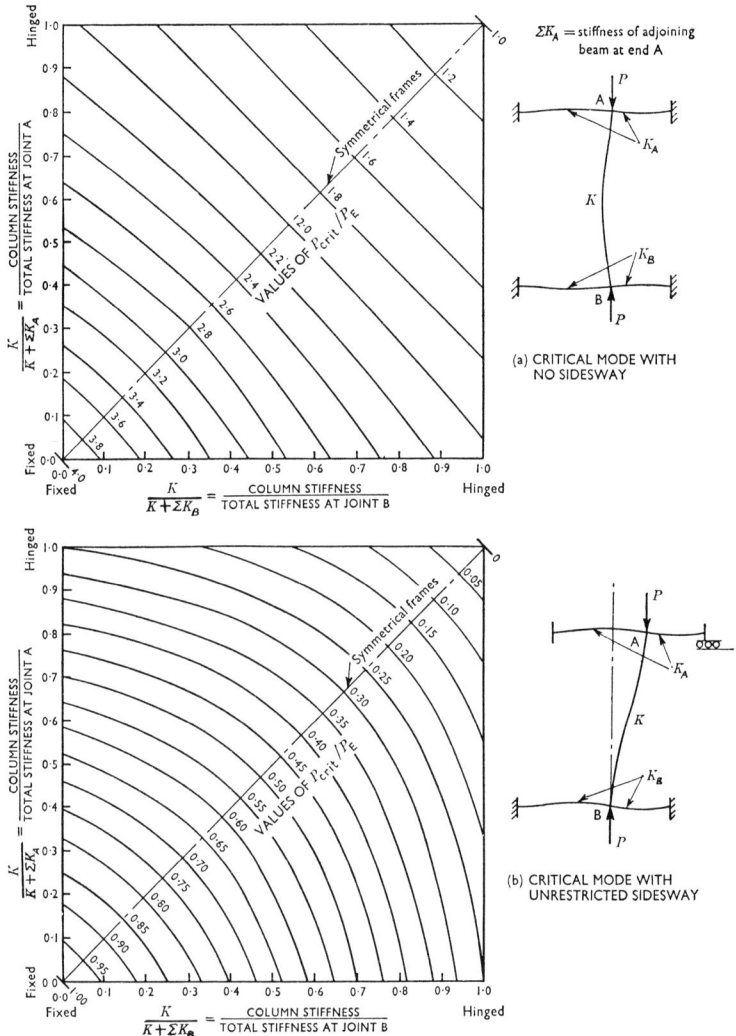

Fig. 3.—Elastic critical loads of simple unsymmetrical beam-and-column structures

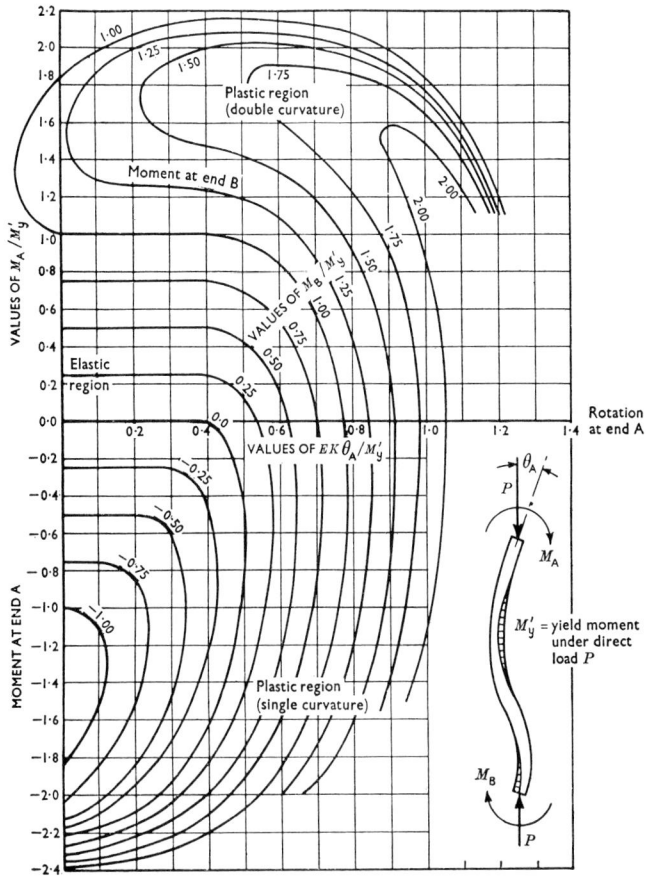

Fig. 4.—Moment-rotation properties of rectangular stanchions at the Euler load, including effects of plastic zones

(Additional lines representing tensile-plastic zones are omitted for clarity.)
The chart is valid for $P/P_E = 1.0$

deterioration of critical loads following the formation of plastic hinges is later discussed.

Research into the effects of plasticity—"Collapse design"

19. Simple plastic design has by now been well tried out by designers on certain kinds of structures. An adequate account has been given of the research work carried out under the guidance of Baker[3] and the recent position was discussed at the 1956 symposium at Cambridge.[20]

20. The reason why consideration has been given to the elastic critical load is that in multi-storey buildings collapse design is no longer synonymous with simple plastic design. Floor-loading arrangements can give a potentially

dangerous single-curvature bending of stanchions (Fig. 7) not met with in portal frames; designs including side-sway are now more difficult. In this setting the deterioration of frame stability due to plasticity is an almost unexplored subject. Above all, the classical tests on model frames with rectangular stanchions carried out by Baker, Horne and Roderick,[21] demonstrating the remarkable properties of stanchions in sustaining further loading after the first yield (Fig. 5), happen at the same time to be a demonstration of the importance of the enhanced critical load of the system as a result of beam restraint. Stiff high-tensile steel beams were used in those tests purposely to demonstrate the enhanced carrying capacity of stanchions. Paradoxically both Horne[2] and Wood[4] have more recently been obliged to recommend that for rapid design of multi-storey frames, even for the same economical no-sway conditions, the basis of design of the stanchions should be at least close to elastic design, once the beams have been designed very economically.

21. To determine why this is so it is necessary to examine (i) interference effects associated with the elastic critical load, and (ii) the necessity of guarding against torsional instability. Dealing with the first it is illuminating to compare the analyses[21, 22] of the original tests by Baker *et al.* (Fig. 5) with the predicted behaviour of the same frame had it been free to sway sideways (Fig. 6). This latter analysis has been carried out on the Building Research Station differential analyser.[23]

22. The beams were very stiff and the ratio $\frac{K_c}{\Sigma K} = \frac{\text{Column stiffness}}{\text{Total stiffness}}$ at the joint is effectively as low as 0·13 approx. From Fig. 3 the critical loads are $3\cdot48P_E$ and $0\cdot87P_E$ for no-sway and unrestricted sway respectively, i.e. 24 tons and 6 tons, since $P_E = 7\cdot0$ tons approx. By a precise analysis Baker *et al.* found the collapse loads to be $P = 7\cdot9$ tons and $P = 6\cdot82$ tons when the loaded beams were so arranged as to give double-curvature (*DC*) and single-curvature (*SC*) conditions in the stanchion. Towards the end of the *DC* test the stanchion suddenly unwrapped into the shape associated with the critical load for no-sway conditions as in Fig. 3a, but with more pronounced plastic zones. While demonstrating the remarkably beneficial effects of plasticity, this is at the same time compatible with a deterioration of the critical load from the original value of 24 tons (because of the dwindling elastic core of the stanchion), the final collapse being still a failure by instability.

23. If the beams are put on rollers an interesting situation arises (Fig. 6). The new elastic critical load (6 tons) is already less than either of the previous *DC* or *SC* collapse loads, and the behaviour for the same beam-loading arrangements is entirely different. Considering the arrangement with loaded beams on the same side of the stanchion (previously *DC*), and applying the beam loads first, and increasing the direct load afterwards, it is found that yielding theoretically first takes place at the ends of the stanchion just before the direct load P reaches 2·0 tons. When $P = 2\cdot22$ tons only a single plastic zone (in compression) is present at both ends, whilst at $P = 2\cdot34$ tons a tensile-plastic zone just commences. A peak instability (=collapse) load is reached at $P = 2\cdot40$ tons, without complete plastic hinges having formed at the ends.

24. It will be noticed that this time with sway allowed, the *DC* loading induces a configuration corresponding to the lowest mode of buckling right from the start. The stanchion does not unwrap on account of this. But if the beams are loaded on opposite sides of the stanchion (*SC*) then with absolute symmetry

The stability of tall buildings 235

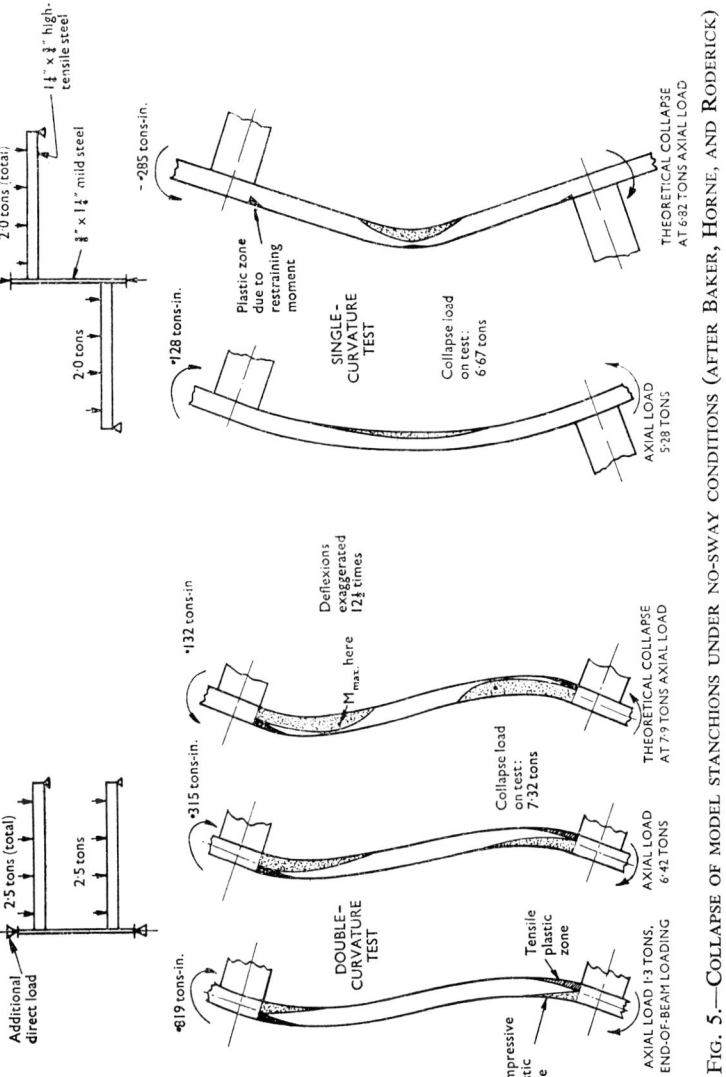

Fig. 5.—Collapse of model stanchions under no-sway conditions (after Baker, Horne, and Roderick)

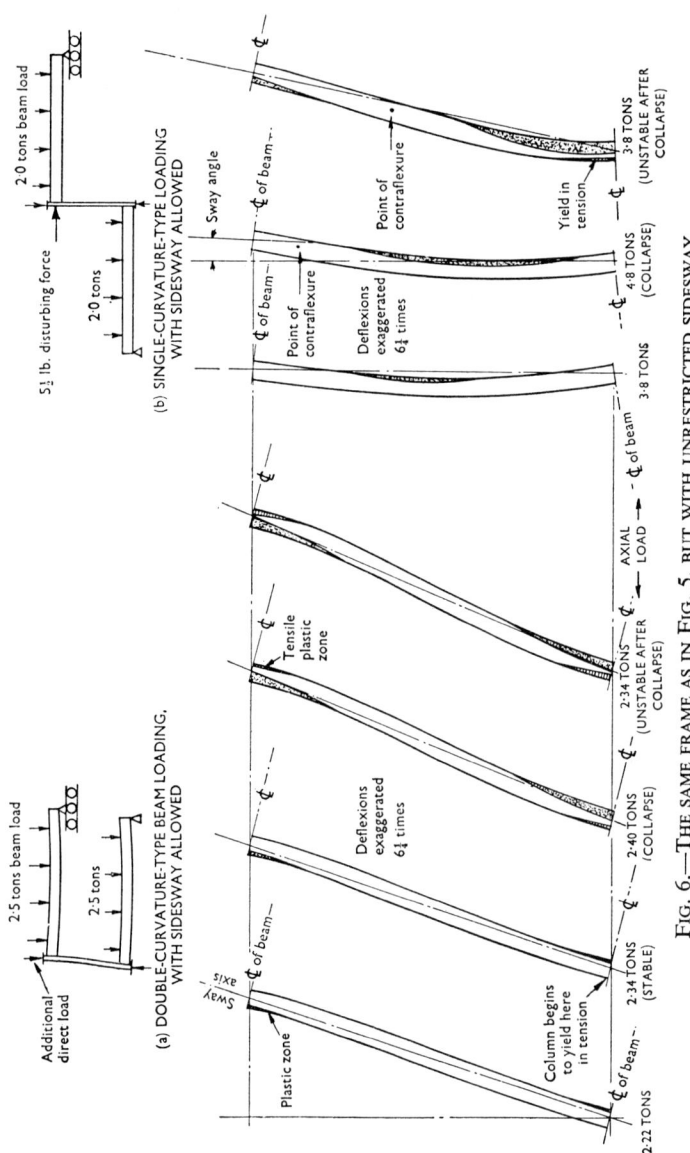

Fig. 6.—The same frame as in Fig. 5, but with unrestricted sidesway
(Results obtained from Building Research Station differential analyser)

the stanchion would remain quite vertical, without side-sway, and the strain history would tend to be identical with that recorded by Baker *et al.* Evidently a very slight disturbing force would cause the column to unwrap, this time into the shape given in Fig. 3b, i.e. into a double-curvature state. The popular belief that double-curvature bending provides the strongest stanchion can be misleading, overlooking the question of frame instability. With a convenient small horizontal disturbing force ($5\frac{1}{2}$ lb., as shown) the early behaviour was almost identical with the no-sway case. But when P was 3·8 tons a thin plastic zone occurred near the centre of the stanchion, the stiffness then deteriorating more rapidly until collapse took place at 4·8 tons. It is not claimed that this unwrapping is traced exactly, but (as summarized in Table 1) it will be noticed that a false sense of stability is always given when the beam-loading arrangement does not coincide with the characteristic elastic critical load configuration.

25. The behaviour of these models brings out the difference between acting and restraining moments on stanchions. Had plastic hinges been allowed to develop in the beams the elastic restraint would have diminished. This has a bearing on the trend of modern design methods.

26. The continued search for rapid simplified methods of design has so far produced two more or less complete (no-sway) methods, when taking into account bending about both axes of stanchions simultaneously. These are:

(i) The Cambridge method using fully plastic beams and elastic stanchions, designated P_xP_y to denote fully plastic beams framing into both axes

TABLE 1.—PARTICULARS RELATING TO THE COLLAPSE OF THE MODEL TEST FRAMES OF FIGS 5 AND 6

(All direct loads quoted in tons)
$P_E = 7·0$ tons (approx.) $P_y = A.f_y = 9·5$ tons (approx.) = "squash" load

Type of beam loading	No sway*				Unrestricted sway†			Type of collapse
	Elastic critical load	Direct load at first yield	Collapse load	Type of collapse	Elastic critical load	Direct load at first yield	Collapse load	
	P_{crit}		P_{coll}		P_{crit}		P_{coll}	
Double-curvature (2·5 tons on beams)	24	0·55	7·9	unwraps	6	2·0	2·4	does not unwrap
Single-curvature (2·0 tons on beams)	24	3·98	6·82	does not unwrap	6	3·14	4·8 (depends on disturbance)	unwraps

* After Baker, Horne, and Roderick
† Results from Building Research Station differential analyser

of the stanchions. This method has been well described elsewhere,[2,3] and is at present undergoing further research.[20] It suffices to note that the beams are of the most economical design that is possible. Owing to the probable absence of elastic restraint the elastic critical load for any stanchion, together with its surrounding beams, might be reduced to the Euler load (with side-sway prevented), and to a first approximation end moments remain acting moments. This is one reason for retaining elastic design of the stanchions. The control of torsional instability, for bending about the strong axis, is another reason.

(ii) A design incorporating stanchions restrained by elastic beams framing into the minor axes of the stanchions (Building Research Station). Although not outright collapse design, this approach has many attractive features. A brief account of the research work leading up to this design method is given below.

Research into rapid methods of evaluating maximum moments in stanchions with elastic restraint

27. It is necessary to find a method of predicting rapidly the maximum moments in elastic stanchions with various beam-loading arrangements, and with due regard to the elastic critical load, if there is to be reasonable prospect of still more advanced elasto-plastic rapid designs. It suffices, for design purposes, to approximate to the elastic critical load, and this can be achieved for no-sway designs by assuming adjacent stanchion lengths, above and below, to be bent into the least stiff configuration of symmetrical single-curvature.

28. The variables must include all values of end restraints and all arrangements of beam loading (the latter expressed as out-of-balance fixed-end moments at the upper and lower joints, C_1 and C_2 respectively). In Fig. 7 several cases are shown of symmetrical *DC* and *SC* bending-moment diagrams for stanchions, and also the special case of one end encastered. As the direct load P increases it will be noticed that the *DC* end moments stay substantially constant and the maximum moments stay at or near the end. The end moments, however, decrease under *SC* conditions, passing through zero moment at the Euler load, and becoming restraining moments at greater direct loads.

29. A voluminous tabulation of all possible solutions must be avoided, but almost instantaneous solutions of the required maximum moment can be obtained using nomograms. A considerable advance has recently been made[4] compared with earlier attempts,[24] in that the maximum moments can now be read directly from a plot of contours of maximum moments (Fig. 8). The technique of finding the required point amid the contours by means of the straight edge is described in this sample nomogram. The values of the (nominal) degree of restraint $K_c/\Sigma K$ at each end of the stanchion, together with the values of C_1 and C_2, decide where the straight edge is placed, and four simultaneous equations involving instability functions are automatically solved at one setting of the straight edge, giving what is tantamount to an exact solution for design purposes.

30. This solution has been made the basis of a new rapid design method incorporating allowances for some degree of plasticity in both beams and stanchions, for torsional instability, and for possible use of stiffening due to composite action. It is described in detail in a recent note issued by the Building

The stability of tall buildings

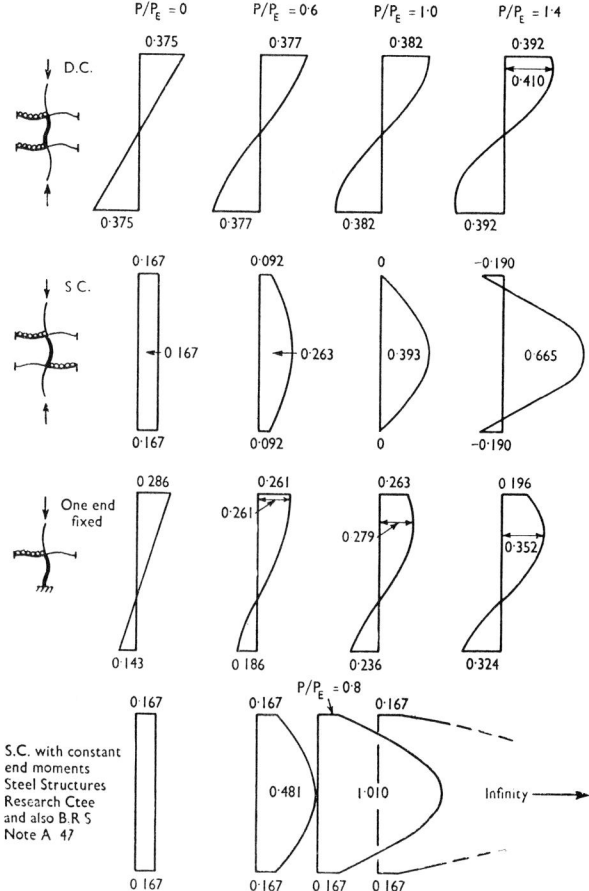

FIG. 7.—EXAMPLES OF BENDING-MOMENT DIAGRAMS WHEN THE STANCHION STIFFNESS IS ONE-QUARTER OF THE TOTAL STIFFNESS AT EACH JOINT

(Moments are expressed in terms of fixed-end moments on beams. No sway allowed.)

Research Station,[4] further suggestions of designers being invited. This method might be suitably modified for high-tensile bolted joints.

31. An advantage of elastically restrained stanchions is the added resistance to torsional instability. Often there is an accumulation of bending moment to one axis of the stanchion, and naturally the designer picks a strong section in that direction, and, if minimum weight is aimed at, a much weaker section in the other direction. An examination of torsional instability, with elastic restraint provided about the minor axes of stanchions has been carried out with the differential analyser, together with tests of destruction, and simple rules have been devised which will permit high slenderness ratios. The importance of

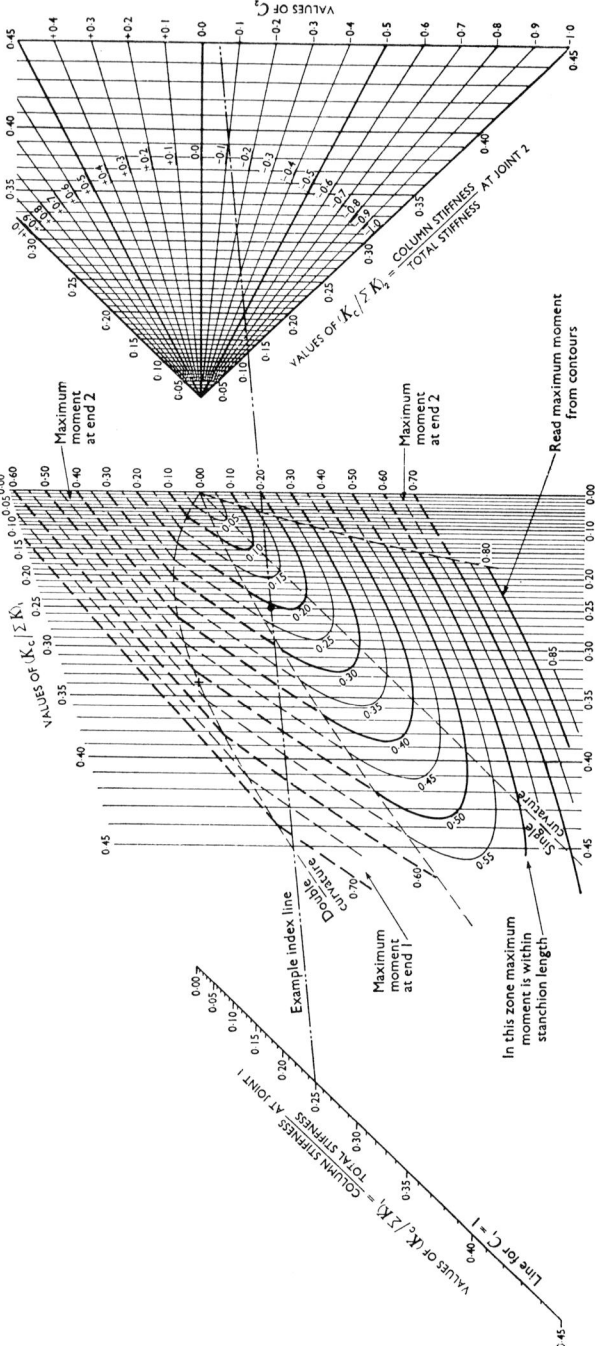

C_1 and C_2 are the out-of-balance fixed-end moments at the two ends of the stanchion, C_1 being the greater value. C_1 is put equal to unity and the values of C_2 and the maximum moment determined are both in units of C_1. Both C_1 and C_2 are of the same sign if they act in the same direction.

In the example $(K_c/\Sigma K)_1 = 0.25$, $(K_c/\Sigma K)_2 = 0.2$, $C_1 = 1$, $C_2 = -0.5$, $M_{max} = 0.21 C_1$ within stanchion length. Innumerable conditions at end 2 on the index line would give the same maximum moment.

Fig. 8.—Typical nomogram for determining maximum moments in stanchions when $P/P_E = 0.4$

The stability of tall buildings 241

restraint is shown pictorially in the torsional behaviour of the model perspex stanchion of Fig. 9 (between pp. 88–89).

Modern studies in frame instability

32. A recent extension of frame analysis dealing with the deterioration, because of plasticity, of the elastic critical load[25, 26, 27] has suggested how the critical load contributes more directly to the actual value of the collapse load. A fully plastic hinge, at constant moment, contributes no more to the stiffness of the structure than does a real hinge, provided that the angular change at the hinge continues in the same direction, once started. Consequently for further displacements, controlled by the remaining elasticity of the various members, there is a new and smaller critical load obtained by substituting, in effect, a real hinge. Naturally how much the critical load drops depends on the exact situation of the hinge.

33. Consider a beam, which is still elastic, carrying a point load (Fig. 10).

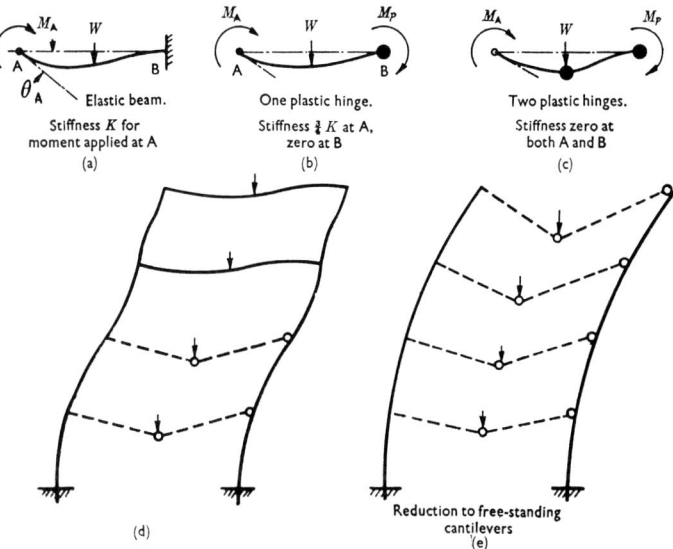

FIG. 10.—STAGES IN THE DETERIORATION OF STIFFNESS OF BEAMS AND OF THE CRITICAL LOADS OF TALL FRAMES DUE TO THE FORMATION OF PLASTIC HINGES

For a unit rotation of joint A, other joints being fixed, the change in the end moment M_A will be proportional to the nominal stiffness K. If a plastic hinge has formed at the far end B, and remains there during further bending, the effective stiffness at end A is reduced, very approximately, to $\frac{3}{4}K$—a result well known to designers—disregarding partial-plastic zones about which comment will be made later. At end B the effective stiffness has become 0, since there can be no further moment distributed to the beam at that end. If an additional hinge forms under the point load (Fig. 10c) thereafter the effective stiffness at end A, for increasing joint rotation, clearly drops to 0. The subsequent contribution of this beam towards the remaining overall frame stiffness has vanished. If this should happen on two consecutive floors (Fig. 10d), the tendency is to convert the original multi-storey frame into a portal frame of great height. If this process is continued (Fig. 10e), effectively free standing cantilevers of great height are sooner or later left (if the soil will stand it). The Author has previously called this[26] a tendency towards "conversion to chimneys". Multi-storey frames provide probably the only case where instability phenomena about the strong axis of a frame may actually be involved. The process can be accumulative and it is proposed to designate the new critical load of the remaining elastic parts of the frame the "deteriorated critical load", at a corresponding load factor λ_{det}. A physical demonstration of this is given in Fig. 2. Here a special model was made with hinges that can be produced at will, by pulling out small pins.

34. The problem now confronting the research worker, and those concerned with future design methods, is that except for secondary effects due to strain hardening, at any stage of loading the load factor that has been reached can never exceed the load factor corresponding to the deteriorated critical load, otherwise the deflexions would increase indefinitely. However, it must be pointed out that any analyses or tests carried out in this subject must relate to representative designs; unless this is observed it is easy to produce evidence for or against catastrophic collapse merely by tinkering about with stiffness ratios. Again a sense of proportion must be retained, since this phenomenon only starts after considerable yielding.

35. The most intriguing point of all is that the theoretical deteriorated critical load of fully developed collapse mechanisms involving side-sway happens to be 0, to a first order of approximation. For a complete mechanism denotes the disappearance, at the last hinge to form, of the remaining elastic core. As it stands this is merely a *reductio ad absurdum*. However, the consequences of this paradox can be followed by reference to a particular case (Fig. 19).

36. The deteriorated critical loads corresponding to various possible combinations of plastic hinges are included in the figure. At first sight the astonishing feature is the wide range of critical loads involved. Since this frame is bent about the strong axes of the beams and stanchions the original critical load is high, as much as $\lambda = 12.9$. The instability effects at working loads are therefore insignificant. If two hinges developed, at the leeward ends of the lower beams, the critical load would be dropped to $\lambda_{det} = 6.3$. Two hinges on each of the lower beams would give $\lambda_{det} = 2.3$. Other combinations lead to remarkably low critical loads. Hence it becomes clear that at about four or five hinges, out of the required ten for the mechanism, the deteriorating critical load clashes with the rising (actual) load, and will prevent the ideal collapse load being reached.

37. The higher the original elastic critical load, the nearer the actual collapse will become to the desired plastic collapse, i.e. the higher the ratio λ_{crit}/λ_p, the more reliable λ_p becomes. If this explanation is to be valid the elastic critical load factors of portal frames that have been tested to destruction must be expected to be very high. To quote two examples of portal frames tested at Abington Hall,[28] one fixed-base, one pinned-base, the values of λ_{crit} were 310 and 77 respectively. The pinned portal collapsed at the correct load, and as the value of λ_{det} was still as high as 27 after the first hinge formed, an advanced stage of development of the last hinge would be necessary before the deteriorating critical load could be of any consequence whatever. Moreover, with the

fixed-base portal Baker and Roderick report that the collapse load was increased because of the extra stiffness due to strain hardening, and the present theory is of course quite compatible with these observations. It is obvious that portal frames and multi-storey frames are in quite a different class.

38. Three examples of multi-storey frames failing by frame instability will be given. In reviewing these examples it must be remembered that buildings are three-dimensional, that failure will occur in the weakest direction, and also that the examples are deliberately chosen to show up frame instability and should not be taken as necessarily average cases. The first example demonstrates the most extreme form of instability likely to be encountered in practice if there is no help from the cladding. The second example represents a more likely practical case, which is nevertheless surprising in that instability can develop even when the frame is bent about the strong axis, and shows that reinforced concrete frames are not necessarily free from instability. The mathematical analyses in both cases are involved, a differential analyser having been used in the later stages. It is hoped to give a more detailed account in the future. The third example is a test of a model to destruction.

39. *The first example of frame instability* refers to the collapse of a series of 4-storey frames about their weak axes of bending when connected by beams of 18-ft span, rigidly welded to the stanchions (Fig. 11). In the other (major) direction are heavy beams of 30-ft span, and in this case there is no side-sway of the stanchions about their major axes. With a restricted number of bays in the minor (18-ft) direction there is, however, the possibility of a "pack-of-cards" collapse of the bents when all floors are loaded. Three features contribute towards this kind of collapse:—

(a) Wind loads.
(b) Floor loads collecting to the tie beams (the triangular area of Fig. 11).
(c) Much heavier floor loads simultaneously taken by the stanchions via the strong beams.

This last item (c) is very important, and distinguishes the behaviour of a real (three-dimensional) building from the usual two dimensional frame analysis, these heavy loads particularly weakening the stanchions for bending about their minor axes. No ordinary elastic analysis, or rigid-plastic analysis either, would make the designer readily aware of the impending instability (Fig. 14).

40. The total (live + dead) floor loads were taken as 145 lb/sq. ft (when $\lambda = 1 \cdot 0$) and the wind pressure 10 lb/sq. ft (also when $\lambda = 1 \cdot 0$). The total wind load was divided among eight bents (= seven bays), providing a relatively small sideways disturbance. The same "free" bending moment in the beams was given by substituting, for convenience, a central point load on each tie-beam of 7 tons ($\lambda = 1 \cdot 0$). A minimum weight idealized plastic design, to collapse at $\lambda = 2 \cdot 0$, was attempted in the first place, the stanchions being subsequently increased slightly to allow for initial curvature, individual instability, etc. (making use of Chapter 15 of reference 3). This had the effect of raising the collapse load, disregarding all forms of instability, to $\lambda = 2 \cdot 21$ (see mode of collapse Fig. 11c). The frames were checked to see that they were strong enough for bending about the strong axis.

41. The analytical problem thus reduces to the solution of a typical interior bay. The effect of the elastic critical load ($\lambda_{crit} = 3 \cdot 47$) on the elastic behaviour of the frame can be seen in Fig. 13, which shows the side-sway at each storey.

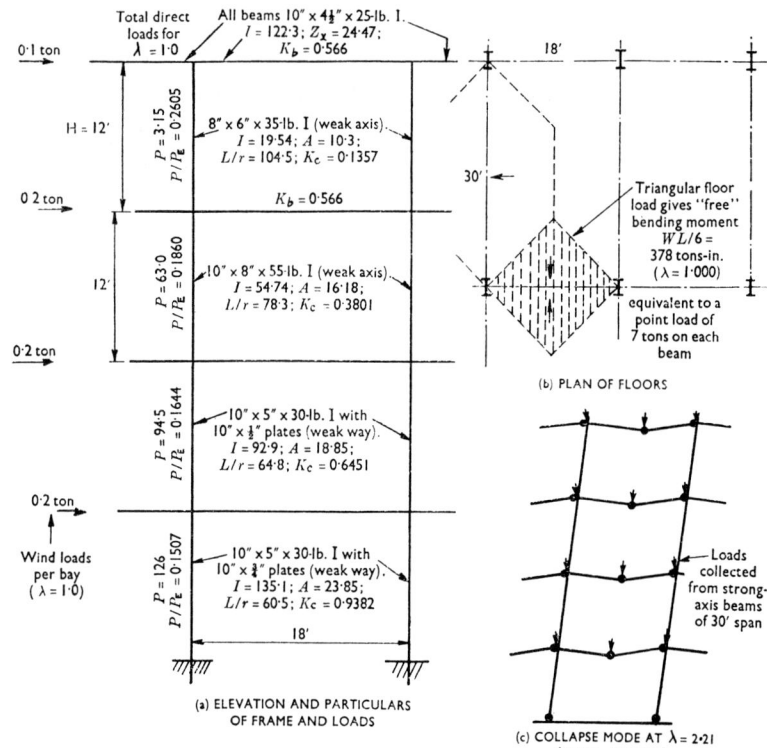

Fig. 11.—Particulars of four-storey multi-bay frame, which collapses about the minor axis of each bent

The corresponding right- and left-hand beam-support moments are seen in Fig. 14. First yield takes place, at $\lambda = 1\cdot50$, of the first-storey beam at the right-hand support.

42. The low form factor of the beams ($M_P/M_y = 428/373$ tons/in. $= 1\cdot15$) is of some importance, implying that the first hinge should have appeared at about $\lambda = 1\cdot70$, and thereafter a new critical load corresponding to $\lambda_{det} = 2\cdot51$ (see Fig. 12) will control the deflexions. But at $\lambda = 1\cdot70$ considerable yielding will clearly also have taken place (Fig. 14) at the second-storey beam support. When completed this second hinge would drop the critical load to $\lambda_{det} = 1\cdot95$. It is clear that no complete mechanism could develop. An additional hinge at the centre of the lowest beam would be most unfortunate ($\lambda_{det} = 1\cdot21$). Moreover, the frame would be sensitive to partial yielding at the base of the stanchion ($\lambda_{det} = 0\cdot75$ for a hinge).

43. It was found possible to produce a refined solution, incorporating the effects of all partial-plastic zones, for the state of the frame at $\lambda = 1\cdot70$, but what was likely to happen if the load factor was increased slightly can be gleaned from Fig. 13. Here the side-sway can be inserted at each storey corresponding to a rigid-link collapse mechanism (after collapse), taking into account the

The stability of tall buildings

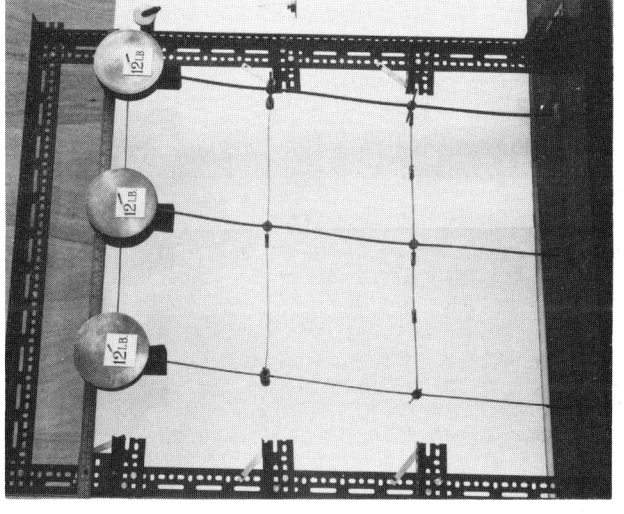

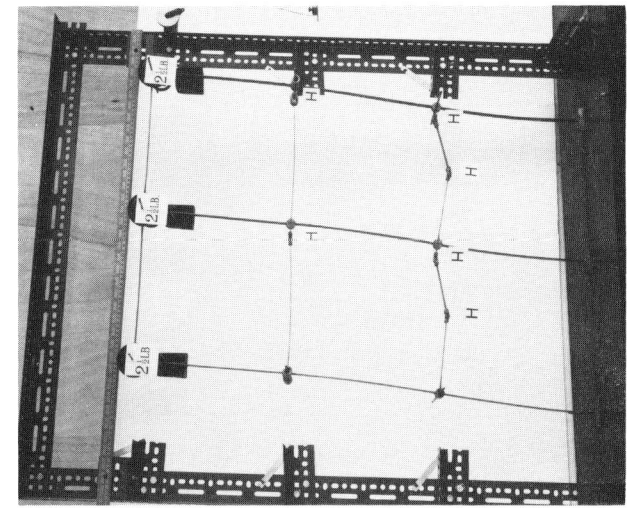

(a) Original elastic frame when free to sway. (With sway prevented, the corresponding critical load exceeds 31 lb. per stanchion.)

(b) Reduction of critical load with hinges as shown.

FIG. 2.—MODEL TO DEMONSTRATE THE DETERIORATION OF ELASTIC CRITICAL LOAD DUE TO FORMATION OF HINGES. CRITICAL LOADS: (a) 12 LB.; (b) 2½ LB. PER STANCHION

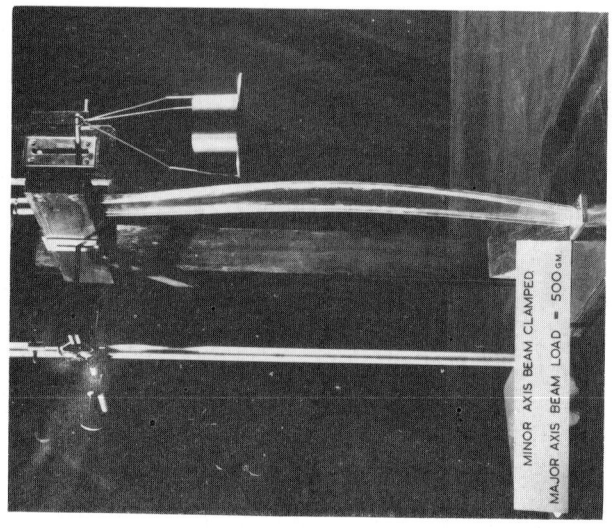

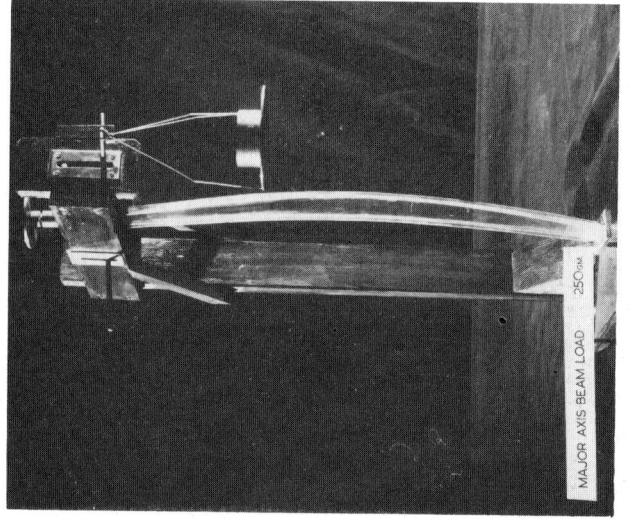

(a) With no restraint about minor axis   (b) With restraint, at top end only, about minor axis

FIG. 9.—TORSIONAL-FLEXURAL BUCKLING LOADS OF A MODEL PERSPEX I-STANCHION, IN PRESENCE OF MAJOR-AXIS BENDING MOMENTS

The stability of tall buildings 247

(a) Just before collapse

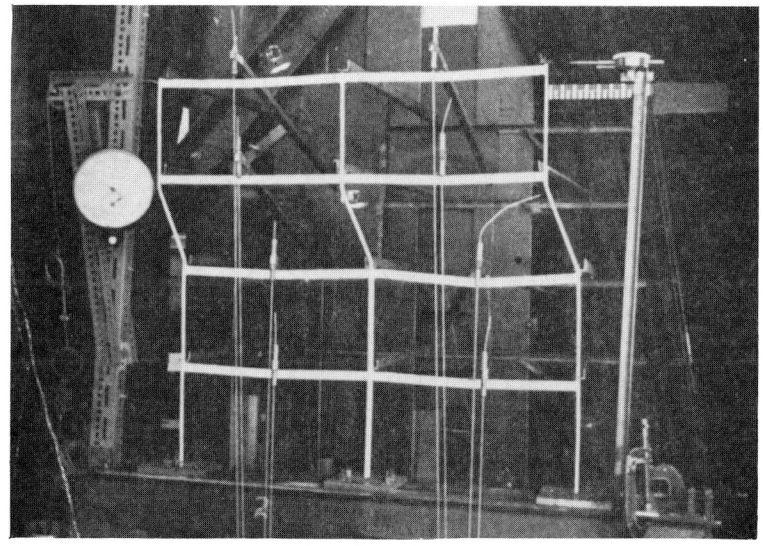

(b) After collapse

FIG. 22.—TEST OF FOUR-STOREY TWO-BAY FRAME

Fig. 23a.—Encased steel frame failed at 20 tons racking load when there was $2\frac{1}{2}$ in. displacement. The corresponding bare frame failed at $9\frac{1}{2}$ tons

Fig 23b.—The composite behaviour of the frame shown above and the brick panel when working in unison

(Note the plastic hinges along the beams. Failure occurred at 53 tons (average of two panels) with $\frac{1}{2}$ in. displacement)

The stability of tall buildings 249

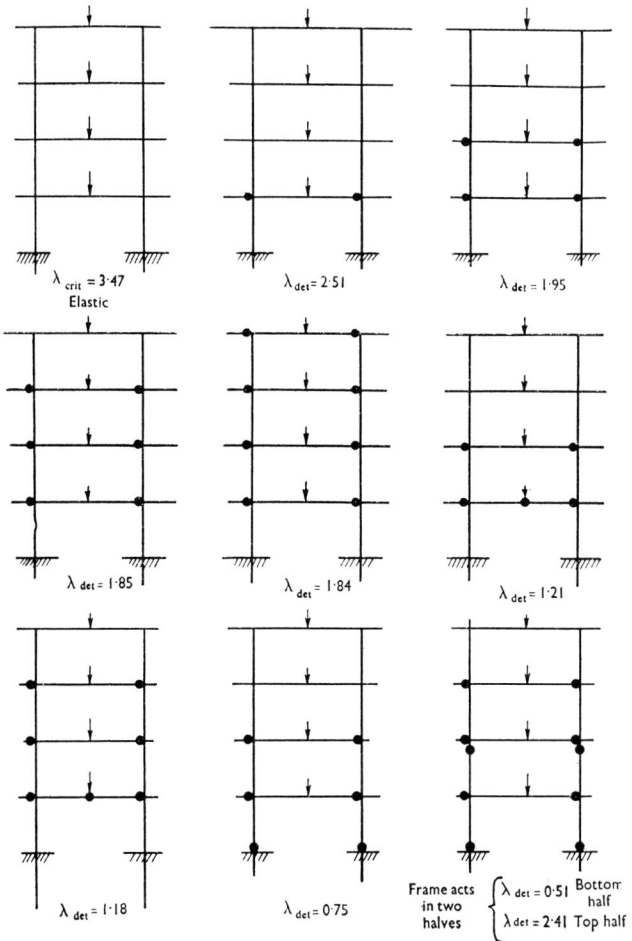

FIG. 12.—FOUR-STOREY MULTI-BAY FRAME. LOAD FACTORS CORRESPONDING TO DETERIORATED CRITICAL LOADS

$P.\Delta$ effects to maintain equilibrium. The problem is to find four curves, one for each storey, leaving the elastic curves at first yield, $\lambda = 1\cdot 50$, presumably asymptotic to the corresponding rigid-link curves, and all four curves attaining a common maximum load factor. Even a preliminary sketch indicates that $\lambda = 1\cdot 80$ is not likely to be attained.

44. Accordingly a new approximate form of analysis was invented for the purpose of following further movements. Moment-rotation characteristics of beams in the elasto-plastic region have been obtained at the Building Research Station (see typical chart, Fig. 15), showing that over a wide range of partial-plastic zones the effective stiffness of a beam is approximately $\tfrac{3}{4}K$ once a hinge has formed at the far end. Certain hinges were therefore assumed to have

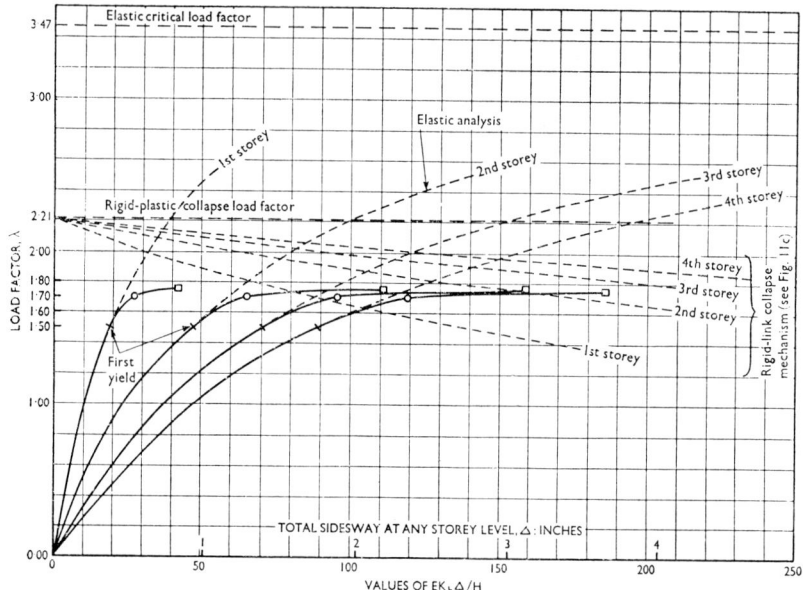

Fig. 13.—Four-storey multi-bay frame. Demonstration of collapse by plotting the side-sway at each storey level

formed at $\lambda = 1\cdot75$, keeping an eye on the critical load all the time, and, having inserted the appropriate beam stiffnesses and fully plastic moments, a revised moment-distribution solution was made for the whole frame. The result is given in Fig. 16. The ringed figures indicate where further yielding would have taken place. Reference to the deteriorated critical load charts however, shows, that any attempt to correct for these plastic zones would be hopeless. In fact hereabouts the analysis appears to run wild. The deflexions and moments shown plotted for $\lambda = 1\cdot75$ (Figs 13 and 14) are definitely "unsafe". It is somewhat meaningless to describe the exact state at collapse, but this takes place between $\lambda = 1\cdot70$ and $1\cdot75$, probably $\lambda = 1\cdot73$.

45. The beam moments of Fig. 14 (left-hand side) indicate a kind of collapse corresponding to a new depressed critical load, a most pronounced "frame" nstability. This suggests that if the deteriorating critical load is regarded as a continuously descending function, for which definite values can be given at certain stages, then collapse takes place when the increasing load and the deteriorating critical load coincide. At states beyond collapse they can cross over. Moreover, this explains why the zero-stiffness concept is preferable for describing the various critical loads for it is now attained at finite displacement. This frame probably has the catastrophic type of collapse mentioned by Bolton,[27] for there is little or no warning (Fig. 14).

46. A more precise estimate of the shape of the frame immediately before collapse ($\lambda = 1\cdot70$) is given in Fig. 17, the deflexions being exaggerated 50 times.

The stability of tall buildings

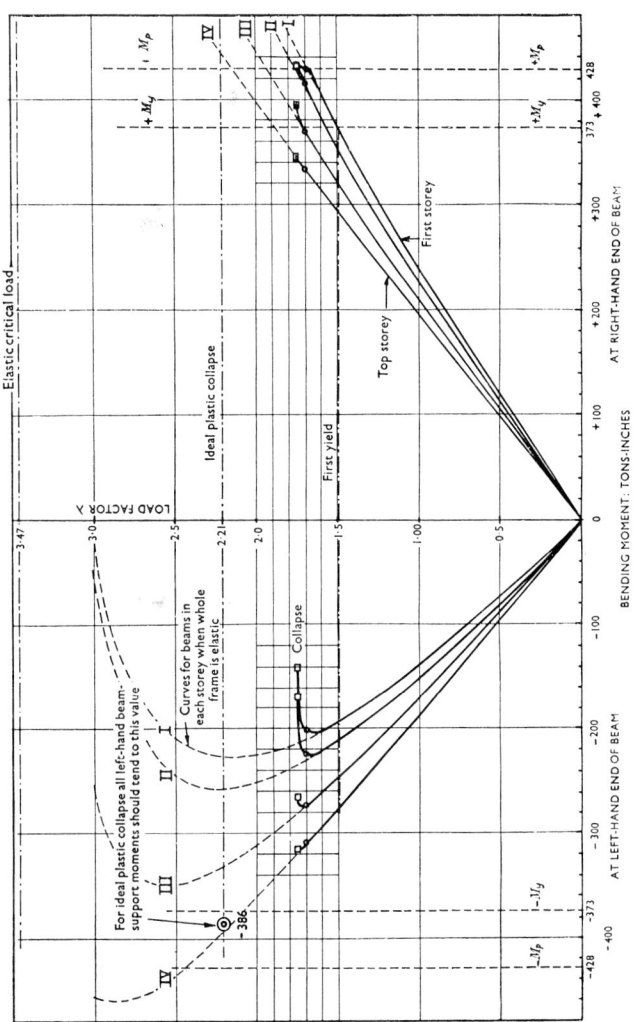

Fig. 14.—Beam moments in multi-storey frame for increasing load factor

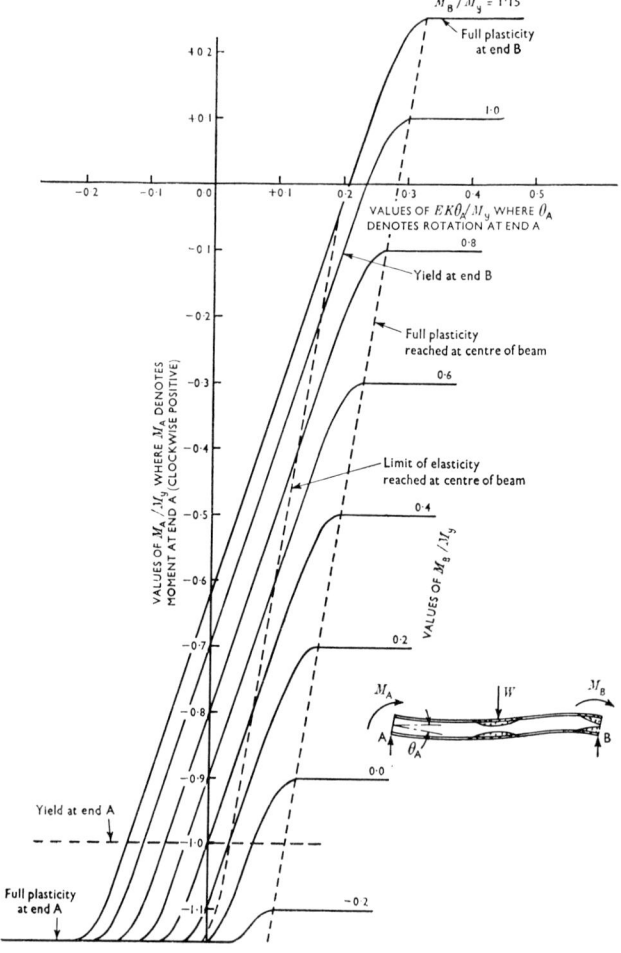

FIG. 15.—Properties of a typical beam, carrying a central point load, bent about major axis, including effects of partial plastic zones

The chart is valid for $\dfrac{WL/4}{M_y}=1\cdot 6$, where L denotes span and M_y the yield moment.

Form factor is 1·15 for bending of beams about major axis

The first hinge is hardly completed and there is only one other significant plastic zone.

47. The above example is a severe case because of the large direct loads fed in at each joint by the strong beams at right-angles to these (weaker) frames. It is clear that anything approaching this state of affairs should be avoided at all costs if the frames really are bare frames.

The stability of tall buildings 253

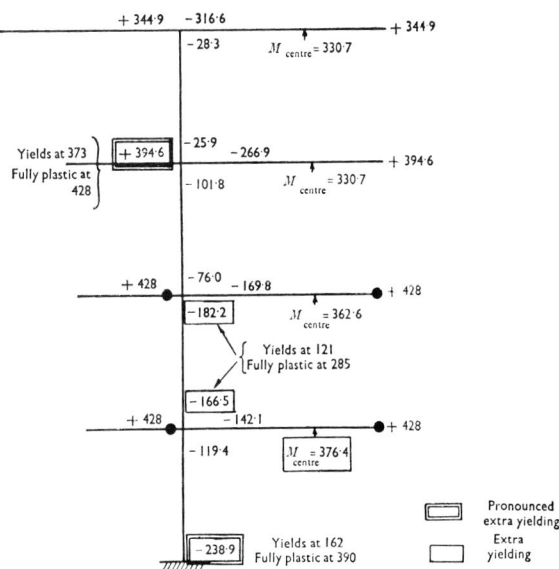

Fig. 16.—End of simplified moment-distribution process for load factor $\lambda = 1\cdot 75$, and indications of further yielding

48. *The second example of frame instability* refers to a 4-storey, single-bay frame (Fig. 18), this time bent about the strong axis. The frame is one of a series of similar bents connected by tie beams, producing a nowadays familiar tall thin building. The intensities of loading for $\lambda = 1$ are (*a*) roof, 125 lb/sq. ft (live+dead), (*b*) other floors, 250 lb/sq. ft, and (*c*) unit wind pressure, $7\frac{1}{2}$ lb/sq. ft. This corresponds to an industrial building with fairly heavy floor loads in a somewhat sheltered situation. It was so chosen to demonstrate instability effects. Heavy wind loads and light vertical loads lessen instability—a feature that can be put to practical advantage, as will be seen.

49. A comprehensive collapse mode was found (Fig. 19a) implying economical design. The stanchion sections allow for torsional instability (see Chapter 15 of reference 3) and are designed for a load factor $\lambda = 2\cdot 0$. Disregarding these subsidiary effects the ideal collapse mechanism could occur at $\lambda = 2\cdot 15$. It was thought that a representative collapse design and subsequent analysis would be given if the bending moments in the main beams could be assumed to be developed by central point loads giving the same total free moment as the floor loads.

50. Deteriorated critical loads (λ_{det}) were calculated in advance (Fig. 19b). The original elastic critical load is high($\lambda_{crit} = 12\cdot 9$) but the range of possible values of λ_{det} is rather startling, remembering that there is no guarantee that yielding does not occur at the base of stanchions.

51. On analysis, first yield occurred at $\lambda = 1\cdot 51$ at the centre of the third storey beam ($M_y = 373$ tons/in., $M_p = 428$, form factor $1\cdot 15$). This hardly altered the rotational stiffness of the whole beam (see Fig. 15), so that elastic analysis was

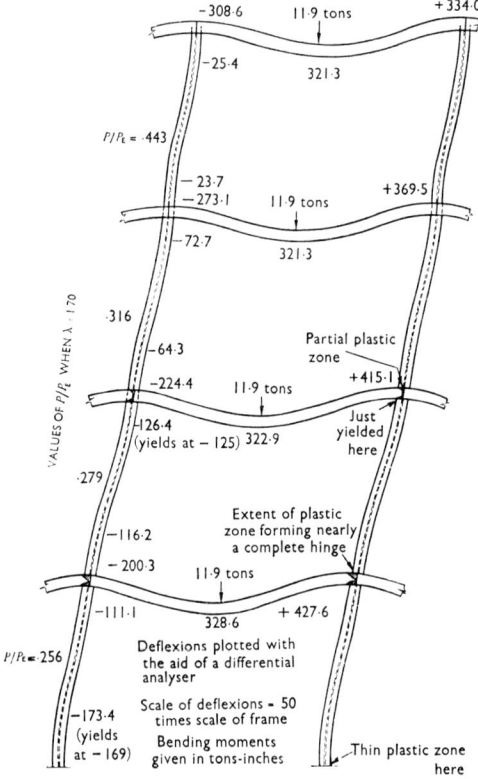

Fig. 17.—State of frame at a load factor of 1·70 just before collapse
(This solution incorporates the effects of partial-plastic zones)

continued tentatively up to $\lambda = 1\cdot70$, showing that a hinge would still not have formed ($M = 419$ at this stage). However, yielding had started at the right-hand end of the first storey beam ($M = 407$) and at the centre of the second-storey beam ($M = 383$). The corresponding deflexions at each storey are plotted in Fig. 20.

52. For $\lambda = 1\cdot80$, it was expected that the first two hinges would appear at the centre of the third storey and the right-hand support of the first storey beams (shown as 1 and 2 in the order of hinge formation, Fig. 19a). The necessity of maintaining constant moments at these two points considerably complicated the analysis, because of lack of symmetry. Only partial-plastic zones appeared at the centre of the first, second, and fourth beams, and very slight yield at the head of the second and third stanchion lengths. Previous experience at correcting for such zones suggested that it would be feasible to go to a higher load factor.

53. Raising λ to $1\cdot90$ promptly resulted in the appearance of hinges Nos 3 and 4 (Fig. 19a), and indicated the likelihood of further plastic zones as shown

The stability of tall buildings

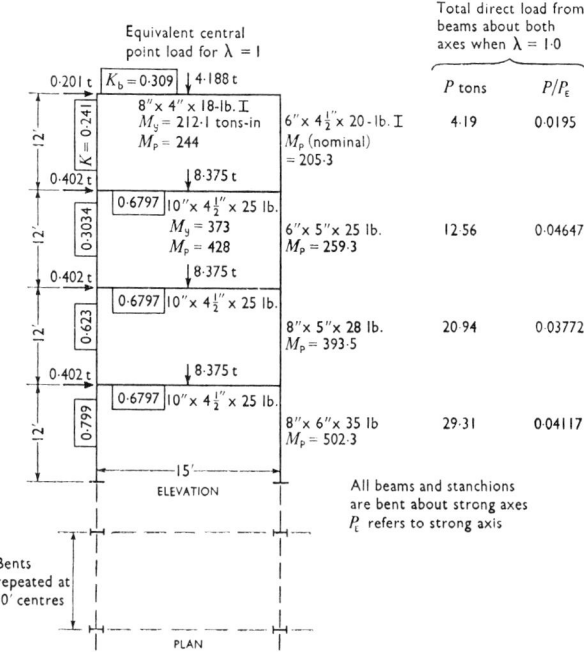

FIG. 18.—DETAILS OF SYMMETRICAL FOUR-STOREY SINGLE-BAY FRAMES

in Fig. 21. It was evident that λ_{det} was then definitely less than 2·30, and most likely less than 2·05. At this critical stage the differential analyser was used to account for partial-plastic zones, the effect being to increase the side-sway as was expected, but curiously enough to delay the formation of other hinges. The extreme sensitivity hereabouts was reflected in the plot of deflexions (Fig. 20), solutions in excess of $\lambda = 1·90$ being highly improbable.

54. Once again instability developed largely as a result of rapidly deteriorating critical loads, in spite of P/P_E values being very small (Fig. 18). Only four complete hinges formed out of the possible ten. The final state of the frame at $\lambda = 1·90$ is shown in detail in Fig. 21. No reversibility of stress was encountered, i.e. plastic zones remained "active". It is unlikely that strain hardening occurred except at the centre of the third-storey beam. There was no possibility of strain hardening in the previous example. In general it is only likely if λ_{crit}/λ_p is very large.

55. *The third example of frame instability* is a demonstration model made from convenient square and rectangular mild-steel sections, in which an attempt was made to keep the beams fairly stiff, and care taken to stabilize the frame against failure out of the plane of bending (Fig. 22, between pp. 88–89), without interfering with the collapse. This model collapses ideally at $\lambda = 2·0$ (for the loading conditions depicted) involving side-sway of both the top two storeys.

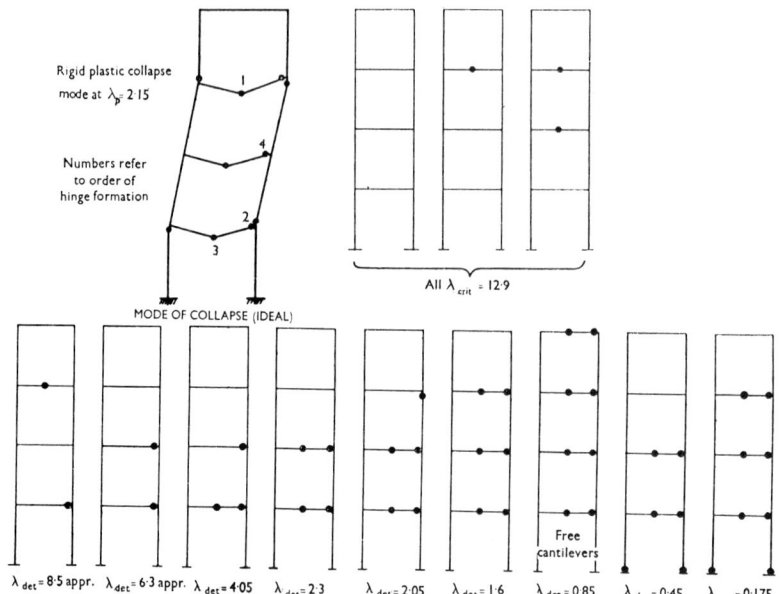

FIG. 19.—IDEAL COLLAPSE LOAD AND CRITICAL LOADS OF MULTI-STOREY SINGLE-BAY FRAME LOADED AS IN FIG. 18

But it failed at $\lambda = 1\cdot69$ on test, and by a different mode involving only translation of the third storey. This mode, discarded as an "upper bound" solution, could not possibly have occurred except that there is an original low elastic critical load, $\lambda_{\text{crit}} = 5\cdot7$ approx., which is associated with a relatively large sidesway at the third storey.

Practical considerations and recommendations

56. It is instructive to tabulate the principal load factors in the three cases above (Table 2). It has been suggested by various authorities, notably Merchant, that a "generalized Rankine" formula $\frac{1}{\lambda} = \frac{1}{\lambda_p} + \frac{1}{\lambda_c}$ might be taken as a convenient starting point for correlating different frames. The Author agrees with this approach, but has previously suggested[29] that local loading arrangements must be included (see Figs 5 and 6). This particular formula gives a fairly good result for the frame bent the strong way but overestimates the deterioration of the collapse load when the elastic critical load is low. There are indications that the part played by plasticity is the more important.

57. Many minor features have admittedly been left out of the above examples, but they are to some extent compensating. Factors which would make matters worse are out-of-straightness of members, locked-up initial stresses, torsional instability (perhaps of the complete frame), local crinkling, slip at joints, alteration in direct loads due to shear in beams, and settlement. There is also the loss of stiffness about the minor axis due to plastic bending about the major

axis—a peculiar form of three-dimensional frame instability discussed in more detail by the author elsewhere.[4] Items improving the collapse load are strain hardening, non-reversibility of stress-strain diagram, and effective shortening of members meeting at a joint.

58. It is probable that the average multi-storey building frame, even unclad, would behave somewhat nearer to plastic theories than is suggested by these

TABLE 2.—LOAD FACTORS APPERTAINING TO COLLAPSE

Example	Ideal plastic collapse λ_p	Elastic critical load λ_c	Actual collapse λ	Remarks
1	2·21	3·47	1·73	Frame weak way
2	2·15	12·9	1·90	Frame strong way
3	2·0	5·7	1·69	Demonstration model

examples, which were rather in the nature of test cases. But collapse designs must incorporate some (variable) correction to account for deteriorating critical loads, and the problem is to find a simple one.

59. There are, however, two important conclusions which can be drawn. First, even a limited appeal to composite action with floors and walls should suffice to avoid the necessity of including instability effects in the design of many frames. The second conclusion is that wherever it is feasible to stop side-sway, this should be done. If bracing and/or composite action with floors and walls is not possible then an attractive alternative would be to design special "wind bents", say every fifth or sixth bent, with floors distributing all the wind loads to

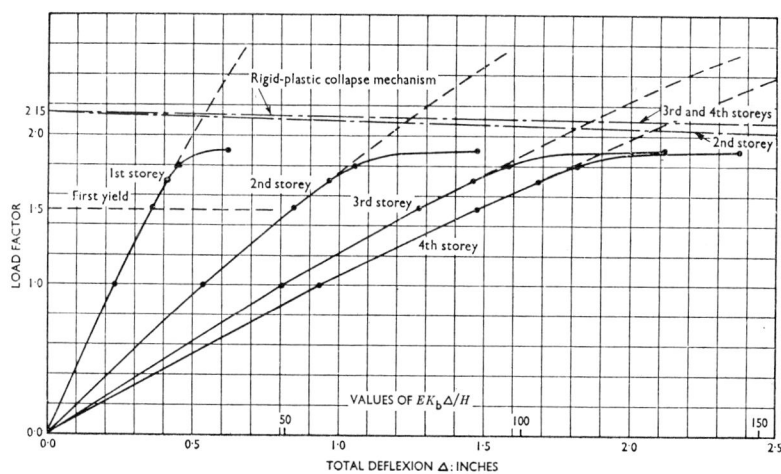

FIG. 20.—DEFLEXIONS OF FOUR-STOREY SINGLE-BAY FRAME UP TO COLLAPSE

these special bents. This considerably increases the stiffness of these special bents, so that they could be designed on a "collapse" basis without much instability. The remainder of the frames could then be designed virtually on a no-sway basis by the methods mentioned earlier in this paper, the overall design being in all probability about the most economical that is possible.

The effects of stiffening due to composite action

60. As far as it affects the stability of multi-storey structures the fundamental contribution of composite action, for which there is now considerable evidence

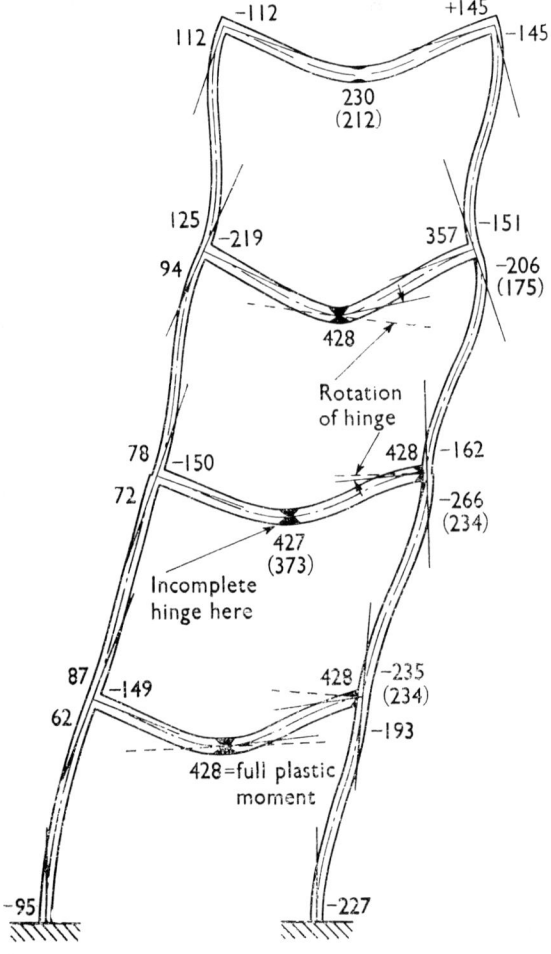

FIG. 21.—CALCULATED FINAL STATE OF FRAME JUST PRIOR TO COLLAPSE AT $\lambda = 1 \cdot 90$. DEFLEXIONS PLOTTED ON 50 TIMES SCALE OF FRAME. CURVATURES DUE TO PLASTIC ZONES OBTAINED USING A DIFFERENTIAL ANALYSER

Final bending moments are given in tons-inches

Figures in brackets refer to moment at first yield, where partial plastic zor from tests on real buildings[7, 30, 31, 32] coupled with a limited number of analytical and experimental reports,[33] falls into three main classes, thus:—

(i) *The alteration of actual loads distributed from floor slabs and walls to beams.* Just as arching action in walls can increase the intensity of load transmitted to a beam near the supports and decrease it at mid-span, so the distribution of loads on beams may be altered by bending and twisting action in adjacent loaded floor slabs. As long as the beams are very stiff the intensity of loading on beams approximates to that assumed in codes of practice. But for slender beams the load intensity is decreased at the centre of the span. There is in fact a continuous decrease in beam bending moments with decrease in relative beam/slab stiffness ratios. All these factors affect the fixed-end moments of loads on beams, the central-span moments of beams, and the induced moments in stanchions, in nearly all cases with marked benefit. For example in a recent test of a reinforced concrete beam in a block of flats carrying a deep brick wall and overhead floors, the measured steel stresses indicated a bending-moment coefficient ranging from $WL/11$ when the bricks were first laid to $WL/27$ when the wall was 5 ft deep, and at a later stage to $WL/132$ after the brickwork was 13 ft high.

(ii) *The increased stiffness of beams in resisting rotation of joints, from combined action with slabs and also encasement.* When couples are applied to the ends of beams the rotational stiffness may easily be doubled because of combined slab action,[33] disregarding the contribution of concrete encasement. Work is in hand to extend this investigation to include T-beam and other membrane effects.

(iii) *The reduction of side-sway by combined wall and frame action in panels.* Fig. 23 shows the difference between the behaviour under racking loads of a complete composite wall panel and the separate behaviour of the frame on its own. In a further test of the $4\frac{1}{2}$-in. brick panel on its own, the panel failure at a load of 13 tons, but with only $\frac{1}{8}$-in. displacement. The failure took the form of a crack, roughly diagonal across the panel. The frame has a plastic failure at the (bolted) joints ($9\frac{1}{2}$ tons racking load for bare frame, 20 tons encased) and the brick wall a diagonal tension failure (13 tons), but the composite mode of failure is a new mode. Compression "bands" in the brickwork are responsible for producing the plastic hinges at points along the beam. Additional information relating to other tests is given in Table 3.

CONCLUDING REMARKS

61. It is obvious that the effects of composite action briefly outlined above would drastically alter the stability analysis of multi-storey frames if only more information was available. As a temporary measure composite action should be capable of bridging the variable gap that exists between the actual collapse loads of elasto-plastic multi-storey bare frames, involving frame instability to some extent, and the collapse loads predicted by simple plastic theories, for only small propping forces are necessary to counteract instability. Admittedly this is not using composite action at its best, but it would seem to provide a reserve of

TABLE 3.—RACKING TESTS ON ENCASED STEEL FRAMES WITH VARIOUS WALL-PANEL INFILLINGS

Type of frame and infill	First visible crack			Ultimate load: tons	Horizontal deformation at ultimate load: inches
	Load: tons	Horizontal deformation Δ_H: inch	Approximate ratio deformation/height: Δ_H/H		
Frame Type 1 Horizontal girders 10 in. × 4½ in. (I 25) Vertical stanchions 10 in. × 8 in. (I 55) (weak way) 6-in. × 4-in. × ½-in. bolted cleat connexions to top and bottom flanges of each beam					
Open bare frame	7	1·0 {First yield}	—	9·3	6·0
Encased frame	14	1·0	1/100	20	2·3
Encased frame with 4½-in. brick panel	35	0·3	1/350	49	2·5
(Repeat test) with 4½-in. brick panel	30	0·28	1/400	56	2·8
Brick-on-edge infilling . . .	21	0·27	1/400	40	2·0
3-in. clinker block	22	0·25	1/450	35	0·8
(Repeat test) 3-in. clinker block .	24	0·28	1/400	36	0·8
3-in. hollow clay block . . .	22	0·40	1/275	30	1·5
13½-in. brick	110	0·26	1/425	135	0·6
4½-in. brick, with door opening .	13	0·11	1/1000	38	2·1
Frame Type 2 (*somewhat stiffer than Type* 1) Horizontal girders 13 in. × 5 in. (I 35) Vertical stanchions 10 in. × 8 in. (I 55) (strong way) 6-in. × 4-in. × ⅝-in. cleat connexions					
Encased frame	17	1·0	1/100	23	2·2
4½-in. brick infilling	37	0·28	1/400	75	1·5

stiffness on which frames might (quite literally) lean. A more economical design method would be to employ deliberately stiffened spine walls or end walls to eliminate side-sway and then make best use of those modern frame design methods which involve no side-sway. The particular frame-design methods being issued by the Building Research Station are specially arranged so that future information relating to composite action can be inserted without changing the form of design.

62. Finally, it is dangerous to attach too much importance to the apparent reliability of the particular load factors that have been associated with semi-rigid and pin-jointed design methods in the past. The elastic critical loads of such frames can be surprisingly low and there must have been considerable help from the cladding of most city buildings. A new statistical and philosophical approach to the whole question of load factors is required. Perhaps the main contribution of a study of composite action is to show (Fig. 23, facing p. 89) that the behaviour of a complete structure is not by any means the same thing as the mere addition of the separate behaviour of component parts. The modern tendency to ever increasing specialization is seen to be somewhat artificial. There is in fact no exact solution to the problem of the stability of complete tall buildings.

Acknowledgements

63. This Paper is published by the permission of the Director of Building Research. The Author wishes to thank the Brixton School of Building for making the models of Figs 2 and 9 (the model frame being a development of those used by Professor Merchant), Mr W. T. Lawton for performing the extensive calculations leading to Fig. 21, Dr M. R. Horne for the method of loading the beams of Fig. 22, and Mr L. G. Simms for carrying out the racking tests.

References

1. J. F. Baker, "The design of steel frames". Struct. Engr, vol. 27 (1949), p. 397.
2. M. R. Horne, "The stanchion problem in frame structures designed according to ultimate carrying capacity". Proc. Instn civ. Engrs, Pt III, vol. 5, p. 105 (Apr. 1956).
3. J. F. Baker, M. R. Horne, and J. Heyman, "The steel skeleton". Vol. 2. Cambridge Univ. Press, 1956.
4. "Rapid design of multi-storey rigid-jointed frames: systematic improvements in stanchion design". Bldg Res. Stn, Note A.58, 1957.
5. M. R. Horne, "Maximum beam moments in welded building frames". Struct. Engr, vol. 28 (1950), p. 109.
6. R. H. Wood, "An economical design of rigid steel frames for multi-storey buildings". Nat. Bldg Studies, Res. Pap. No. 10, H.M.S.O., Lond., 1951.
7. Final report. Steel Structures Res. Cttee. H.M.S.O., Lond., 1936.
8. "Rapid design of multi-storey rigid-jointed frames". Bldg Res. Stn, Note A.47, 1955. A later simplified version of reference 6.
9. A. Berry, "Calculations of stresses in aeroplane wing spars". Trans. roy. Aero. Soc., No. 1 (Nov. 1919), p. 3.
10. B. W. James, "Principal effects of axial load on moment distribution analysis of rigid structures". N.A.C.A. Tech. Note 534, 1935.
11. E. E. Lundquist, "Principles of moment distribution applied to stability of structural members". Proc. 5th Int. Congr. App. Mech., Cambridge, Mass., 1938. Wiley, N.Y., 1939.
12. D. B. Chandler, "The prediction of critical loads of elastic structures". Ph.D. thesis, Manchester Univ., 1955.

13. R. K. Livesley and D. B. Chandler, "Stability functions for structural frameworks". Manchester Univ. Press, 1956.
14. Eric Shepley, "Continuous beam structures". Concrete Publications Ltd, Lond., 1942.
15. R. H. Wood and E. Goodwin, "Degree of fixity methods for certain sway problems". Struct. Engr, vol. 30 (1952), p. 153.
16. L. A. Beaufoy and A. F. S. Diwan, "Analysis of continuous structures by the stiffness factors method". Quart. J. Mech., vol. 2 (1949), p. 263.
17. H. G. Allen, "The estimation of the critical loads of certain frameworks". Struct. Engr, vol. 35 (1957), p. 135.
18. W. Merchant, R. B. L. Smith, and R. E. Bowles, "Critical loads of tall building frames". Pt 1: Struct. Engr, vol. 33 (1955), p. 84; Pt 2: vol. 34 (1956), p. 284; Pt 3: vol. 34 (1956), p. 324.
19. A. Bolton and J. B. B. Owen, "Stability of axially loaded continuous beams". J. roy. Aero. Soc., vol. 59 (1955), p. 848.
20. Symposium on the plastic theory of structures; Cambridge, Sept. 1956. Brit. Weld. J., vol. 3 (1956), p. 331, and vol. 4 (1957), p. 1.
21. J. F. Baker, M. R. Horne, and J. W. Roderick, "The behaviour of continuous stanchions". Proc. roy. Soc., Series A, vol. 198 (1949), p. 493.
22. J. W. Roderick and M. R. Horne, "The behaviour of a ductile stanchion length when loaded to collapse". Rep. FE 1/11, Brit. Weld. Res. Ass., 1948.
23. "The Building Research Station open days for civil and structural engineers". Struct. Engr, vol. 33 (1955), p. 329.
24. R. H. Wood, "A derivation of maximum stanchion moments in multi-storey frames by means of monograms". Struct. Engr, vol. 31 (1953), p. 316.
25. See reference 2. Discussion by R. H. Wood, p. 149.
26. W. Merchant, "Frame instability in the plastic range". Symposium on the plastic theory of structures; Cambridge, Sept. 1956. Brit. Weld. J., vol. 3 (1956), p. 366. Discussion by R. H. Wood, p. 26.
27. A. Bolton, "Structural frameworks". Ph.D., thesis, Manchester Univ., 1957.
28. J. F. Baker and J. W. Roderick, "Test on full-scale portal frames". Brit. Weld. Ass., Rep. FE 1/26, 1956.
29. R. H. Wood, "The absolute and the intuitive in dimensional analysis". Euler Soc. Note No. 8. Cement & Concr. Ass., 1955.
30. R. H. Wood and R. J. Mainstone, "Stress measurements in the steel frame of the new Government offices, Whitehall Gardens". Conf. on the correlation between calculated and observed stresses and displacements in structures, Instn civ. Engrs, 1955.
31. J. Harrop, "Analysis of stresses in steel-framed and reinforced concrete structures". Ph.D. thesis, Leeds Univ., 1956.
32. A. J. Ockleston, "Load tests on a three-storey reinforced concrete building in Johannesburg". Struct. Engr, vol. 33 (1955), p. 304.
33. R. H. Wood, "Studies in composite construction". Pt 1, Nat. Bldg Studies Res. Pap. No. 13, H.M.S.O., Lond., 1952: Pt 2, Res. Pap. No. 22, H.M.S.O., Lond., 1955.

The Paper, which was received on 4 October, 1957, is accompanied by eight photographs and nineteen sheets of drawings from which the half-tone page plates, and the Figures in the text have been prepared.

Written discussion on this Paper should be forwarded to reach the Institution before 15 November, 1958, and will be published in or after March 1959. Contributions should not exceed 1,200 words.—SEC.

12

NATIONAL BUILDING STUDIES

Research Paper No. 13

DEPARTMENT OF SCIENTIFIC AND
INDUSTRIAL RESEARCH

(BUILDING RESEARCH STATION)

STUDIES IN COMPOSITE CONSTRUCTION

Part I. The Composite Action of Brick Panel Walls Supported on Reinforced Concrete Beams

BY

R. H. WOOD, Ph.D., B.Sc., A.M.I.C.E., A.M.I.Mech.E., A.M.I.Struct.E.

(*Building Research Station*)

LONDON

HER MAJESTY'S STATIONERY OFFICE

1952

STUDIES IN COMPOSITE CONSTRUCTION

Part I. The Composite Action of Brick Panel Walls Supported on Reinforced Concrete Beams

SCOPE OF THE INVESTIGATIONS

INTRODUCTION

THE tests described in this paper were carried out originally to determine the most economical design of foundation beams which are required to support the brick walls of houses, schools, etc., when the whole of the building is carried on short bored piles spaced at regular intervals. The necessity for short bored piles arises from the shrinkage of the soil which takes place on clay sites, and has been described elsewhere.[1] It is clear however that the results of the tests have a more extensive application, for example in the design of lintels carrying brick walls, and in the general study of composite action in multi-storey buildings where beams carry brick or stone walls.

In the design of building frames it is common practice to ignore the composite effects of the floors and walls which, however, materially alter the loads which are actually transmitted to the beams. The stiffening effects of floors and walls are being studied by the Building Research Station both in the laboratory and in the field and the evidence so obtained, including that from the preliminary tests described in this paper, points quite conclusively to the importance of composite action with walls. The considerable stiffening action which can be attributed to floors is outside the scope of the present paper.

It is customary in practice to design beams and lintels carrying brickwork so as to be capable of supporting a triangular load of bricks, where the base of the triangle is the span of the beam, provided that the remainder of the brickwork is adequately supported. If the bricks themselves carry any superimposed load above the apex of the triangle it is not at all clear what proportion if any of this extra load should be taken into account, and it is frequently ignored.

The bending movements induced in the beam obviously depend in some manner on the relative stiffnesses of the beam and the brick panel. The greater the stiffness of the beam, the more load is transmitted to the beam at the centre of the span. With a very flexible beam a considerable degree of " arching " can be expected to take place and in the limit the panel may tend to become self-supporting. Again, the term " relative stiffness " is difficult to define, for in common parlance the stiffness of the beam is assessed in terms of the moment of inertia and the length, and pre-supposes that bending stresses are the governing feature. However, in the case of a composite panel, the " beam " may exhibit a marked tendency to act as reinforcement for the whole panel since—by so doing—a greater overall stiffness is achieved and a smaller amount of work is done by the applied loads. Such a phenomenon clearly depends on the degree of bond or shearing forces that can be developed between beam and panel, particularly near the ends.

There is a tendency for the arching in the brickwork to transmit to the ends of the beam forces that develop extra restraint or " fixity " at the ends of the beam, and the beam stresses may be further relieved by friction at the supports. At the centre of the beam the underside will of course be in tension, but it is by no means certain that the top of the beam will be in compression and in practically all the tests to be described tensile forces were found. There is, in general, so much reduction of stress that it is unlikely that the " no-tension-in-concrete " theory will apply any longer, and the concrete itself may even be acting as " reinforcement " for the composite panel.

When there are openings in the brickwork the stress patterns are considerably modified, and several types of openings have been investigated. The position of the opening in relation to the span is of some importance.

In a steel or reinforced concrete framed building it seems that eventually it will be neither rational nor economical to design the frame as though it were independent of the walls, floors, etc. Where there are any substantial brick panels or stiff floors the modern tendency to reduce the frame stiffness, although safe, often leads to an elaborate analysis of the frame, arbitrarily isolated from the remainder of the building, in which the predicted behaviour of the frame under either elastic of " plastic " conditions (or a combination of the two as in shake-down problems) may be far removed from the truth. It will be shown that in some cases the brick panels, even with openings, are capable of " holding up " the supporting beam under conditions of self-weight or of superimposed loads, and in due course a rational design would take account of the real behaviour of the " beam " in such cases.

The laboratory tests here described are now being supplemented by tests on actual buildings. The scope of this note is restricted to producing simple design rules which are safe and economical for reinforced concrete beams supporting permanent brick walls, with and without openings. The design recommendation should be applicable to lintels and no doubt to any beam of a reinforced concrete frame which carries brick walls, provided that most of the load is due to the weight of the wall itself and the superimposed load on the wall. Since the experimental work was especially directed in the first instance to the determination of suitable dimensions for reinforced concrete beams carrying house walls and spanning short bored piles, the span of the beam chosen for the tests was determined by practical considerations, but the results have been analysed in such a way as to permit other spans to be used.

The stresses which occur during bricklaying, when the mortar is still wet and little arching can develop, may attain a high proportion of the total working stresses; consequently they have been investigated in some detail.

DETAILS OF THE EXPERIMENTS

At the commencement, a suitable estimate of a supporting beam was made for an effective span of 10 ft. 6 in., based upon traditional methods which assume a triangular load of bricks as in Fig. 1. The weight of a triangle of 9 in. brickwork (11 in. cavity wall) was calculated and all additional superimposed loads neglected. A somewhat nominal, highly stressed beam such as is shown in Fig. 2 was adopted, since it was anticipated that the stresses appertaining to the triangular loading would never, in fact, be realized. It was subsequently found that this assumption was justified.

In no case did this beam fail to deal with working loads, and in most cases with considerable overload. The dimensions 12 in. by 7 in. were chosen so as to accommodate an 11 inch cavity wall, and the reinforcement consisted of four $\frac{5}{8}$ in. dia. bars—together with light shear reinforcement. This beam is subsequently referred to as the " light " beam.

The general arrangement of the walls and supporting beams can be seen in Plate 3(A) and Fig. 3. The walls, in general, were made 8 feet high and the superimposed loads on the brickwork were applied by means of hydraulic jacks at the

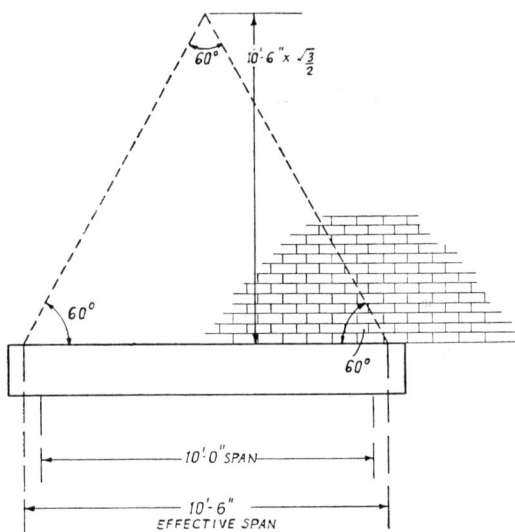

Fig. 1. *Commonly assumed design loading for reinforced concrete supporting beams*

top of the wall to simulate the load which is transmitted in practice to the brickwork from the upper storey or storeys. The load from the jacks was distributed by means of a system of steel beams and rollers. Scaffolding was necessary both for loading the jacks and for providing a certain degree of lateral stability for the walls. The reaction from the jacks was taken by a system of tie rods. The total load which could be applied by the jacks was 28·8 tons and the average brick panel tested weighed about 4 tons. The total load, therefore, carried by the supports and beam reached a maximum of about 33 tons.

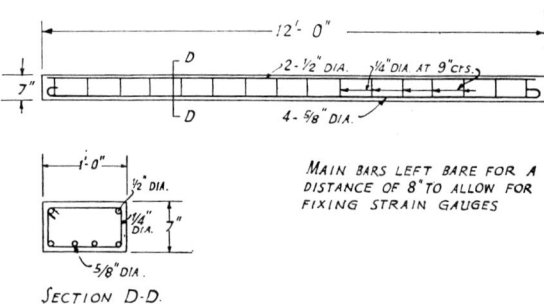

Fig. 2. *Details of reinforced concrete beam*

In the special case of house foundations the nominal load was expected to be about 1 ton per ft. run, for a two-storey house. With a span of 10 feet, if the beam safely withstood a total load of 33 tons this would represent a factor of safety of at least 3·3. For the general case of composite design, however, it should be noted that a purely nominal " light " beam was under test and the stresses in the beam were analysed accordingly, the real factor of safety being fixed by the resultant steel stresses for whatever load it is desired to carry.

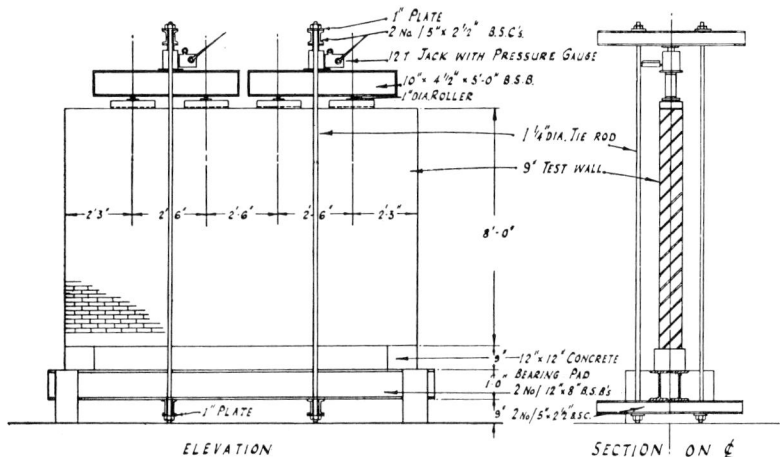

FIG. 3. *Loading arrangement*

For measuring strains in the steel reinforcement the " Maihak " gauge was used (Plate 2), the concrete being cut away locally for the purpose. These gauges (4 in number) were mounted by the device shown in the photograph and the mean of four readings was taken. This instrument has been described in detail elsewhere[2] and is of the vibrating-wire type, similar to the " Acoustic " gauge[3]. The strain in the working wire, stretched between the gauge points of the instrument, was measured by synchronising the frequency of vibration of the working wire against a reference wire which was strained by means of a screw and micrometer head. Both wires were plucked electromagnetically and the subsequent vibrations, which were recorded on a cathode ray oscillograph, were such that a stationary figure was produced when the two wires were in unison. The strain was then determined from a previous calibration of the reference wire.

Strains were measured in the bottom courses of the brickwork in order to enable an estimate to be made of the loads transmitted to the beams. At first electrical resistance strain gauges were used, as shown in Plates 1 and 3(A). The bricks were dried out with a battery of infra-red lamps and carefully polished and cleaned with acetone before fixing the gauges with " Durofix ". Since the direction of the principal stresses was not known at the commencement, rectangular rosettes were employed. These were later dispensed with when it was found that sufficient information was given by vertical gauges alone. Afterwards check-results were obtained by the use of roller-mirror extensometers of one-inch gauge length. Wooden blocks for holding these instruments in position are shown in Plate 3(A).

Deflections of the beams and walls were measured with dial gauges. Subsidiary tests were carried out on samples of bricks to determine Young's Modulus and Poisson's Ratio.

CALIBRATION OF SUPPORTING BEAM

During the tests the beam stresses were found to be so low, and no sign of any damage could be detected, that the same beam could conveniently be used for many tests. It became apparent, however, that the " no-tension-in-concrete " theory, by which the beam was designed, would hardly apply. Accordingly an identical beam, cast at the same time, was subjected to a series of calibration tests, in which a careful record of the steel stresses, concrete stresses and deflections of the beam was made, the system of loading approximating very closely to uniform distribution. The beam was simply supported on a 10 ft. 3 in. span, and as far as possible the history of the beam under calibration was made the same as that of the supporting beam taking the steel stresses as the criterion (Fig. 6). Thus, for example, the steel stresses in practice were very largely within the cycle of loading OA which was repeated several times, and here the stress distribution in the beam is nearer to the theoretical distribution taking tension in the concrete. At higher loads in the calibration test the " no-tension " theory applies approximately, but only in one of the wall-and-beam tests, viz. that in which there was a doorway near a support, did the steel stress exceed those included in the cycle OA, and then only as the result of considerable overload.

TYPES OF WALL TESTED

The walls tested were:—

(a) 9 in. solid English-bonded wall without beam support.

(b) 9 in. solid English-bonded wall with beam support.

(c) 11 in. cavity with beam support.

(d) 11 in. cavity wall with central window opening, on supporting beam.

(e) 11 in. cavity wall with central door opening, on supporting beam.

(f) 11 in. cavity wall with door opening at one end of span, on supporting beam.

In one instance, case (c), a damp-proof course was provided between the beam and brickwork.

TESTS ON WALLS WITHOUT OPENINGS

SOLID 9 IN. WALL WITHOUT SUPPORTING BEAM

The first wall to be tested was an unsupported wall as this gives some indication of the degree to which a supporting beam is necessary. The brickwork was laid on temporary supports which were removed at the end of the curing period. English bond was used and a control was made of the proportions and consistence of the mortar used. Beyond this no particular care was taken to produce brickwork of exceptional quality, the aim being to imitate practical conditions as far as possible. Small attachments for the dial deflection gauges were built into the third and fifth courses from the base of the wall.

The mortar used was a 1:2:9 by volume cement-lime-sand mix giving a compressive strength range ranging from 484 to 807 lb. per sq. in. (taken over several samples), and a briquette (tensile) strength of 80–100 lb. per sq. in.

The central deflections of the wall for a series of superimposed loads applied at the top of the wall are given in Fig. 4. In the case of a deep panel it is of course difficult to define what is meant by " central deflection " since all points along the central vertical section do not undergo quite the same displacement. This point is neglected in the usual theory of bending of slender beams. In this instance the deflections were taken near the base of the wall (3rd and 5th courses) so that they could be compared with the deflections of a supporting beam.

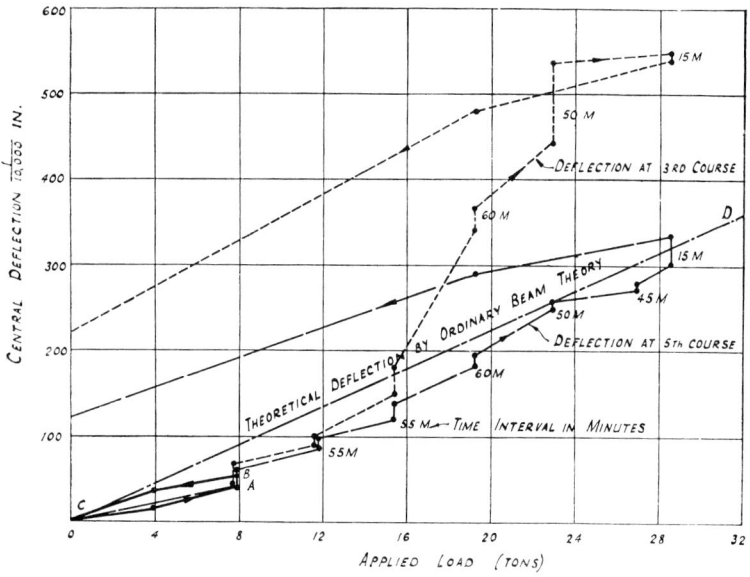

FIG. 4. *Deflection under load of 9 in. solid wall without supporting beam*

The cycle OABC (Fig. 4) for 8 tons maximum applied load was obtained from the readings of the dial gauge 5 courses from the base. The load of 8 tons was kept on for $1\frac{1}{2}$ hours and some " creep " can be detected. As the deflections were very small the creep over an extended period was subject to interference from temperature effects. For normal working loads the bricks exhibited a considerable degree of elasticity. The line OD represents the deflection curve which would be obtained by the theory of simple bending using a value of $E = 0.8 \times 10^6$ lb. per sq. in. for the bricks (determined by test) and taking the effective span as 10 ft. 6 in. and total depth of brickwork as 8 ft. 10 in.

In the special case of a house wall the total working load to be expected on a 10 ft. span is roughly 10 tons. Excluding the weight of brickwork (4 tons) the additional applied load is about 6 tons. At this load the deflection of the wall is very small indeed, being approximately 0·004 in. At half working load (3 tons) the deflection of the wall was therefore of the order 0·002 in. whereas from the special calibration test of the proposed supporting beam such a distributed load produced a deflection of the beam of about 0·180 in., or 90 times as large.

The system of cracking which developed in the wall at larger loads is shown in Fig. 5. Considered as a freely supported beam the whole panel would have failed at a total load of about 12 tons, but it will be seen that the unaided brickwork successfully withstood a load of 32 tons, the cracking, of course, having extended. The sytem of cracking points to the existence of an effective arch inside the panel itself, the brickwork under the arch being virtually suspended from the arch. The function of any supporting beam would be to prevent brickwork falling away and to prevent spreading of the arch which in this test was held by friction together with the shear strength of any mortar bed at the supports.

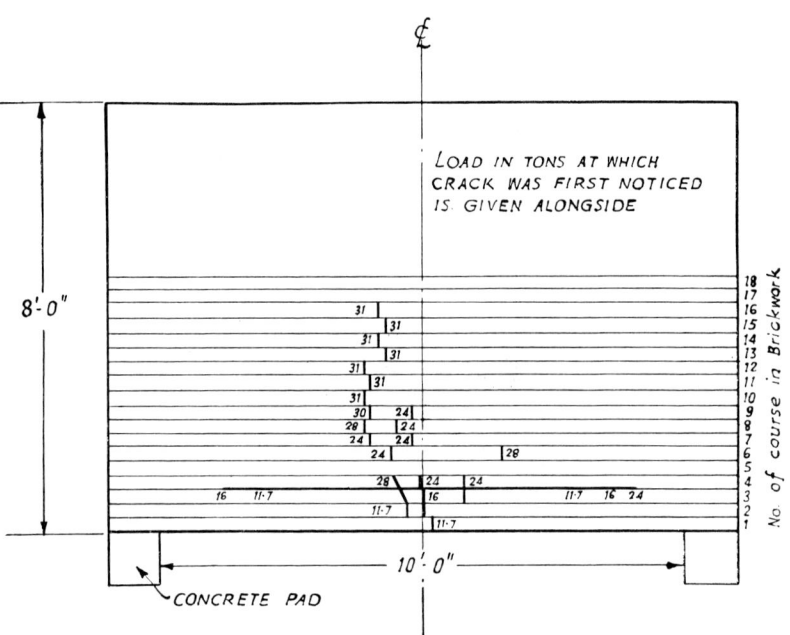

FIG. 5. *Crack formation in brick panel—solid 9 in. wall without supporting beam*

The deflection at the third course was greater than that at the fifth course, because of horizontal cracking. Although no bricks fell away during the test, it is to be expected they would do so after a time.

Since the deflections of the unsupported brick panel were exceedingly small compared with those of the beam, it was decided to use such a nominal supporting beam for the remainder of the tests.

SOLID 9 IN. WALL ON SUPPORTING BEAM

The beam was first used on a 10 ft. clear span to support a solid 9 in. wall approximately 8 ft. high. The beam was supported at each end on concrete bearing blocks which simulated the caps of the foundation piles in the case of house construction.

An opportunity was taken in this test to record the beam stresses during the bricklaying. At the completion of 5 courses the steel stress at the centre of the beam was found to be 700 lb. per sq. in. It was noticed that, because of deflection, the beam had a tendency to separate from the outer edges of the supporting blocks and to " ride " on the inner edges giving an equivalent span at the higher loads of about 10 ft. 3 in. Consequently this was the span on which the specially calibrated beam was tested, and for that beam a steel stress of 700 lb. per sq. in. corresponded to a uniformly distributed load of about 0·23 tons (see Fig. 6 line OA). But the actual weight of 5 courses of bricks on this span is about 0·58 tons. Taking the steel

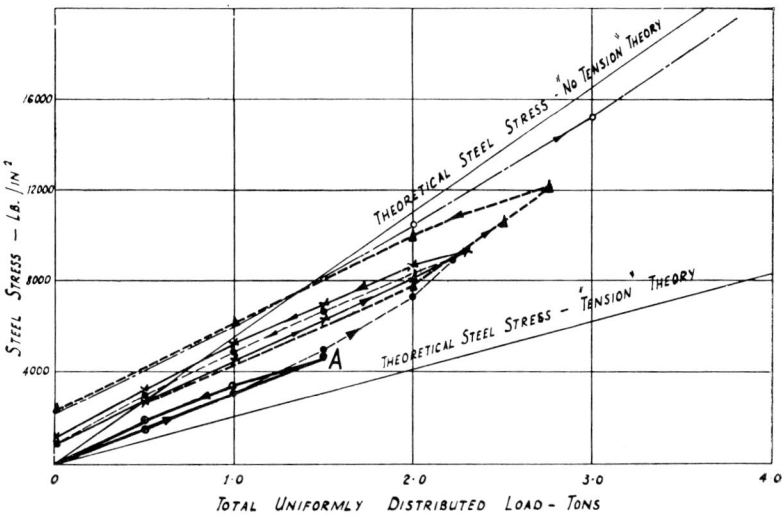

FIG. 6. *Calibration of supporting beam*

stress as the criterion we conclude that the steel stress at the centre of the supporting beam would be given by adopting an equivalent moment of

$$\frac{WL}{8}\left(\frac{0\cdot 23}{0\cdot 58}\right) = \frac{WL}{20}$$

for a correspondingly freely supported beam.

The following table gives a summary of the beam stresses and moments during bricklaying.

TABLE I—BEAM STRESSES DURING BRICKLAYING

No. of Courses	Stress in steel lb./in.²	Equivalent Distributed load tons	Actual brick load tons	Equivalent moment at centre
5	700	0·23	0·58	$\dfrac{WL}{20}$
10	1380	0·44	1·16	$\dfrac{WL}{21}$
15	2060	0·67	1·74	$\dfrac{WL}{21}$
20	2480	0·79	2·32	$\dfrac{WL}{23·5}$
25	2620	0·83	2·90	$\dfrac{WL}{27·8}$

As there can be no appreciable " arching " with unset mortar over only 5 courses, the low central moment must evidently be due to fixity at the supports. The beam was bedded down on cement mortar to the 12 in. by 12 in. concrete blocks, and calculations showed that the supports were capable of developing almost full fixing moments at the commencement of bricklaying. The central bending moments are therefore of the correct order. It is interesting to note that they decrease considerably as the number of courses increases; this no doubt points to the onset of arching in the brickwork. The steel stresses eventually become about 3,000 lb. per sq. in. When extra loads were applied at the top of the panels the beam could no longer be considered encastered, for cracks developed along the bedding plane.

The measured deflections during bricklaying agree substantially with the above conclusions, as is shown in Table II, where actual and theoretical deflections (assuming complete fixity of supports) are good.

TABLE II—BEAM DEFLECTION DURING BRICKLAYING

No. of courses	Measured central deflection (inches)	Theoretical deflection— no tension in concrete (inches)	Theoretical deflection— with tension in concrete (inches)
5	·010	·0135	·005
10	·020	·027	·0105
15	·026	·041	·016
20	·027	·0545	·021
25	·028	·068	·0265

At the commencement of bricklaying the actual deflection lies midway between the two theoretical values, which is to be expected in view of the results of the test on the control beam (Fig. 6). After a certain number of courses have been laid there is practically no further increase in measured deflection.

The steel stresses were measured also for repeated cycles of loading. It is proposed to derive an equivalent bending moment for each range of load such that when used in conjunction with a freely supported identical beam the same steel stresses would result as were found in the composite beam and panel under test.

With a superimposed load of 6 tons (giving 10 tons total, including the weight of the wall) the steel stress due to the superimposed load was found to be 500 lb. per sq. in. This corresponds to a distributed load of 0·175 tons on a freely supported beam (Fig. 6), and consequently an equivalent central bending moment of

$$\frac{WL}{8} \times \frac{0\cdot175}{6} = \frac{WL}{274}$$

Corresponding results at higher loads are given in Table III.

TABLE III—EQUIVALENT BEAM MOMENTS WITH 9 IN. SOLID WALL

Applied load* tons	Steel stress† (measured) lb./in.²	Equivalent bending moment
6	500	$\dfrac{WL}{274}$
15	1440	$\dfrac{WL}{248}$
18·5	2160	$\dfrac{WL}{218}$
25·8	4920	$\dfrac{WL}{130}$

*Not including dead weight of wall = 4 tons.
†Due to superimposed load only.

It will be noticed that these equivalent bending moments are remarkably small. A very considerable degree of overload was necessary (about $2\frac{1}{2}:1$) before the steel stresses exceeded those found during bricklaying.

CAVITY (11 IN.) WALL ON BEAM (WITH DAMP-PROOF COURSE)

In the special case of house foundations it was thought that the presence of a damp-proof course might be detrimental to the bond between bricks and concrete which to some extent is essential for composite action. Accordingly a bituminous damp-proof course was interposed between beam and panel in the test of a cavity wall, such as is seen in Plate I.

Repeated loadings on this wall panel resulted in the following average recorded steel stresses and equivalent live load moments.

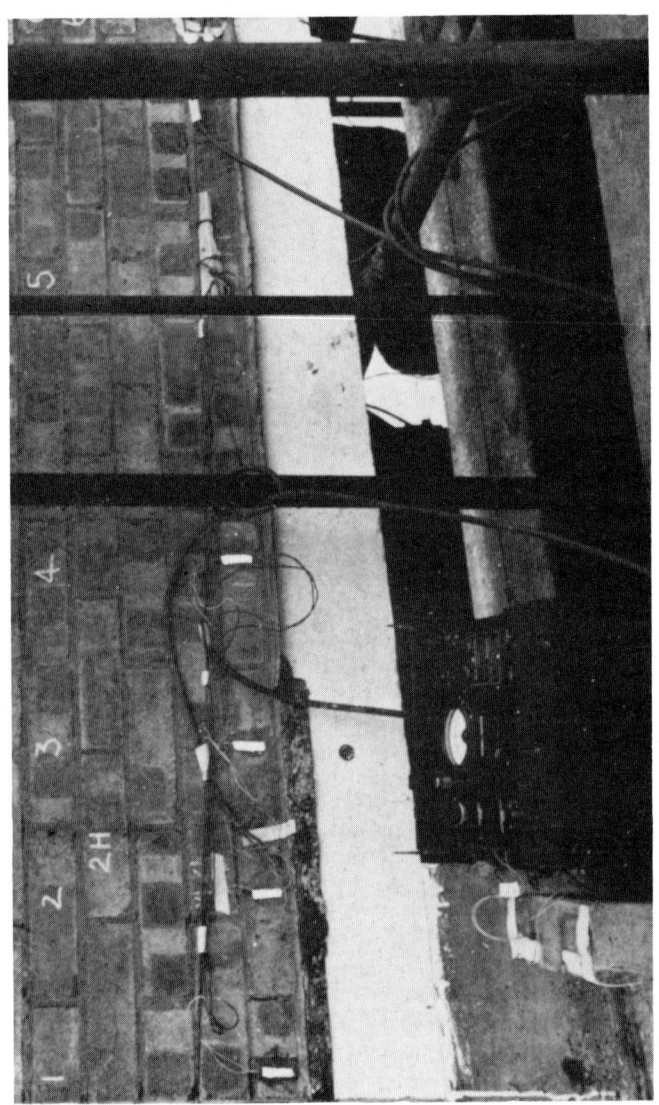

A series of resistance strain gauges at the base of a cavity wall for measuring the load transmitted to the beam

PLATE 1

Showing the method of attachment of the "Maihak" strain gauges to the main reinforcement of the supporting beam

PLATE 2

(A). *Testing a cavity wall with window opening.*

(B). *Cavity wall with central door opening*

PLATE 3

The composite action of walls on beams

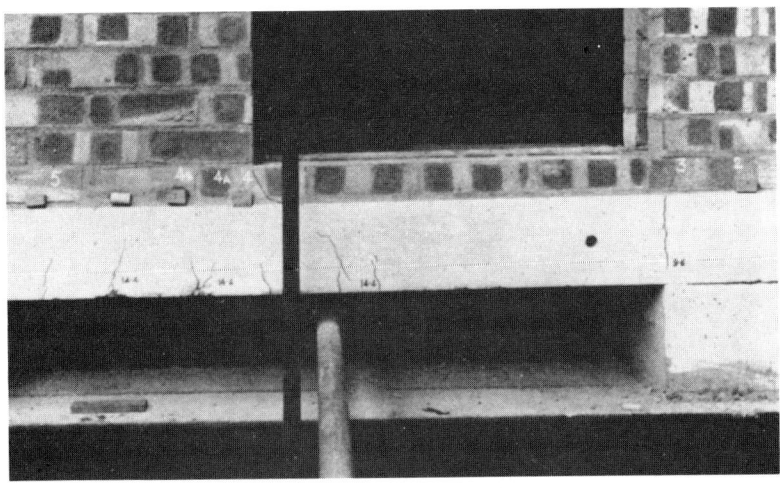

(A). *Cavity wall with door opening near support.*

(B). *Test of cavity wall with door opening near support*

PLATE 4

TABLE IV.—EQUIVALENT BEAM MOMENTS WITH 11 IN. CAVITY WALL

Live load tons	Steel stress lb./in.^{2}	Equivalent distributed live load from Fig. 6 tons	Equivalent bending moment
6	160	·05	$\dfrac{WL}{960}$
9·6	280	·09	$\dfrac{WL}{850}$
19·2	760	·24	$\dfrac{WL}{640}$
28·8	1480	·47	$\dfrac{WL}{490}$

Here the steel stresses are remarkably low and as this test was conducted out of doors in a period of intense cold weather (February 1947) it was decided to wait until a general thaw had set in and then repeat the test; the following results were then obtained:

TABLE IV.(A)—EQUIVALENT BEAM MOMENTS WITH 11 IN. CAVITY WALL

Live load tons	Steel stress	Equivalent load	Equivalent bending moment
6	188	·06	$\dfrac{WL}{800}$
9·6	328	·105	$\dfrac{WL}{730}$
19·2	1120	·36	$\dfrac{WL}{430}$
28·8	2160	·69	$\dfrac{WL}{330}$

It will be noticed that the stiffness of the cavity wall dropped appreciably after thawing out (throwing larger loads on to the beam) but that the stiffness is still considerably greater than the solid 9 in. wall of the previous test. This feature is quite surprising, and is probably due to the type of bond used. In the case of the cavity wall all the bricks were stretchers, and the use of these probably caused a greater degree of arching.

At this stage it was decided to investigate the degree of arching in the panel as a means of relieving the beam of load. Accordingly, resistance gauges were fixed in position at a series of points on the lowest course of bricks, just above the top of the beam, the aim being to obtain the vertical intensity of stress, which is the stress contributing principally to the beam bending moment.

The approximate distribution of vertical stress at the base of the wall, for various superimposed loads, is shown in Fig. 7, and is based on a pre-determined modulus of elasticity for the bricks. This shows a remarkable increase in vertical stress over the supports and for a short distance along the beam. Towards the centre, where

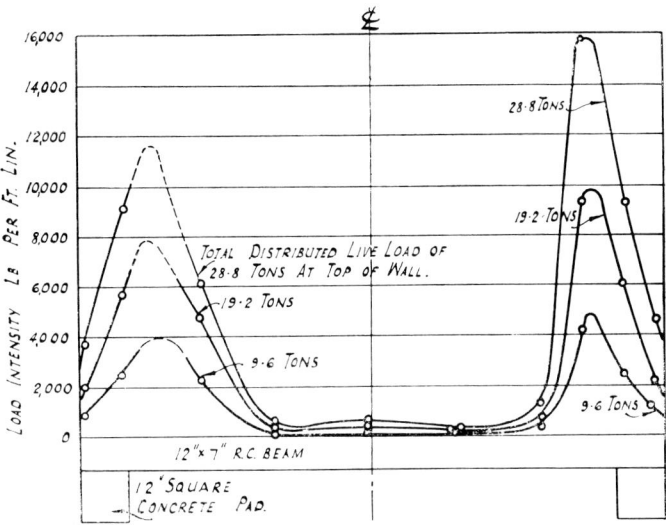

Fig. 7. *Load distribution on beam, measured by resistance gauges on first course of brickwork. Cavity wall on damp-proof course with 12 in. by 7 in. support-beam*

the beam deflects the most, the vertical stress is reduced almost to zero. If the wall had been perfectly homogeneous and the loading had been applied without any eccentricity, the stress distribution in Fig. 7 would have been symmetrical about the middle of the span, and the area under the curve would have corresponded to the applied load. In practical tests, with a rather variable material, exact correspondence is not to be expected, although it might have been improved if additional gauges could have been used to measure strains on both sides of each leaf of the cavity wall. Qualitatively the arching of the wall is clearly demonstrated.

By this time cracks had commenced under load at the bedding plane of the beam and concrete supports, extending from the outside of the concrete supporting blocks to within 3 in. of the inside edge of the support. This means that the reactions became very nearly point loads corresponding to a span of about 10 ft. 3 in. Combining such reactions with the load intensity curves of Fig. 7 it is found that the central bending moment is of the order of $WL/50$.

This low value for the moment indicates that only a small proportion of the load is effectively borne by bending of the beam, the major part being supported by the wall as a result of composite action. Full composite action was apparently developed, since an investigation of the distribution of concrete stresses in the beam indicated that the top of the beam was in *tension* under load and the " neutral axis ", if it could still be so called, was in the brickwork, about 6 inches above the concrete.

It will be noticed in Table IV. (A) that, for the maximum live load applied in this test, the equivalent bending moment, deduced from the steel stress in the beam, was only WL/330.

It is clear that there must be other beneficial effects of composite action, over and above the redistribution of vertical forces which has been noted in Fig. 7. Presumably the chief of these effects is the horizontal restraint at the supports, due to either friction or direct bond with the supports, coupled with horizontal shears at the wall-beam interface. Tensile stresses in the concrete over the whole cross-section of the beam may also have helped considerably to relieve the steel stress.

It is interesting to compare the maximum steel stress recorded in the test with that corresponding to the use of the simplified design procedure, based on a theoretical analysis of composite action, outlined on page 21. The total depth of the panel was about 8 ft. 7 in. and, according to the design method, the moment arm would be assumed to be two thirds of this, i.e. 5 ft. 9 in. For a load of 28·8 tons on a span of 10 ft. 3 in., the steel stress (taken as uniform) is estimated to be 8,900 lb. per sq. in., using the total area of steel, top and bottom. The maximum steel stress recorded in the test was, however, only 2,160 lb. per sq. in. so that the design method is very conservative for this particular example.

CAVITY WALLS WITH WINDOW AND DOOR OPENINGS

Cavity walls were supported on the " light " beam and tests were carried out with three types of openings as follows:—

(i) central window opening (Plate 3(A)).

(ii) door openings at mid span (Plate 3(B)).

(iii) door opening near a support (Plate 4).

The maximum steel stresses in the beams have been grouped together in the table below and in each case the equivalent bending moment on a free, simply supported beam is given which, according to the tests on the calibrated control beam, would result in the same steel stress.

From the figures in the last line of the table it will be seen that here, for the first time, the steel stresses have exceeded the value at A in Fig. 6. There is some hysteresis here, but the history of the specially calibrated beam was kept as close as possible to the beam in question and the corresponding distributed loads should be reasonably accurate for practical purposes.

In each case the weight of the brickwork itself was an additional load of about 3 tons. In particular for a beam of 10 ft. span supporting house walls, the maximum total load on the beam represents a considerable overload for each type of wall. Except in the last case of a door opening near the supports the steel stresses were remarkably low, even with the overload.

In each case the equivalent " free " bending moments increased slightly with increase of load. This non-linear effect shows that there was a gradual redistribution or transference of load from the brick panel to the beam but at no stage, even after cracks had appeared in some cases, did the composite action appear to show any signs of breaking down. It should be noted that there was no reinforcement in the brickwork. A point of considerable interest was the consistently smaller degree of composite action when the door opening was next to the support as is shown by the equivalent " free " moment increasing to a value of about WL/50. The reason

TABLE V.—TESTS OF CAVITY WALLS ON SUPPORTING BEAMS WITH VARIOUS OPENINGS IN THE WALLS

Type of wall	Applied load tons (not including weight of wall)	Measured steel stress lb./in.^{2}	Equivalent distributed live load (tons)	Equivalent bending moment
Wall with central window opening with lintel at top of window ..	6	280	0·09	$\dfrac{WL}{530}$
	12	608	0·19	$\dfrac{WL}{510}$
	20	1120	0·36	$\dfrac{WL}{440}$
	28	1694	0·54	$\dfrac{WL}{415}$
Wall with door opening at mid span	6	480	0·15	$\dfrac{WL}{320}$
	19·2	1540	0·49	$\dfrac{WL}{310}$
	28·8	2700	0·86	$\dfrac{WL}{270}$
Wall with door opening near to beam support	6	3000	0·95	$\dfrac{WL}{50}$
	9·6	5310	1·6	$\dfrac{WL}{48}$
	14·4	11,160	2·25 to 2·60	$\dfrac{WL}{51}$ to $\dfrac{WL}{44}$

for this can now be seen by a study of the vertical stress intensity in the bottom course of the brickwork in each case.

With the central window opening, the approximate vertical stress distribution in the bottom course of bricks, as measured by resistance gauges on one face, is shown in Fig. 8. It will be noticed that the arching effect has spread further over the supports, no doubt because of the relatively large size of the window opening. In support of this, a certain degree of compression was found in the lintel over the window, in addition to the usual bending stresses. At a later date a check on the strain distribution at the maximum load was carried out using roller-mirror extensometers, with the result shown in the figure. Up to a point very near to the outside edge the results are almost identical, a difference being found in the stress at the outside edge itself. As this is a point at which cracking frequently occurred it is not certain that the bricks were in exactly the same condition in the repeat tests.

In this case the line of reaction at the supports could not be determined by inspection so that bending moments could not be deduced from Fig. 8. Nevertheless it is easy to see that these vertical stresses would give rise to bending moments of a very small order.

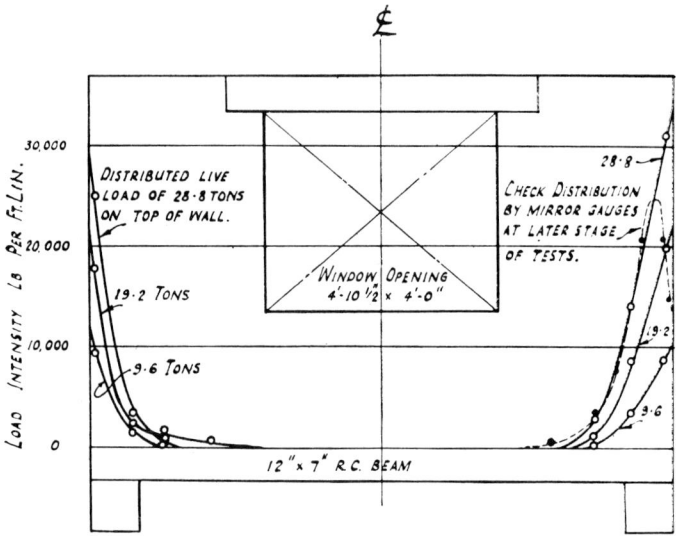

FIG. 8. *Load distribution on beam, measured by resistance gauges on first course of brickwork. Cavity wall with window opening*

With the door opening at midspan a sketch of the layout is shown in Fig. 9 together with the distribution of vertical stress for one half of the span in the bottom course of bricks as determined by roller-mirror gauges. As the door opening has a smaller width than the window opening the spread of the arch is smaller, and the vertical stress distribution is nearer to that of a plain wall. The lintel and any

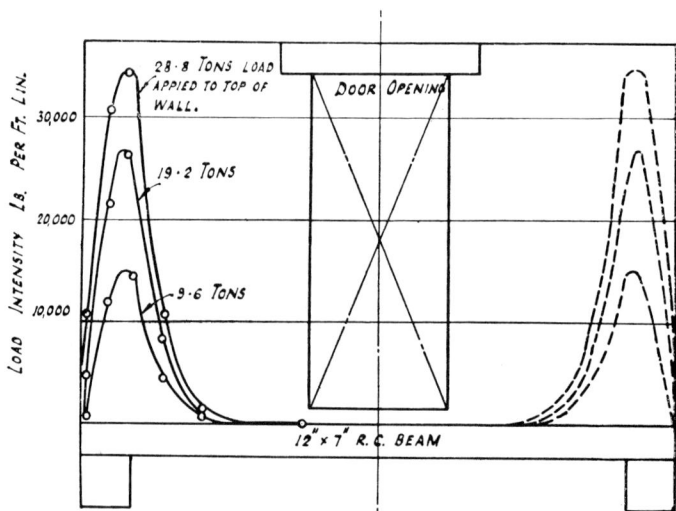

FIG. 9. *Load distribution on beam as measured by roller mirror gauges on first course of brickwork. Cavity wall with central door opening*

brickwork over the door complete the "arch". The steel stresses are slightly higher than those in the previous test on a wall with a window opening, probably because the combined depth of beam and brickwork at the centre of the span, beneath the opening, is considerably smaller.

Undoubtedly the worst case occurs when the door opening is placed at ¼ span or thereabouts (Plate 4). Here the vertical stress distribution in the bricks at the base of the panel is quite remarkable (Fig. 10). As direct transmission of the load to the right hand support is impossible because of the doorway, the left hand part of the panel develops an arch system of its own, and there is a sharp peak of stress down the left hand side of the door for a matter of one brick width only. This peak is of high intensity and is the reason why, for the first time, steel stresses of some magnitude were recorded. Some direct arching to the left hand support takes place, and the load from the lintel is transmitted down the right hand side of the doorway.

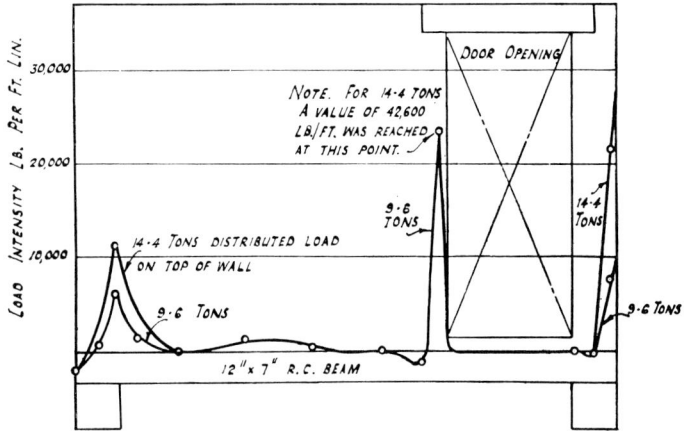

FIG. 10. *Load distribution on beam, measured by roller mirror extensometers Cavity wall with offset door opening*

The behaviour of the beam itself (Plate 4(A)) agrees with the type of stress distribution in the bricks. The beam developed certain cracks at a 50 per cent overload, but could not be considered to have failed. The cracks in Plate 4(A) have been exaggerated for photographic purposes, and the first crack to appear was at the right hand support, hogging moments being produced at the supports by the nature of the brick stress distribution curves. Some cracks can be noticed immediately under the peak in the brick stress curves.

In spite of the fact that the presence of a doorway in this position gives rise to a point-load effect partway along the span, the tabulated equivalent bending moments still demonstrate the existence of a considerable composite action.

THE STRESS DISTRIBUTIONS IN DEEP WALLS
USING THE "RELAXATION" TECHNIQUE

A fairly clear idea of the behaviour of loaded panels on supporting beams can be obtained from a study in the first instance of a panel without a supporting beam. The behaviour of the bricks during test demonstrated a remarkable degree of

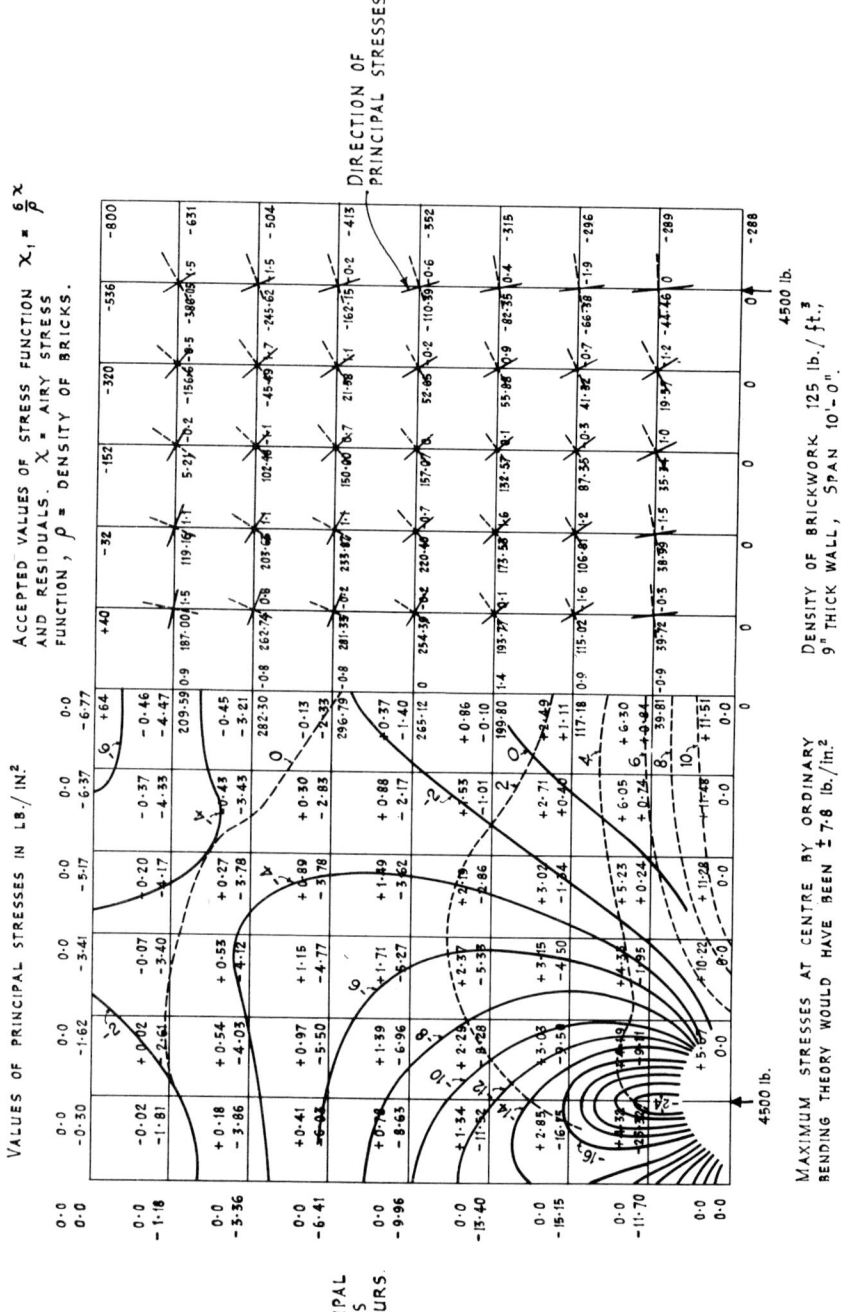

FIG. II. STRESSES IN BRICK PANEL DUE TO GRAVITY.

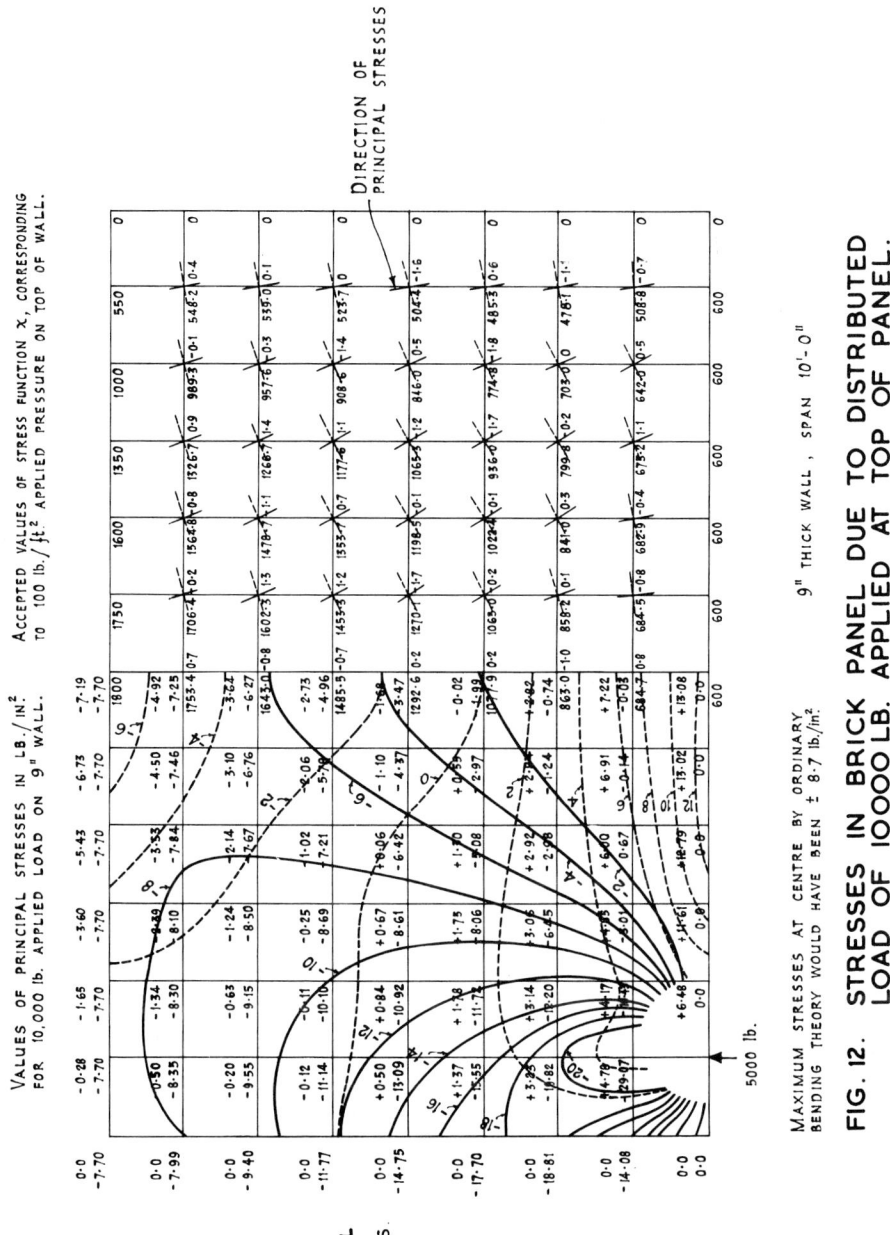

FIG. 12. STRESSES IN BRICK PANEL DUE TO DISTRIBUTED LOAD OF 10000 LB. APPLIED AT TOP OF PANEL. 9" THICK WALL, SPAN 10'-0".

elasticity: creep, if it occurred at all, was only of the same order as temperature movements. In spite of heavy loads only small stresses were recorded in the panels, and the assumption of a perfectly elastic medium up to the point of tensile cracking should be fairly satisfactory.

It is proposed to examine the case of a brick panel without any supporting beam, having line load reactions at the supports. The proportions of the panel are given in Figs. 11 and 12, where the shape of the panel roughly coincides with those which have been tested. There are two cases to consider:

(a) when a uniform superimposed load is applied to the top of the wall;
(b) when the weight of the wall is treated as a " body " force.

The stress distribution may be determined by evaluating the Airy Stress Function χ at numerous points in the panel such that at each point the biharmonic equation

$$\frac{\partial^4 \chi}{\partial x^4} + \frac{2\partial^4 \chi}{\partial x^2 \partial y^2} + \frac{\partial^4 \chi}{\partial y^4} = 0 \quad \ldots \quad (1)$$

is satisfied, together with the appropriate boundary conditions.

The stresses are then determined from

$$\widehat{xx} = \frac{\partial^2 \chi}{\partial y^2} + \rho\Omega = \text{direct stress in x direction} \quad \ldots \quad (2)$$

$$\widehat{yy} = \frac{\partial^2 \chi}{\partial x^2} + \rho\Omega = \text{direct stress in y direction} \quad \ldots \quad (3)$$

$$\widehat{xy} = -\frac{\partial^2 \chi}{\partial x \partial y} = \text{shear stress} \quad \ldots \quad (4)$$

Here ρ refers to the density of the bricks and Ω is the body force potential due to gravity which is zero in case (a).

From the known boundary conditions the values of the stress function χ can be set up all round the boundary by a process of integration and the problem is then to determine values of χ at internal points. This was achieved by means of the " relaxation " process developed by Southwell, using a fine " mesh " and working numerically in terms of the finite difference approximation to equation (1).

The solution of case (b) is given in Fig. 11, the right hand half showing accepted values of χ and the " residuals ", together with the direction of the principal stresses. These clearly demonstrate the arch effect.

The left-hand half shows contours of principal stresses, again demonstrating the arch and the rising stresses near the supports.

An interesting feature is the degree of stress imposed on the vertical edge of the brickwork (which was observed in some of the experiments) and it should be remembered that the simple bending theory would assume zero stresses all along this edge.

Fig. 12 gives similar results for the distributed load at the top of the wall. This stress pattern is very similar to that in the previous case. In both cases it will be noticed that the maximum horizontal tension at the centre section due to bending is much greater than the maximum compression; this shows the departure from the simple bending theory. The bending stresses are plotted in Fig. 13 and show that the " neutral axis " is below the centre line, and the stress distribution curve has a characteristic double curvature. Nevertheless, it can be proved that the internal " moment arm " is still very close to the Navier-value of two-thirds of the depth.

The composite action of walls on beams

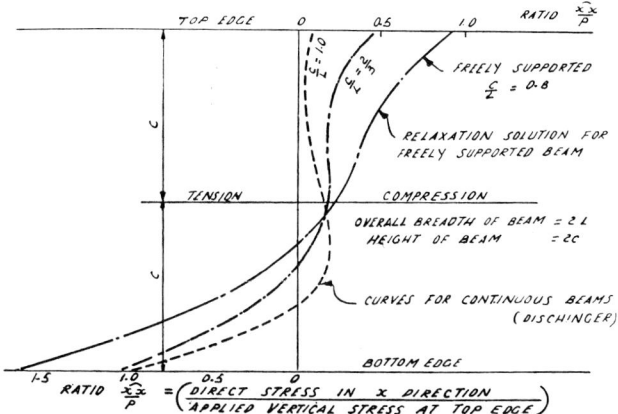

FIG. 13. *Bending stresses on the centre line of a deep beam*

USING FOURIER SERIES

In 1932 Dischinger made a " Contribution to the theory of wall-like girders "[4], in which he examined the stress distribution for deep walls resting on multiple supports in the form of continuous beams. Dischinger found that above a certain ratio of depth to span (about $\frac{2}{3}$) the " moment arm " at the centre of the beam was no longer dependent upon the depth but changed fairly abruptly to that value of the moment arm given by an infinitely deep beam, viz. about 0·47 times the span (see Fig. 15). Dischinger mentions some unpublished results referring to simply supported beams but quite recently (1947) Durant and Garwood[5] have solved this case for deep simply supported beams. In this case it appears impossible to satisfy all the boundary conditions simultaneously when using Fourier Series, and the solution of Durant and Garwood is to that extent an approximation. Thus the solution gives stresses on the free (vertical) edges of the wall or deep beam which are not identically zero, but which add up to total zero forces and total zero moment, and according to the authors are sufficiently small to be ignored.

With very deep beams some quite remarkable stress patterns occur. In Fig. 14 the result of Durant's and Garwood's analysis is given for a panel with a height equal to three times the width, resting on narrow supports, and it will be noticed that the distribution of stress is very nearly the same as that for an infinitely deep beam by the same theory. The " moment arm " tends to a limiting value clearly dependent upon the span only and not on the depth. Practically the whole of the beam action takes place in the bottom of the panel, and the top of the panel is virtually acting as a column supporting the superimposed load. This can best be seen if we turn the panel upside down, and regard the support reactions as loads on a " column ". By the well known principle of St. Venant we should then expect the local effects to have died out in about $1\frac{1}{2}$ column widths. The same kind of phenomenon was previously noted by Dischinger (loc. cit.) for continuous walls. Fig. 13 shows the centre-span stress distribution in two deep continuous wall-beams, with depth to span ratios of 1·0 and 2/3, compared with the stress distribution for the freely supported single-panel wall (ratio 0·8) which has been previously studied in this paper by relaxation methods. It will be noticed that the maximum bending

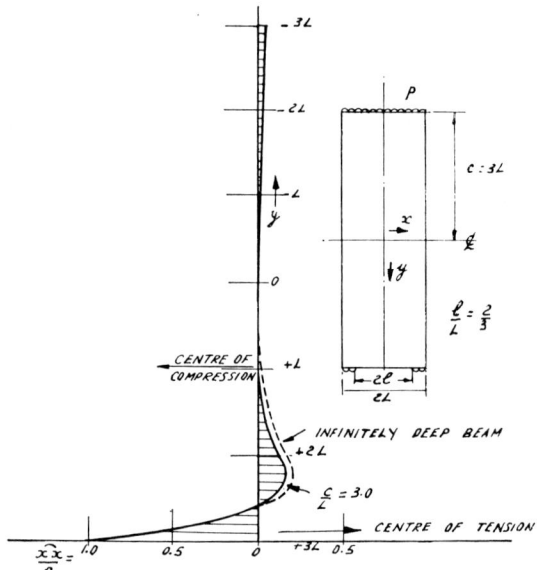

FIG. 14. *Horizontal stresses on centre line of freely supported deep beam (Durant and Garwood)*

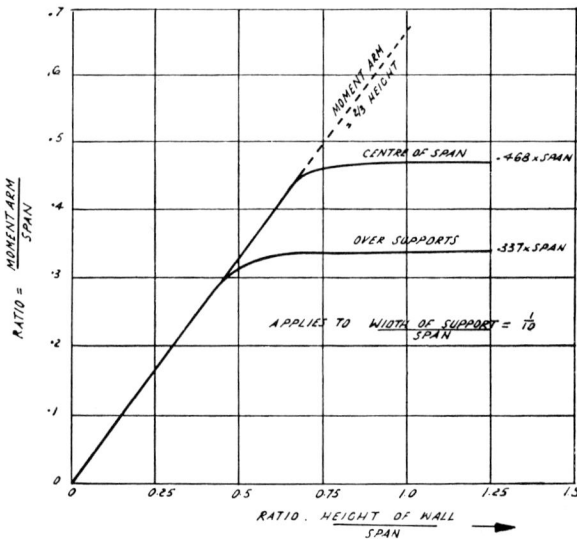

FIG. 15. *Values of the moment arm for distributed loads on a continuous deep wall resting on several supports (Dischinger)*

stress $\widehat{xx}$ tends to become equal to the applied vertical stress p, which is a peculiarity of deep continuous walls, and that the solution of Durant and Garwood tends to the same result for simply supported wall panels (Fig. 14). Recent work, however, at the Building Research Station by W. T. Lawton, employing relaxation methods, has shown that for simply supported wall panels the limiting maximum stress $\widehat{xx}$ for a very deep wall is some 40 per cent higher than that given by Durant and Garwood. The discrepancy is due to the fact that the edge stresses in the Fourier analysis are no longer negligible in a deep wall.

For an infinitely deep panel the compression at the top approaches zero and it will be seen from Fig. 14 that the centre of compression is not easily located. We can, however, obtain it by adapting the solution of Durant and Garwood, which is given for uniformly distributed loads as:*

$$\text{Horizontal Stress } \widehat{xx} = ky + \sum_{m=1}^{\infty} \frac{h_m \cos \xi}{\sinh^2 (2\beta) - 4\beta^2} \Big\{ \sinh 2\beta \, (u \cosh v - \sinh v) - 2\beta \, (v \cosh u - \sinh u) \Big\} \quad . \ (5)$$

where $\xi = \dfrac{m\pi x}{L}, \ \beta = \dfrac{m\pi c}{L}, \ \eta = \dfrac{m\pi y}{L}, \ u = \beta - \eta, \ v = \beta + \eta$

$h_m = -\dfrac{2p}{m\pi} \left(\dfrac{L}{L-l}\right) \sin \dfrac{m\pi l}{L}$ and $k = \dfrac{L^2 p}{4c^3} \cdot \dfrac{L}{(L-l)} \cdot \left(\dfrac{l}{L} - \dfrac{l^3}{L^3}\right)$

It can now be shown that in the neighbourhood of the bottom edge of an infinitely deep beam the horizontal stress is

$$\widehat{xx} = \Sigma h_m \cos \xi (u-1) e^u,$$

from which the curve of Fig. 14 has been plotted. The neutral axis lies at $0 \cdot 40L$ from the bottom edge for $\dfrac{l}{L} = \dfrac{2}{3}$.

With a total depth of wall of $2c$ the total compression can be found from the value of $\int \widehat{xx} \, dy$ between the neutral axis and the top edge (the wall being considered of unit thickness). Thus we eventually find that as $c \to \infty$

$$\int_{+c-0\cdot 40L}^{-c} \widehat{xx} \, dy = \sum_{m=1}^{\infty} \frac{2pL}{m^2 \pi^2} \left(\frac{L}{L-l}\right) \cdot \epsilon e^{-\epsilon} \sin \frac{m\pi l}{L} \quad . \ . \ (6)$$

where $\epsilon = 0 \cdot 4 m\pi$

Similarly the moment of the compression area about the bottom edge can be shown to be $(c \to \infty)$:—

$$\int_{+c-0\cdot 40L}^{-c} \widehat{xx} \, (c-y) \, dy = \frac{5}{81} pL^2 + \sum_{m=1}^{\infty} \frac{2pL^2}{m^3 \pi^3} \left(\frac{L}{L-l}\right) \cdot (1+\epsilon+\epsilon^2) \, e^{-\epsilon} \sin \frac{m\pi l}{L}$$
$$. \ . \ . \ (7)$$

From (6) and (7) the "centre of compression" can be determined, from which it is found that the "moment arm" of this infinitely deep wall approaches a value almost equal to the span, depending on the width of the reaction at the supports. By satisfying more rigidly the conditions on the vertical edges, Lawton, however,

*The symbols refer to panel shown in Fig. 14.

finds that the limiting moment arm is approximately equal to 0.7 × span. Accordingly, it is proposed to adopt a conservative design rule, after the lead of Dischinger, as follows :—

" In the design of freely supported deep walls, the steel reinforcement may be designed using Navier's moment arm of 2/3 × depth, with a limit of 0.7 times the span of the wall ".

It should be remembered that in actual practice there will be other features such as friction at the supports and cracking of the concrete, which will increase the apparent moment arm.

For deep walls the reinforcement may be considered effective up to $\frac{1}{10}$ span from the lower edge ; for a shallow panel, up to $\frac{1}{4}$ depth.

For continuous beams Dischinger found the limiting moment arm at the centre to be about 0·47 × span and that at the supports about 0·34 × span, depending on the support width, and the effective positions of reinforcement are also given (loc. cit.).

The question of what working steel stresses should be used with the above design rule is discussed later.

RECOMMENDED METHODS OF DESIGN

From the previous discussion it will be obvious that both the " tension " and " compression " reinforcement of the supporting beam used in the tests lie in the tensile zone of the panel as a whole when openings are not present. Hence we may make use of this form of analysis for more economical design or alternatively use the " equivalent bending moments " given in the analysis of steel stresses providing in that case the ratio of beam depth to span is approximately equal to that used in the tests ($\frac{1}{15}$ to $\frac{1}{20}$).

From the foregoing tables it appears that the beam reinforcement may be designed for a bending moment of $WL/50$ where there are door or window openings near to the supports or for $WL/100$ either for plain walls or when door and window openings only occur at the centre of the span. These recommendations are conservative since they apply to overload conditions.

Thus referring to the 9 in. solid wall, 8 ft. high which has previously been tested on an equivalent span of about 123 in. using the 12 in. by 7 in. " light " beam (referred to in Table III.):—

With 25·8 tons distributed load the bending moment $WL/8$ would be 890,000 lb. in.

Since $\frac{2}{3}$ × depth does not exceed 0·7 × span we may use a moment arm of $\frac{2}{3}$ × 103 in = 69 in.

The total steel area is:

$$4/\tfrac{5}{8} \text{ dia.} = 1\cdot 23 \text{ sq. in. bottom reinforcement}$$
$$2/\tfrac{1}{2} \text{ dia.} = \underline{0\cdot 39} \text{ sq. in. top reinforcement}$$
$$1\cdot 62$$

Hence the design steel stress $= \dfrac{890{,}000}{1\cdot 62 \times 69} = 7970$ lb. per sq. in.

The recorded steel stress was less than this, being 4920 lb. per sq. in., showing that the design procedure is still somewhat conservative.

This additional factor of safety accrues from:
(a) horizontal friction at the supports reducing the tension at the beam level, and therefore apparently increasing the moment arm,
(b) direct aid from tension in concrete acting as " reinforcement " for the complete panel,
(c) some degree of fixing moment developed near the supports,
(d) an increase of the moment arm by virtue of strain energy considerations as a result of minute cracking in the brickwork.

Had we used the more elementary design method we should have obtained for the same beam a stress based on a bending moment of $WL/100 = 71,000$ lb. in., whence

$$\text{steel stress} = \frac{71,000}{0.88 \times 6 \times 1.23} \text{ approx.} = 10,900 \text{ lb. per sq. in.,}$$

which is obviously conservative.

It will generally be found that the first method is more economical for panels without openings.

The simple method of designing lintels in common use is to take a triangular load of bricks (Fig 1, bending moment $WL/6$) and ignore any superimposed load.

Considering a 9 in. thick wall, span 10 ft. 6 in. and height equal to 10 ft. 6 in. $\times \frac{\sqrt{3}}{2}$ (Fig. 1) then the bending moment based on the *total* load would be as follows, for various superimposed loads:—

Superimposed load	0	10 Tons	20 Tons	30 Tons
Bending moment	$\frac{WL}{12}$	$\frac{WL}{42}$	$\frac{WL}{72}$	$\frac{WL}{102}$

Thus at low loads the design is most uneconomical. At high loads there are insufficient data to indicate whether or not the simple design method is actually safe. It must also be remembered that with a single panel there is little justification for this simple (triangular) method for there is no apparent external support for the bricks outside the triangle.

For walls of greater depth, for this same span, the bending moment deduced from the simple method is unchanged, since the increase in depth does not increase the arch included in the triangle shown in Fig. 1. For a particular total load, the actual maximum stresses in the beam will, however, be less for this case than for the wall previously considered, since the additional depth of brickwork increases the internal moment arm of the composite panel, up to a limit approximately equal to $0.7 \times$ span.

PRESENT LIMITATIONS AND FUTURE RESEARCH WORK

The proposed methods outlined above should at this stage of the work be subjected to certain limitations and conditions as follows:
(a) The " limiting moment-arm method " clearly applies only to the case of panels without openings or to that part of a panel below an opening.
(b) Whereas both the suggested design moment-coefficients and the " limiting moment-arm method " are quite safe with deep walls (in the former case with openings in addition) it is not intended to use either method for panels of considerably smaller depth than those tested. It will have been noticed that neither the brick stresses nor the concrete stresses nor the bond between

bricks and concrete are at all critical in deep panels, for there the "arch" is of such depth that shear stresses become small in magnitude, and no particular precaution need be taken except that of good workmanship. When the panel is small in depth, however, the moment arm becomes practically coincident with that of the simple beam theory, and in such a case, i.e. with a shallow "arch", shear stresses would become a matter of considerable importance. Until further tests are carried out it is proposed to restrict the application of the above design rules to panels where the depth is not less than 0·6 × span.

(c) Since a very considerable proportion of steel is saved by composite design it is not necessary, nor is it desirable at this stage, to use the normal working stresses for steel in reinforced concrete. As an additional factor of safety it is recommended that the design stresses should not exceed the following :—

7 tons per sq. in. where the beam is propped up during the bricklaying process;

5 tons per sq. in. where the beam is unsupported during the bricklaying process.

Under these conditions the whole of the dead weight may then be reckoned as superimposed load. As experience is gained it will no doubt be possible, using the same underlying theory, to raise the design stresses.

(d) In the present report the superimposed loads were applied at the top of the walls, and the results are immediately applicable to house walls. With a perfectly elastic system the stress "patterns" so obtained can be shown to apply also to distributed loads at beam level, provided an overall vertical tensile stress is superimposed over the whole panel statically equal to the distributed load at beam level. The nett effect is to relieve the panel of vertical compression above the "arch" whilst introducing a tensile stress below the "arch", the stress pattern being otherwise unaltered. Consequently the results obtained *do not apply to lower beam loading unless tensile connectors can be placed between wall and beam.* Where this is not possible the beam can be designed to take floor loads (at that level) independently as a beam, whilst wall loads and superimposed loads may be taken by composite action.

(e) No attempt must be made at this stage to design beams incorporating standard rolled steel sections by the above methods. The Building Research Station is studying the problem of composite action for walls and floors carried by encased steel beams under both laboratory and field conditions.

(f) Where door or window openings occur near the supports it is advisable to include a nominal shear reinforcement in the beam (or otherwise to provide an additional support).

(g) Frequent use is now made of clinker blocks for the interior of a cavity wall. Although the tests were carried out with bricks the stresses so obtained were so low as to suggest that the design methods should be suitable for clinker blocks, provided these comply with British Standard Specification.

(h) Some light reinforcement may be necessary in the walls to take care of shrinkage and also for continuity in the case of continuous walls.

(i) When using the suggested design moment coefficients the ratio of the beam depth to the span should be approximately that used in the tests ($\frac{1}{15}$ to $\frac{1}{20}$).

CONCLUSIONS

(a) Throughout the tests, remarkably low stresses have been recorded in the supporting beams. For a wall about 8 ft. in height supported on a reinforced concrete beam of light construction, spanning 10 ft., the maximum steel stress due to the construction of the wall was about 3000 lb. per sq. in. The addition of a superimposed load of 6 tons, uniformly distributed along the length of the wall, resulted in an increase in steel stress which exceeded 500 lb. per sq. in. in only one test, in which there was a door opening near to one end of the span. In this latter test the steel stress due to the superimposed load of 6 tons was 3000 lb. per sq. in.

(b) It has been shown that these low stresses are largely due to arching effects in the brickwork, the bricks in fact forming a composite beam of a much greater depth than the supporting beam. The provision of a damp-proof course did not prevent this latter effect taking place. In some instances the whole of the beam including the top (compression) reinforcement, was found to be in tension.

(c) During bricklaying, the degree of composite action is small, particularly for the first few courses, and the stresses reached during the construction of the wall may often be greater than those arising subsequently from superimposed loading. The tests indicated that after about 30 courses of brickwork have been laid, the stresses due to additional courses are similar to those due to a superimposed load, distributed on the completed wall, equal to the weight of the additional brickwork.

If the beam is propped during bricklaying and until the mortar has hardened, the maximum stresses due to the weight of the wall may be assumed to be the same as for an equal load applied to the top of the wall.

(d) There are two feasible methods for calculating the amount of steel reinforcement required in the supporting beam. The first consists of taking equivalent bending moments on a freely supported beam, and tables of these equivalent moments have been given. From these it appears that a design moment of $WL/50$ based on total load for brick panels where there are door or window openings near the supports, and $WL/100$ for panels where door and window openings are absent or occur at mid-span, will give a safe design. An additional factor of safety is present since the beam design will be based on a " no tension in concrete " theory. For the most part the equivalent free-span moments during test ranged between $WL/960$ to $WL/130$. When using this method the ratio of beam depth to span should be approximately that used in the tests ($\frac{1}{15}$ to $\frac{1}{20}$).

The second method is based on a more rational analysis of a reinforced panel as a whole and is applicable at this stage only to panels without openings or to that part of a panel below an opening. In most cases this method is more economical. This rational method is referred to as the " limiting moment-arm method " and calculations show that a limiting moment-arm approximately equal to $0.7 \times$ span occurs in very deep panels; otherwise the Navier moment-arm of $\frac{2}{3} \times$ depth is permissible.

In neither case need the concrete or brick stresses be evaluated, nor are deflections of any consequence.

(e) The " arching " effects observed in the tests occurred even though the " tie " was provided by only small amounts of steel reinforcement. It appears therefore that the addition of a little reinforcement to brick panels in the traditional type of house could be of great value, particularly in strengthening the wall against the effects of unequal settlement.

(f) The greatest advantage from composite action of wall and beam is obtained when door and window openings are kept away from the supports and preferably placed at the centre of the span.

(*g*) It appears to be desirable to provide some top steel in the beam over the supports, where hogging moments develop due to the nature of the composite stress distribution. A nominal shear reinforcement should also be provided close to the supports when door and window openings occur other than at mid span.

(*h*) The present practice of designing lintels and beams to support a triangular weight of brickwork is very uneconomical with low superimposed loads.

(*j*) With a reasonable quality of brickwork no shear connectors appear to be necessary when the depth of the panel is above the minimum for which these design rules apply (viz. 0·6 × span). If the live loads are not applied at the top of the wall, but instead at beam level, *tensile* connectors should be arranged where feasible; otherwise composite action cannot be relied upon. It is intended to carry out further research work with applied loads at beam level.

(*k*) The design rules are not applicable at the present moment to structural steelwork, but further tests are in progress.

(*l*) The above design rules are conservative when compared with the test results and there are many features which give rise to additional margins of safety. As a further precaution it is proposed to limit the steel stresses to 7 tons per sq. in. if the beam is supported when the wall is being built and to 5 tons per sq. in. if it is unsupported during construction.

REFERENCES

(1). WARD, W. H. House foundations. *R.I.B.A. J.*, February, 1947.
(2). BAKER, J. F. An investigation of the stress distribution in a number of three storey steel building frames. *2nd Report of the Steel Structures Research Committee* (p. 242), 1934.
(3). JERRETT, R. S. The acoustic strain gauge. *J. of Sci. Instruments*, 1945, **22**, No. 2.
(4). DISCHINGER, F. Beitrag zur Theorie der Halbscheibe und des wandartigen Balkens. *International Association for Bridge and Structural Engineering*, 1932, **1**, 69-93.
(5). DURANT, N. J., and F. GARWOOD. Stresses in a deep beam. *Ministry of Works Technical Note No. 78*, October, 1947.

INDEX

Arching
  in brickwork, 259,264

Beams
  Fourier analysis, 282-290
  composite, 124,138
  continuous, 138-139
  reinforced concrete, 137
  simply supported, 141
Brick walls
  on beams, 264 et seq
  without openings, 268-280
  with openings, 280-283
Buckling
  lateral-torsional, 23
  of centrally loaded columns, 31

Cladding
  calculation procedures for, 115-117
  functional requirements of, 110
  indented profiled sheets, 110
  roof sheeting, 112,113
  roof decking, 115
  sandwich panels, 27
  structural benefits of, 128-131
  types of, 109
Cold-formed profiles
  roof sheeting, 112,113
  roof decking, 115
  wall cladding, 114,115
  trapezoidal, 113,116-117

Columns
  bent in single curvature, 20,23
  bent in double curvature, 20,23
  composite, 137-138
  design of, 5-6,227
  deterioration of critical loads, 20
  effective lengths, 5
  initial lack of straightness, 2,5
  material properties, 2
  Perry-Robertson formula, 2-3,5,9
  simple design, 5
  stability, 3,9,31
Composite
  beams, 124
  brick walls on beams, 264 et seq
  columns, 137-138
  dry floor decks, 124
  floor decks, 120-121,126
  infilled frame, 259
Concrete
  ductility of, 139-141
  slabs, 145 et seq.
Connections
  beams, 63
  beam-to-column, 58,76
  characteristics, 58
  choice of, 75-77
  cleat type, 77,78
  design of, 75-108
  end-plate type, 75-108
  moment-rotation curves, 61,82-83,98-99
  pinned, 57
  prying action, 79,80,102

rigid, 57
semi-rigid, 5,58-59,66,82
semi-rigid, partial strength, 84,106-107
restraint, 63-66

Decking
  cold-formed 109-134
  composite floor, 120,124
Deep beams
  stress distribution, 283 et seq
Deteriorated
  critical loads, 12-15,20
Ductility
  concrete, 139-141
  composite beams, 141
  shear connectors, 139-141

Effective lengths
  columns, 5
Elastic theory
  second order, 39-40
Elastic buckling load
  columns, 4
Elasto plastic
  frame design, 230 et seq
End-plate connections
  design of, 84-99
  failure modes, 90-98
  moment capacity, 98
  rotational stiffness, 98-99
Engesser's double modulus theory 3
Equilibrium-bifurcation theory 31,32
Equilibrium method 195-219

Fire
  resistance of composite floors, 126
Frames
  analysis of, 34,35,73 et seq
  braced, 29
  continuous, 2,33
  design of, 19

no-sway, 5
pin-jointed, 57
portal, 23
rigid, 57
simple design, 5
stability of, 10-20,30,225 et seq
swagebeam, 129
tests, 66-68

Geometric
  imperfections, 31,36,256
  properties, 2

Joints
  rigid, 29
  semi-rigid, 29,58-59,66

Loads
  elastic critical, 232
  point, 291

Nodal forces
  application to frameworks, 208-210
  application to slabs, 211-213,215-217
  examples of, 197-207

Plastic hinge theory
  application of, 44-48,241 et seq
  examples of, 48-54
  first order, 38-39,40
  second order, 38
  simplified second order, 38,40
Plastic flow rules 175-193
  choice of, 180-183
Profiled sheeting
  calculation procedures, 116-117
  capacity of, 117,119
  tests, 117-120

Index

Rankine-Merchant
  formula, 16-18,33,35,256
Reinforced concrete
  beams, 137
  slabs, 145 - et seq.
Residual stresses
  patterns for rolled sections, 37

Semi-rigid
  action, 58-59,61,66
  connections, 5,29
Shear connectors
  ductility of, 139-141
  partial, 139-141
Slabs
  design of r.c, 145-174

Stability
  columns, 3
  frames, 10-20
Stanchions elastic design
  SSRC Design Method, 227
Steel studies
  frameless, 128
Stressed arch building, 131
Strip method
  extensions to, 162-173
  Hillerborg's, 145-174

Ultimate strength theory, 36

Wood-Armer reinforcement rules, 149
Yield criteria, 183-186